Liengme's Guide to Excel® 2016 for Scientists and Engineers

Liengme's Guide to Excel® 2016 for Scientists and Engineers

Windows and Mac

Bernard Liengme

Keith Hekman

ELSEVIER

ACADEMIC PRESS
An imprint of Elsevier

Academic Press is an imprint of Elsevier
125 London Wall, London EC2Y 5AS, United Kingdom
525 B Street, Suite 1650, San Diego, CA 92101, United States
50 Hampshire Street, 5th Floor, Cambridge, MA 02139, United States
The Boulevard, Langford Lane, Kidlington, Oxford OX5 1GB, United Kingdom

Library of Congress Cataloging-in-Publication Data
A catalog record for this book is available from the Library of Congress

British Library Cataloguing-in-Publication Data
A catalogue record for this book is available from the British Library

ISBN 978-0-12-818249-9

For information on all Academic Press publications visit our
website at https://www.elsevier.com/books-and-journals

Publisher: Katey Birtcher
Acquisition Editor: Steve Merken
Editorial Project Manager: Katerina Zaliva
Production Project Manager: Anitha Sivaraj
Cover Designer: Christian J. Bilbow

Typeset by SPi Global, India

Working together
to grow libraries in
developing countries

www.elsevier.com • www.bookaid.org

Contents

Preface

This book is for people in technical fields, students and professionals alike. Its aim is to show the usefulness of Microsoft® Excel in solving a wide range of numerical problems. Excel does not compete with the major league symbolic mathematical environments such as Mathematica, Mathcad, Maple, and the like. Rather it complements them. Excel is more readily available and is easier to learn. Furthermore, it generally has better graphing features and ways of handling large datasets.

The examples have been taken from a range of disciplines but require no specialized knowledge, so the reader is invited to try them all. Do not be put off by an exercise that is not in your area of interest. Each exercise is designed to introduce and explain an Excel feature. The two modeling chapters will help you learn how to develop worksheets for a variety of problems. This is very much a practical book designed to show how to get results. The problem sets at the ends of the chapters are part of the learning process and should be attempted. Many of the questions are answered in the last chapter. The *Guide* is suitable for use as a textbook in a course on scientific computer applications, a supplementary text in a numerical methods course, or a self-study book. Professionals may find Excel useful to solve one-off problems rather than writing and debugging a program, or for prototyping and debugging complex programs. A few topics are not covered by the *Guide*, such as database functions and making presentation worksheets. These are fully covered in Excel books targeted at the business community, and the techniques are applicable to any field.

After using the *Guide* for several years in my classes, I was honored to have the opportunity by Dr. Liengme to update the *Guide to Excel 2016* and grateful for his improvement suggestions. Based on my experience with students' needs, I added support for Apple OS commands where applicable in this version. Also, I created videos of me doing the exercises in the earlier chapters, as many students find that it is easier to watch where to click, than to read instructions. In addition, the opportunity has been taken to add new exercises and problems.

I wish to thank Dr. Marta Maroń for her assistance with this version of the book. Finally, I would like to thank my wife Rana for her patience and support, making this version of the book possible.

I welcome e-mailed comments and corrections, and will try to respond to them as soon as I can. Please check my website and the *Guide*'s companion webpage https://www.elsevier.com/books-and-journals/book-companion/9780128182499 for supplementary material.

I hope you enjoy learning to "excel."

S.D.G.
khekman@calbaptist.edu   **Keith Hekman**
https://calbaptist.edu/faculty-directory/profileview?id=209

Chapter

1

Welcome to Microsoft Excel 2016

When Microsoft Excel is started you are presented with a window similar to that in Fig. 1.1. From there you may (i) select from the left panel a recently opened workbook, (ii) click on Open Other Workbook, or (iii) click on the icon Blank Workbook to start a new project. For a Mac, it is similar to Fig. 1.2. Recent workbooks and other workbooks have their own separate tab. Note that while in Word we speak of a *document*, in Excel we use the term *workbook*. In either case, we are referring to a file. In this book we shall not explore using OneDrive, Microsoft's online cloud storage, so we can ignore the *Sign In* option in the top right corner (top left on a Mac).

When we open a new workbook we have a window showing the Excel Interface. Fig. 1.3 is a screen capture from the author's computer with the Excel window was "restored down" to occupy about half of the monitor screen. The Excel window on your computer may differ slightly depending on your monitor size and resolution. For a Mac, the window is similar to Fig. 1.4.

It is helpful to know the correct name for the various parts of the window. This makes using the Help facility more productive and aids in conversing with other users. As a new term is introduced it is displayed in italics, and the reader should try to remember the meaning of such terms.

Liengme's Guide to Excel 2016 for Scientists and Engineers. https://doi.org/10.1016/B978-0-12-818249-9.00001-7

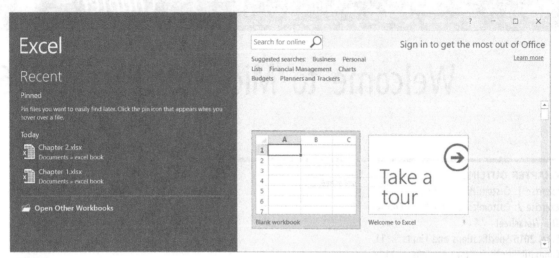

■ FIG. 1.1

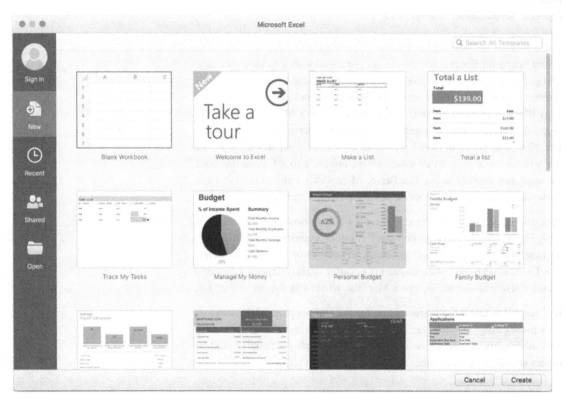

■ FIG. 1.2

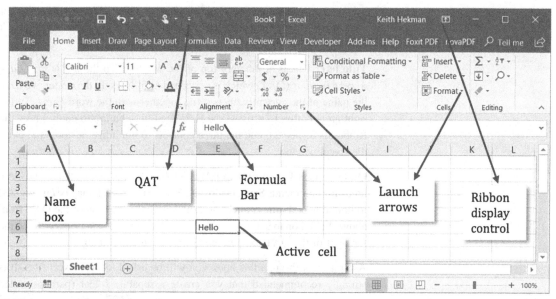

FIG. 1.3

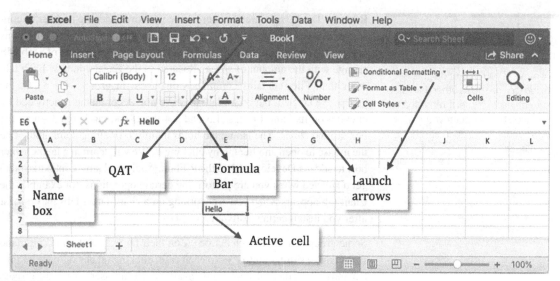

FIG. 1.4

It is recommended that you read this chapter while seated at the computer and experiment as you read it. Remember that pressing the $\boxed{\text{Esc}}$ key will back you out of an action you do not wish to pursue.

Title bar: This is at the very top of the window. To the left is the *Quick Access Toolbar* (QAT) which is described as follows. In the center, we have the name of the currently opened file together with the word *Excel*. To the right are a button to activate the Help facility; a button to control how the Ribbon is displayed; and the three controls to minimize, restore, and close the Excel window.

Quick Access Toolbar (QAT): When Excel 2016 is first installed, the QAT holds the commands Save, Undo, and Redo. However, it may be customized to hold others. Furthermore, one can change the location of the QAT from above the ribbon to below the ribbon. Click on the *launcher button* ▼ at the far right of the QAT to open the QAT customizing dialog.

As you work through this book you will be asked to save Excel files. It is strongly recommended that you create a separate folder (perhaps called *Excel Practice*) in which to keep these. The first 🖫 tool on the QAT, displaying an icon of a floppy disk (something no one uses anymore!), will open the File Explorer where you can make folders and save files.

Warning regarding Undo: Excel keeps a single undo stack. This means that if you issue an undo command you may undo changes made to worksheets other than the currently active one. If more than one workbook is open you may even undo an action in another workbook.

Ribbon: The *Ribbon* stretches across the window under the title bar. It consists of a number of *tabs* (File, Home, Insert, Page Layout, etc.). The Ribbons in Figs. 1.3 and 1.4 have the Home tab selected. The appearance of a tab will change with the amount of space allocated to the Excel window. Each tab, other than File, contains commands displayed in *groups*. On a PC, the groups are labeled, but on a Mac, they are not. A command is activated by clicking on its icon. In Figs. 1.3 and 1.4 the Home tab is open—note the box around *Home*. The Home tab holds mainly formatting commands. Use the mouse to open another tab by clicking it. We will learn in a later chapter how to add the Developer tab to the Ribbon. Additional tabs (contextual tabs) get displayed when you are performing certain operations; for example, the Chart tab appears when you are working on a chart. Other tabs may appear after you install certain software.

Some groups have a launch button $\boxed{\searrow}$ on their far right and some command icons have a similar button ▼. In each case, clicking on one of these buttons expands the choice of commands available to the user. We shall discuss these on an as-needed basis.

On Excel for a Mac, in addition to the ribbon, some commands are accessed from drop-down menus at the top of the screen.

File tab: This tab gives the user access to the so-called *back stage* to do things like open, save, or print a file. It also gives us access to the Options dialog where we can customize certain Excel features. We will look at this in later chapters. Excel for a Mac has a drop-down file menu instead of the file tab.

Title Bar Tools: On a PC, to the far right of the Title Bar we have four icons.

Ribbon Control: By default, the Ribbon displays tabs, their groups, and commands. The Ribbon Control tool gives us the Excel of both tabs and command, showing only the tab, or having the Ribbon auto-hide. The second two options are useful when the user needs to see more of the working area of the window. There is no Ribbon Control tool on a Mac.

Minimize, Maximize, and Close: The last three buttons are familiar to all users of Microsoft products and need no further explanation.

Help facility: On a PC, Quick access to help can be gotten by typing in the box. Additional help by clicking on the help tab to access the help ribbon on both a PC and a Mac. The help menu can also be accessed by pressing the F1 key.

Formula Bar and Name Box: Just under the Ribbon is the Formula Bar with the Name Box to the left. In Figs. 1.3 and 1.4 the Name Box is displaying E6. You will notice that both the E column and the six row headings are highlighted and that the cell at the intersection of this column and row is picked out by a border. We call E6 the *active cell*, and we say that the Name Box displays the reference (or address) of the active cell. When the active cell contains a literal (text or number), the Formula Bar also displays the same thing, but when the cell holds a formula then the formula bar displays the actual formula while the cell generally displays the result of that formula. Quick experiment: type B4 in the name box and press Enter↵; note how this takes you to cell B4.

Worksheet window: The worksheet window occupies most of the Excel space. A workbook (that is to say a single Excel file) may contain *worksheets* and *chart sheets* (collectively called *sheets*); we will concentrate on worksheets for now. A worksheet is divided into *rows* (horizontally) and *columns* (vertically); the intersection of a row and a column is called a *cell*.

Sheet tabs: Below the worksheet window we have tools to navigate from sheet to sheet and to scroll a sheet horizontally. By default, Excel 2016 opens a new workbook with one worksheet; this number can be changed in the Options setting. To the left of the first sheet tab ⊕ are arrows

Note: It is becoming common to talk about tabs when worksheets are meant. This is very poor practice since it can cause confusion and will not benefit a user searching in Help.

for navigating from sheet to sheets, but merely clicking a sheet tab is the most rapid way. To the right of the last sheet tab is a tool to insert a new worksheet. To the right of the sheet tabs is the horizontal scroll tool; the vertical scroll tool is on the right side of the worksheet. We will see later how to rename sheets. If your mouse has a wheel you can use it to scroll up and down a worksheet.

Status bar: At the very bottom of the Excel window we have the *status bar*. To the left is the mode indicator. When you move to a cell this displays *READY*; when you start typing it becomes *ENTER*; if you double-click a cell (or press the F2 key) it becomes *EDIT*. Other status conditions like *POINTING* and *Copy/Paste* we will discuss later. We will ignore the second tool (Macro Recorder) for now. To the right, we have *Page View* buttons that let us display the worksheet in different ways: Normal, Page Layout, and Page Break Preview: more on this topic later. Finally, there is the *Zoom* tool that enlarges/reduces the display. You can also change the magnification of the worksheet by rotating the mouse wheel while holding down the Ctrl key.

If we experiment with the *Page View* buttons, we may notice that the worksheet gets vertical and horizontal dotted lines. These show how much will fit on a printed page. On a PC, right-clicking the status bar brings up a dialog box that allows you to customize the status bar. For a Mac, click on the status bar while holding down the Ctrl key to bring up the dialog box. We will show more features of the status bar in Exercise 3.

EXERCISE 1: CUSTOMIZING THE QAT

Any of the Excel commands can be reached by opening the appropriate tab and locating the command within one of the tab groups. If there is an operation that you perform frequently it is convenient to be able to access it from the Quick Access Toolbar which explains its name. As a demonstration, we will add the File Open command to the QAT.

(a) Start Excel and let the mouse pointer hover over the QAT launch button ⩧ which is always the last item on the QAT. A screen tip box will open with the text *Customize Quick Access Toolbar*.
(b) Now click on the launch button to open the dialog shown in Fig. 1.5. On this dialog, we see the more commonly needed commands. To add one of the common items to the QAT just click on it to bring up a check mark. Correspondingly, click on an item with a checkmark to remove it. The dialog closes immediately so it must be reopened to make further selections.

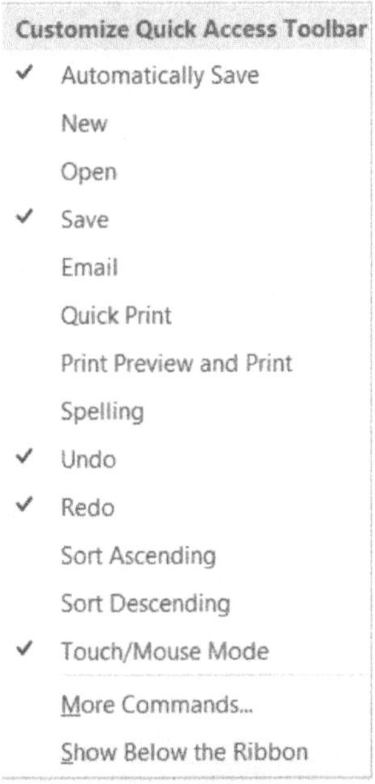

Customize Quick Access Toolbar

✓ Automatically Save

New

Open

✓ Save

Email

Quick Print

Print Preview and Print

Spelling

✓ Undo

✓ Redo

Sort Ascending

Sort Descending

✓ Touch/Mouse Mode

M̲ore Commands...

S̲how Below the Ribbon

■ **FIG. 1.5**

(c) If the command you need is not shown, then click on *More Commands...* to bring up a second dialog—Fig. 1.6. To add a command to the QAT, select an item in the left panel and click the Add button. To remove a command, click it in the right panel and click the Remove button. Locate the Copy command and add it to the QAT. Close the dialog by clicking the OK button (or Cancel button to correct a mistake)

Note: The procedure earlier shows how to add any command to the QAT but there is a much simpler method for commands that are already on the Ribbon. On a PC just right-click (on a Mac hold the Ctrl key and click) the command icon and select *Add to Quick Access*

(d) There is little merit in having the Copy command on the QAT since there is a very convenient shortcut (Ctrl+C) for this purpose. On a PC right-click (on a Mac hold the Ctrl key and click) on the Copy command on the QAT (it looks like two sheets of paper) and use the Remove command in the popup menu.

(e) It is sometimes said, tongue in cheek, that there are always three ways of doing the same thing in Excel! To demonstrate that this is not too great an exaggeration, if you are using a PC, open the File tab, on the left side locate and click on Options. This opens a dialog box; click on Quick Access Toolbar in the left panel. This again brings us to the dialog shown in Fig. 1.6. Close the dialog by clicking the Cancel button. If you are using a Mac, click on *Excel* in the top menu. From the menu, click on *Preferences...* and in the *Authoring* section, click on *Ribbon &*

Note: If a Print command is needed on the QAT it is recommended that one uses Print Preview and Print rather than Quick Print. This lessens the risk of wasting paper at home or mistakenly printing confidential material in an office setting

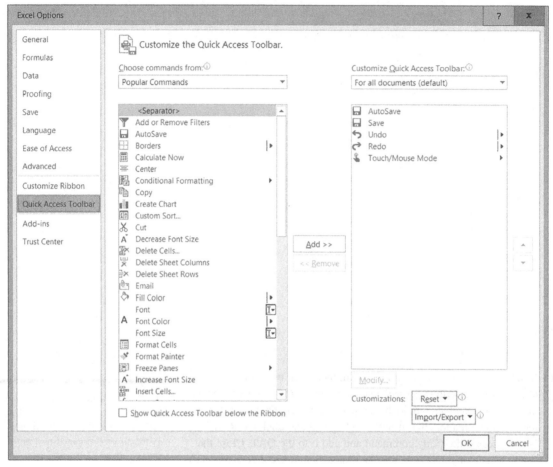

■ FIG. 1.6

Toolbar. This will bring up the Ribbon and Toolbar dialog. Click on the *Quick Access Toolbar* option similar to Fig. 1.6. Close the dialog by clicking the Cancel button.

As you become more familiar with Excel we will condense the second and third sentences in the above to the simple instruction: Use *File / Options / Quick Access Toolbar* (*Excel / Preferences… / Ribbon & Toolbar / Quick Access Toolbar* on a Mac).

EXERCISE 2: CUSTOMIZING THE RIBBON CONTROL

The Ribbon may be customized using the steps discussed in (c) earlier, but this is not a suitable topic for this book. Rather we shall change how the entire Ribbon is displayed, not what commands are displayed on it. (This cannot be done on a Mac)

(a) To the right on the Title Bar is the Ribbon Control tool ⌃. Click on it and experiment with the three options.

(b) In your own words state how the Ribbon behaves with (i) Auto Hide, (ii) Show Tabs, and (iii) Show Tabs and Commands. What are the advantages/disadvantages of each setting?

THE WORKSHEET

The worksheet window is the heart of the Excel application. It is here that we enter and work with data. It is helpful to learn some terms.

Columns and rows: A worksheet is divided vertically into columns and horizontally into rows. The intersection of a column and row forms a cell. At the top of the worksheet we have the column headers (the letters A, B, C, etc.) and to the left the row headers (the numbers 1,2,3, etc.). The last column is XFD (there are 16,384 columns); the last row is numbered 1,048,576; thus a single sheet has some 17 billion cells. Your computer would need to have a very large amount of memory if you planned to fill every cell.

Cell: A cell is the unit on the worksheet; it may be empty, or it may hold data. Generally, cells are outlined by gridlines. However, it is possible to request Excel not to display gridlines for a particular worksheet. Note that gridlines are not printed unless otherwise specified in *Page Layout / Sheet Options* (*Page Layout* on a Mac).

Active cell: If you click on a single cell on the worksheet, it is displayed with a solid border. We call this the *active cell*. The reference (such as A1) of the active cell is displayed in the Name Box. The correct term for the combination of column letter and row number (as in A1) is *reference,* but *address* is acceptable. What is not acceptable is *name* since this has a very special meaning in Excel. It is possible to configure Excel to use another reference system in which the top left cell is referred to as R1C1 but we shall not be concerned with that method. As noted before, the Name Box displays the reference of the active cell.

Range: A range is a group of contiguous cells. The shaded areas (see Fig. 1.7) B2:B109, D2:G2, D5:F9, and H8 are examples of ranges. Technically a single cell is also a range—it is a range consisting of just one cell. We refer to a range using the addresses of the top left cell and the bottom right cell separated by a colon.

Data and Formulas: A cell may contain either data or a formula. Data and formulas are frequently entered by typing in the cell. You can complete (*commit*) your entry in a number of ways: pressing the Enter [Enter↵] key; pressing one of the arrow keys ([↓], [←],[→],[↑]) or the Tab key [Tab⇆]; or clicking the checkmark (✓) to the left of the formula bar. There is another

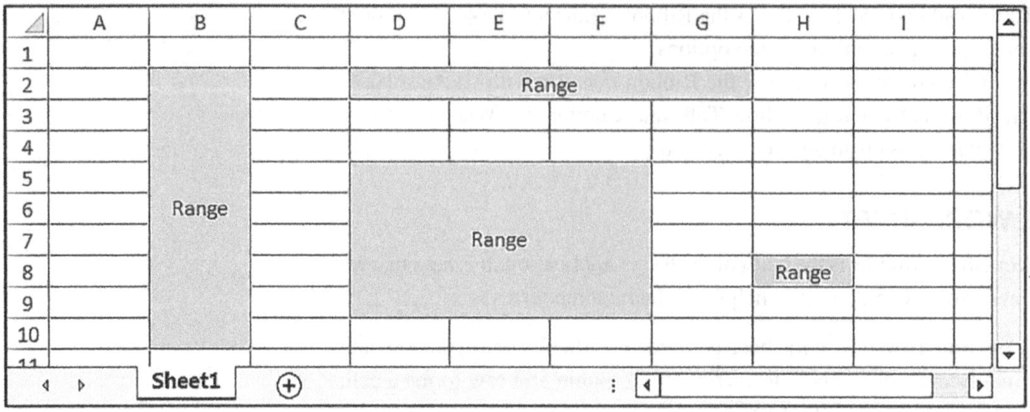

FIG. 1.7

method—clicking on another cell—but this is a very poor habit to pick up since the result when entering a formula is generally not what you want! The Enter⏎ key generally takes you down to the cell one below, but we can change this with an option setting to move one to the right. Data and formulas can also be placed in cells by copying (or cutting) them from other cells and then using the Paste command. The source cells can be in the same worksheet or in another worksheet, or even from another workbook.

Data: The data we enter into a cell can be one of four types. It could be text (such as the word Experiment), a number (123.45), a date (1/1/2013), or a Boolean constant (TRUE or FALSE). Later we shall see that a date is actually a number with a special format applied.

Formulas: A formula always begins with an equal sign (=) followed by a combination of constants and cell references (examples: =2*1.2345, =2*A2). It may also contain one or more functions (Examples: =SUM (A1:A10), =4*MAX(A1:A5)+2). A formula normally displays a value in the cell; this can be any one of the data types listed before. So the cell containing the formula may display a value such as 6.28318, but when it is the active cell the formula bar may display the formula =2*PI(). If the formula fails, it may display (we say it returns) an error value. We start to use formulas in Chapter 2.

Formatting: This is the term used to describe changing how the value in a cell is displayed. We may format a cell to alter the font (typeface, size, color) and to add a border or a fill color. By far the most important aspect of this topic relates to numbers. In a newly opened worksheet, every cell is formatted in what is called General. If I type 1.23456789 into a cell I may see 1.234568 because the combination of column width and font size allows just seven digits and the decimal. So my entry is rounded. The Formula Bar will display the actual number 1.23456789. We may widen the cell to display more digits. This rounding occurs only with real numbers (number having decimal parts) and not with integer numbers. If I type 1234567890, Excel will widen the cell, but when more digits are used, as in 123456789012, Excel displays it in scientific notation as $1.234567E+11$ (meaning 1.234567×10^{11}). Had the column been formatted to a narrow width beforehand, the result would show with fewer digits. We will see later that we may change the format of a number. What is important to remember is that changing the format does not alter the actual stored value. We examine this in a later exercise, but it is good to learn early that there are *stored* values and *displayed* values.

EXCEL 2016 SPECIFICATIONS AND LIMITS

As with the vast majority of Windows mathematical applications, Excel is limited to a 15 decimal precision. That means you cannot store a 16-digit credit card number as a number in Excel but since it is not actually a number (you never perform any arithmetic operations on a credit card number) the solution is to store it as text by typing a single quote (apostrophe) before the number. More importantly, Excel appears to get some math slightly wrong; we examine this topic when we look at the IEEE 756 convention in Chapter 2.

On a PC, if you type `specifications` in the Tell Me query box you are shown all sorts of details like what is the size of the largest number that can be stored? ($9.99999999999999E+307$), or what is the maximum width of a column? (255 characters). Likewise, you could find that the maximum number of worksheets in a workbook is limited by the size of the computer memory. The specifications for Excel 2016 are also available at https://support.office.com/en-us/article/excel-specifications-and-limits-1672b34d-7043-467e-8e27-269d656771c3.

COMPATIBILITY WITH OTHER VERSIONS

Some people are still using Office 2003 so we need to know how to communicate with them. Starting with Office 2007, Microsoft made a radical change to the format of Office files; even the extensions changed. So an Excel 2003 user cannot open workbook generated in one of the newer versions (e.g., an Excel file with the extension XLSX) unless the Microsoft Compatibility pack is installed on the computer. Alternatively, an Excel 2016 user may save a workbook in the older file format making a file with the extension XLS. However, neither of these methods will help if the original Excel 2016 file

Average: 48.8   Count: 10   Sum: 488

■ **FIG. 1.8**

made use of new features (e.g., advanced conditional formatting) or functions introduced since Excel 2003 (e.g., UNICODE). One can open an old Excel file (extension XLS) in Excel 2016. The title bar will include the phrase *Compatibility Mode*. The reader should also be aware that Excel 2010, Excel 2013, and Excel 2016 introduced new features and functions. So, opening an Excel file in a previous version can result in problems.

EXERCISE 3: THE STATUS BAR

Open an Excel workbook and on Sheet1 in A1:A10 type some numbers. Now using the mouse select A1:A10. The status bar should display to the right something like Fig. 1.8. If this is not shown, right-click the Status Bar to bring up its customization dialog and put check marks in the required boxes with the mouse.

Note also that one can have the status bar display Caps when c is on. The reader may wish to experiment with other options on the status bar.

PROBLEMS

If you like puzzle solving, try these problems. We will be covering the topics in subsequent chapters, but you may enjoy the challenge.

1. Type your name in any cell. Make it bold and italic. Can you find how to remove bold or italic?
2. In cell D1 enter this =TODAY( ) and press the [Enter←] key. It should show the current date. Maybe it displays something like 15/3/2013 (or 3/15/2013 if your Windows Regional setting specifies the American date format); can you change it to March 15, 2013? Hint: look in the *Home / Number* group
3. Copy the cell with your name. Paste it in another cell. Copy the cell with the date. Note the "ant track" running around the cell you copied. If you double-click an empty cell, the track disappears and you can no longer paste. You have been using the Windows clipboard. Now click the *Clipboard* launcher on the Home tab (far left). This opens the Office Clipboard, which can hold more than one item. Experiment with it.
4. In A5 type the formula =22/7 and press [Enter←]. This gives an approximate value for π. Can you discover how to make this display with eight decimal places?

5. Type some numbers in the cells D1 to D5—later we will give this instruction as "put numbers in D1:D5." Click D6—or, in technical terms, make D6 the *active cell*. Look for the Σ icon (it is in *Home / Editing (Home* on a Mac)). Click it to see what happens.

APPENDIX: SUPPLEMENTARY MATERIAL

Supplementary material related to this chapter can be found on the accompanying CD or online at https://doi.org/10.1016/B978-0-12-818249-9.00001-7.

Chapter **2**

Basic Operations

CHAPTER OUTLINE

This book is about problem-solving so we shall spend little time on the preparation of presentation-worthy worksheets. We will give some information on how to make a worksheet more readable, but the emphasis is on mathematical operations. The topics in this chapter include:

- Entering numbers, including fractions and percentages
- Simple formulas such as =A1+B1+C1
- Range finders (colored borders showing what cells are used in a formula)
- Arithmetic operators +, −, *, /, and ^
- The Evaluate Formula tool
- Error values such as #DIV/0! and #VALUE!
- Copying with commands and shortcuts
- Formatting numbers
- The difference between stored and displayed values
- Round-off errors resulting from the IEEE 754 standard.

Liengme's Guide to Excel 2016 for Scientists and Engineers. https://doi.org/10.1016/B978-0-12-818249-9.00002-9

	A	B	C	D	E
1	a	b	c	sum	product
2	1	3	4	8	12
3	4	5	6	15	120
4	5	7	9	21	315
5	6	8	3	17	144

■ FIG. 2.1

If you are familiar with an earlier version of Microsoft Excel, you may be tempted to skip this chapter. You are urged to at least read the Exercises to find out about Excel 2016 features.

EXERCISE 1: SIMPLE ARITHMETIC

Imagine that from time to time you are given some data consisting of rows of three numbers and you are asked to find the sum and product of each triple. Of course, this could be done with a simple calculator, but a spreadsheet offers three advantages: we can reuse our spreadsheet from day to day, we can see the values we have entered, and we can make a neat printout of the results. Our completed spreadsheet will look like Fig. 2.1.

(a) In cells A1 to E1, enter the text shown in Fig. 2.1. In A2:C3, enter the numbers shown. You will note that as you enter the text it is left aligned in a cell while numbers are right aligned.

(b) Use the mouse to select A1:E1. On the Home tab, click the right alignment command in the Alignment group; it is the third command in the second row of this group.

(c) Unless we have used a spreadsheet before we might be tempted to type =1+3+4 in cell D2. Try this (remembering to press [Enter←] when finished) and it will give the correct answer, but this totally ignores the main idea behind a worksheet. We should not have to retype data. Wherever possible, formulas should refer to cell values. Click on D2 and tap the [Del] key to remove this formula.

(d) Now type =A2+B2+C2 and click the check mark to the left of the formula bar when the formula is complete. Notice that, as you type, the status bar displays ENTER, but once the checkmark tool ✓ to the right of the Formula Bar (or the [Enter←] key) is used to commit the formula, it shows READY.

(e) Next, we see another way to build a formula. In D3 type an equal sign (=), but rather than typing A3, click the A3 cell. Now type + and continue building the formula with this pointing method. The status

◢	A	B	C	D	E
1	a	b	c	sum	product
2		1	2	4	=A2+B2+C2

■ FIG. 2.2

will alternate between POINT and ENTER. Note how the cells take on a colored border that matches the colors of the cell references in the formula—Fig. 2.2. Again click the checkmark to commit the formula once it is finished. In this case, pointing has little advantage over typing, but in other cases, it has some advantages. Pointing helps to ensure we reference the correct cell in a complex worksheet, and it is very useful to reference a cell on another sheet that could be in another workbook.

Of course, we do not have to rebuild the formula for every cell. We can copy from one cell to another. Here we look at two ways of doing this, and a little later we will see a third (and the fastest) method.

(f) The first method uses the Copy and Paste commands located on the Clipboard group of the Home tab (far left)—see Fig. 2.3. With D3 as the active cell, click the Copy command. Select D4:D5 and use the Paste command (this is the larger icon on the group). Note how the cell we copied (D3) has a mobile dotted border. While this "ant track" is visible, the contents of the cell are still on the Clipboard and may be copied to any other range or cell. The ant track disappears as soon as you start to edit any cell but can also be removed by pressing Esc.

On a PC, if you allow the mouse pointer to hover over the commands in the Clipboard group, screen tips pop up to tell the purpose and the shortcut keystrokes for each command. So Copy is Ctrl+C and Paste is Ctrl+V.

(g) Delete D3:D5 and repeat the copy-and-paste action using the Ctrl+C and Ctrl+V shortcuts. If you accidentally delete D2 use Ctrl+Z to undo the action.

(h) Delete D3:D5 again in preparation for another way to copy D2 down to D5. Move to D2 and note that the active cell border has a small solid square in the lower right corner; this is the fill handle. Carefully move the mouse pointer until it is over the fill handle—the pointer changes from an open cross to a solid cross. Hold down the left mouse button and drag the solid cross down to D5. In step (k) we shall see yet another method of filling a range.

For the final stage in this Exercise, we look at another approach to building formulas. Rather than typing the formula in the cell, we will

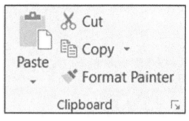

■ FIG. 2.3

Note: The Esc key may be used whenever you want to cancel what you are doing. Also if you start to edit a cell and wish to terminate the operation without making any changes you can use the **✗** tool located between the Name Box and the Formula Bar.

Note: It is customary to show shortcuts with capital letters like Ctrl+C but it is not necessary to hold the ⇧ Shift key to generate a capital letter when using a shortcut.

type it in the formula bar. There is an advantage to doing this when the formula is long, but we shall do it here for demonstration purposes.

(i) Make E2 the active cell. In the formula bar type =. Now complete the formula to be =A2*B2*C2 either by typing or by pointing. Commit the formula with either ⎡Enter↵⎤ or the checkmark on the formula bar. Note that the multiplication operator is an asterisk (*).

(j) Lastly, we will fill in cells E3: E5. With E2 as the active cell move the mouse pointer over the fill handle (watch for the change from open to solid cross) and double-click the fill handle. The formula from E2 is copied down to E5. This AutoFill feature can be used with vertical tables (data arranged in columns) but not with horizontal tables. It can be used to renew formulas when you make a change to the top cell.

(k) Double click on any cell in the range D2:E5. Note the status bar displays Edit. But more importantly, observe the colored borders around the cells in the corresponding cells in columns A, B, and C. Excel uses these range finders to pictorially show you which cells a formula refers to.

(l) In A6:C6 type some numbers. When you complete the last entry Excel will automatically add the formulas in D6 and E6. This is an example of Auto Extend. It requires that the table has at least four rows of entries above the current row. For more information on this topic, see http://support.microsoft.com/kb/231002.

(m) In the QAT, click the Save command (picture of a floppy disc) and save the file as Chap2.xlsx.

> Note: If we had numbers in A1: A100 to be summed we would be ill advised to start with =A1+A2.... Rather we would use the SUM() function to be discussed later. There is also the PRODUCT() function to find continued multiplications.

EXERCISE 2: THE MATHEMATICAL OPERATORS

The following table lists the arithmetic operators, their symbols, examples, and order of precedence.

You may not be accustomed to treating the % symbol as an operator; essentially, it means to divide the preceding number by 100. Exponentiation, of course, means raising a number to a certain power.

Excel evaluates formulas left to right using the order of precedence. So =60-20/5 will yield 56, not 8 since the division precedes the subtraction. We can override the order of precedence by the use of parentheses. Thus =(60-20)/5 gives 8 because everything within parentheses is done first.

Open the workbook (the Chapter2.xlsx file) made in Exercise 1 by opening the File tab on the Ribbon and locating Chap2.xlsx in the Recent Document

Operation	Symbol	Examples	Order of precedence
Negation	−	−1 and −A1	1
Percentage	%	5%	2
Exponentiation	^	2^3, A1^3, A1^B1	3
Multiplication	*	2*3, 2*A1, A1*B1	4
Division	/	2/3, A1/3, A1/B1	
Addition	+	2+3, A1+3, A1 + B1	5
Subtraction	−	3−2, A1 − B1	

list (Fig. 2.4). If it does not appear, click on Browse in the left panel to go to your Documents folder. On a Mac, if the file does not appear in the recent documents list, click on open to bring up the open file dialog in Fig. 2.5 and navigate to your file location.

(a) Click the Sheet2 tab (just above the status bar) to open a new worksheet or, if there is no Sheet2 tab, use the Insert Worksheet tool, which is to the right of the last tab in the sheet tab list.

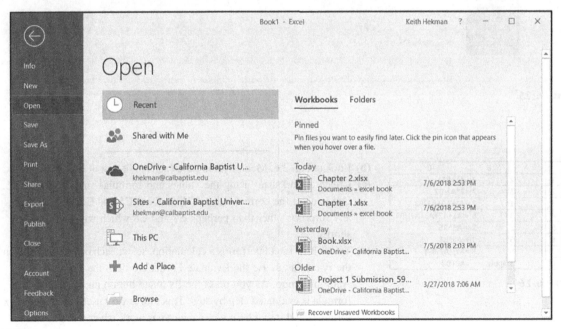

■ FIG. 2.4

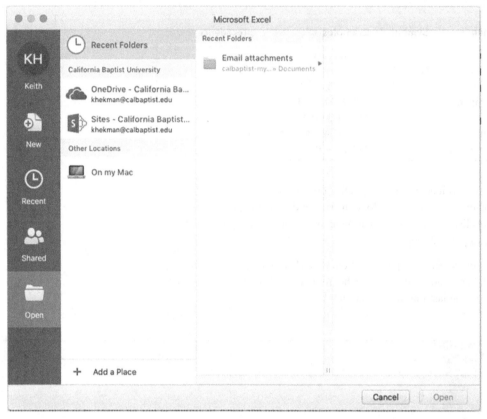

■ FIG. 2.5

	A	B	C
1	2	-3	=A1+B1
2	1	2	=2*A2-B2
3	-1	2	=A1+A2+A3/B3
4	3	5	=(A1+A2+A3+A4)/B4
5	5	2	=-A5^B5
6	-5	2	=-A6^B6
7	5	5	=A7/(A7-B7)
8	5	apple	=A8+B8

■ FIG. 2.6

(b) Look at Fig. 2.6. Mentally compute each result—or at least the first six.

(c) Create a worksheet using the values and formulas shown in Fig. 2.6. Did you get the correct values? How about C5 and C6? Were there any surprises other than perhaps C7 and C8 which will be discussed shortly?

(d) To see how Excel performs a calculation, select each cell in turn and on the Formula tab use the Evaluate Formula tool in the Formula Auditing group. As you press the Evaluate button on the dialog, the formula is evaluated step by step. This tool is very useful with complex formulas and it can help you debug your work when you get an unexpected result—see Exercise 7. You will note that $-5\,\char94\,2$ is evaluated as $(-5)\char94\,2$ and not $-(5)\,\char94\,2$ since negation precedes exponentiation.

(e) Save the workbook. You may wish to experiment with the shortcut [Ctrl]+S to do this.

We have not explained the results #DIV/0! in C7 and #VALUE! in C8. These are examples of *error values*. In the first evaluation step of the formula in C7, we get =5/0. Division by zero is said by mathematicians to be "undefined" so Excel tells us this is an error value. A green triangle in the upper-left corner of a cell indicates an error in the formula in the cell. The Trace Error button appears when you select the cell. Click the button's arrow for a list of options (refer to Fig. 2.7). Experiment with this for yourself. You will find that Ignore Error removes the green triangle but this reappears if you double-click the cell and press the [Enter↵] key without making changes to the formula.

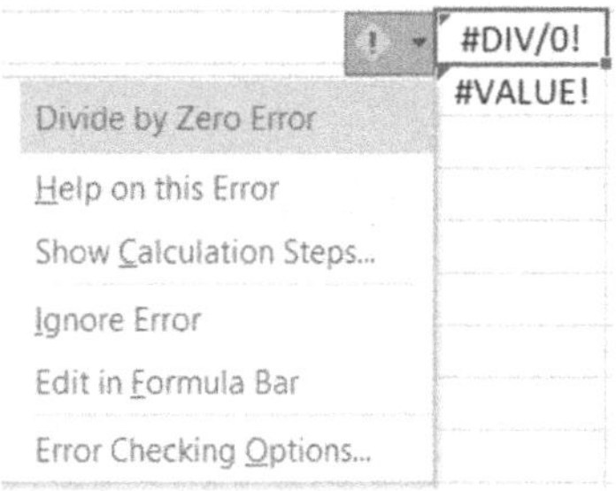

■ FIG. 2.7

The #VALUE! error in C8 occurs because we are using the wrong data type: B8 contains a text value that is incompatible with the addition operator. We will soon discover that the SUM function ignores nonnumeric data, so it can be used in circumstances when we want to add values in a range, ignoring any text that happens to be in the range. Excel has many shortcuts.

In addition to #DIV/0! and #VALUE!, other error values are #REF!, #NUM!, #NAME?, and #N/A. We will discuss them as we proceed but for now, note that each begins with a number (hash or pound) symbol. A worksheet displaying an error value has a mistake in it and needs attention except that #N/A is often acceptable and is taken to mean "not applicable" or "not available." Later we shall see how Conditional Formatting may be used to hide error values.

In Chapter 5 we meet some conditional functions that enable us to avoid error values such as #DIV/0! and #N/A in many circumstances.

EXERCISE 3: FORMATTING (DISPLAYED AND STORED VALUES)

This is a very simple Exercise, but it is most important for an Excel user to know the difference between a displayed and a stored value (Fig. 2.8).

(a) Open your Chap2.xlsx workbook and move to Sheet3 or insert Sheet3 if necessary. Type the text shown in the rows 1 and 3 of Fig. 2.8. Right align the entries in A3:C3 as we did in Exercise 1.

	A	B	C
1	Displayed and Stored Values		
2			
3	N	2+N	2*N
4	1.249	3.249	2.498
5	1.25	3.25	2.498

■ FIG. 2.8

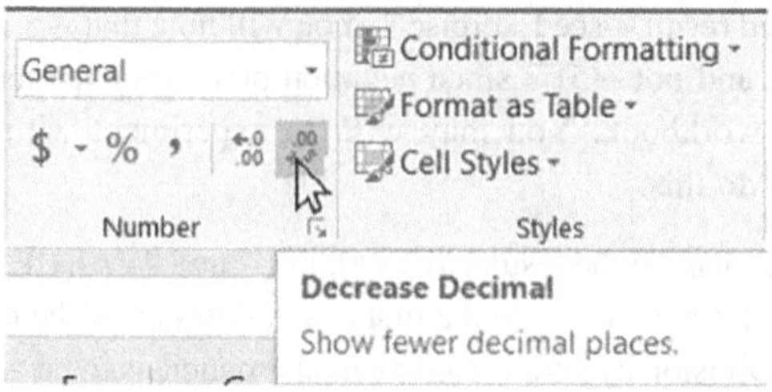

■ **FIG. 2.9**

(b) In A4 and A5 type the value 1.249. With A5 as the active cell, locate the Decrease Decimal tool in the Number group of the Home tab (refer to Fig. 2.9). Click this once to have A5 show 1.25. We say that 1.25 is the displayed value. Note that the Formula Bar still reports 1.249; this is the stored value.

(c) In B4 and C4 enter the formulas =2+A4 and =2*A4, respectively.

(d) In B5 and C5 enter the formulas =2+A5 and =2*A5, respectively. For the purpose of this exercise, please do not copy them from the row above but type them in. By typing in the formula in B5, excel took the number of displayed decimal places from A5.

(e) Save the workbook.

Row 4 has no surprises, but look again at row 5. We know that A5 has the value 1.249 and that it was formatted to display only two decimal places. The addition of 2 gives 3.249, while multiplication gives 2.498. C5 displays the expected value, but B5 has a rounded result. It is a feature of Excel that a cell with a simple formula (such as =A1 which has no arithmetic operator or =A1+2 and =B1-3 with just the addition/subtraction operators) inherits the format of the referenced cell. Had we copied the formula from B4 to B5, the cell would have displayed 3.249.

Note that if you move to A5 the Formula Bar displays the stored value of 1.249 rather than the formatted value of 1.25. If we had wanted the formula to treat 1.249 as 1.25, then we could have used the ROUND function as shown later. Alternatively, we can have Excel treat all numbers to have the precision of the displayed values. This can be helpful in some financial accounting work but can lead to some confusion in other cases, so we shall not pursue this feature.

EXERCISE 4: WORKING WITH FRACTIONS

Most of us work with decimal numbers, but there are still occasions when we would like to do some arithmetic with fractions. In this exercise, we shall learn how to enter a number like 2 ¼ and how to have a number such as 14.6667 displayed as 14 10/15.

	A	B	C	D	E	F
1	Working with Fractions					
2						
3	Adding numbers to give an answer to the nearest 1/2					
4	4 7/8	3/4	6		11 1/2	
5						
6	Displaying and adding numbers to give answer in fifteenths					
7	3 3/15	6 5/15	5 2/15		14 10/15	

■ **FIG. 2.10**

(a) Enter the text shown in A1, A3, and A6 of Fig. 2.10.

(b) Enter the numbers in A4:C4. The number in A4 is entered by typing the 4 followed by a space and then 7/8. Note that the formula bar displays 4.875. The ¾ is entered as 0 3/4 and Excel helpfully omits the zero. If you enter only 3/4, Excel will be over helpful and think you mean a date (3 Apr or 4 Mar of the current year depending on the date format in Regional setting). If C4 has previously held a fractional value before you enter the 6, then the value will be displayed with spaces following it.

(c) In E4 enter =A4+B4+C4 either by typing or using the pointing method mentioned earlier. The result will be displayed as 11 5/8. With E4 as the active cell, look at the Number group on the Home tab; it is displaying Fraction; Excel has formatted this cell to reflect the format of the cells being added.

(d) Since we were set the task of having the result in E4 displayed to the nearest ½, we need to format the cell. Use *Home / Number* launcher (*Format / Cells* on a Mac) to open up the Formatting dialog (Fig. 2.11) from where we may select *As Halves* ½ in the Fraction category and click the OK button.

(e) Enter the values in A7:C7 and copy the formula E4 to E7.

(f) The result in E7 is actually 14.66667, but it displays as 14 ½ because the copy action copied both the formula and the format of E5. If you repeat the instructions of step (d) you will find that the only fractions Excel offers are halves, quarters, sixteenths, tenths, and hundredths. Are we out of luck in wanting fifteenths? No; all we need to do is move to the Custom category in the Format Cells dialog and in the Type box replace #?/2 by #?/15. Note there is a space after the # symbol. Save the workbook.

EXERCISE 5: A PRACTICAL WORKSHEET

In this Exercise, we demonstrate a practical worksheet. An electrical engineer wishes to compute the effective resistance of four resistors in parallel; refer to Fig. 2.12 for a diagram of what is meant by this and for the equation

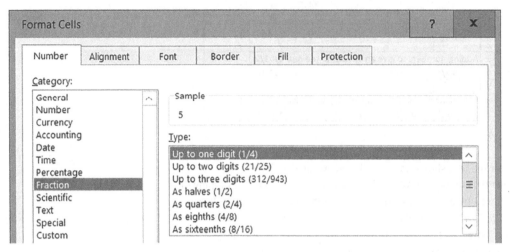

■ FIG. 2.11

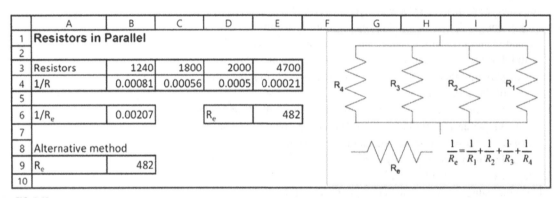

■ FIG. 2.12

used to compute the answer. You are not expected to make the diagram! Also, ignore the fact that gridlines are not seen and there are borders around some cells; we will find how to do this shortly.

(a) Open Chap2.xlsx and use the Insert Sheet command (last item on the sheet tab list) to create Sheet4.

(b) Enter the text shown in A1, A3, A6, D6, A8, and A9. Enter the values shown in B3:E3.

(c) In B4 enter the formula =1/B3 and copy it across to E4 by dragging the fill handle.

(d) In B6 we compute 1/R1+1/R2+1/R3+1/R4 using the formula =B4+C4+D4+E4. You may wish to compose this using the pointing method. We will see in Chapter 5 how the use of functions can make the worksheet more useful.

(e) In E6 we find the reciprocal of the sum of reciprocals with =1/B6 to give us R_e.

(f) In B9 enter the formula =1/(1/B3+1/C3+1/D3+1/E3) to demonstrate a shorter method.

(g) Use the Decrease Decimal tool on the *Home / Numbers* group to display E6 and B9 with no decimal places.

It is dangerous to rely on the results of any computer program (including an Excel worksheet), which has not been tested. Try your worksheet with some simple values such as four resistors of 2Ω or four of 100Ω. Does your worksheet agree with the results you computed in your head? This does not constitute a total validation of the worksheet but it gives us more confidence in its results.

Does the worksheet have any limitations? Clearly, it cannot be used for more than four resistors, but that is not a serious drawback from a practical point of view. How about fewer than four?

(h) Move to E3 and press the [Del] key. Oh dear, our worksheet displays a number of #DIV/0! error values. The blank value in E3 is treated as a zero value. Excel cannot compute the formula in E4 (=1/E3); this generates the first #DIV/0! error. The error is carried over to B6 which tries to use the result in E4, and then to E7 which tries to use the B6 value. B9 also attempts to compute 1/E3 and generates its own #DIV/0! error. Let's think about the physical meaning of removing a resistor. It does not mean inserting a resistor of 0Ω; that would be a short circuit. Rather, it means replacing R_4 by a very large resistance since air is a nonconductor. So we might solve our problem with a large number such as $1\,M\Omega$.

> Note: When a mathematical formula references a blank cell, the blank is treated as a zero.

(i) Enter values of 2 for the first three resistors and 1E6 (you may be familiar with this notation meaning 1×10^6 from your hand calculator) for the last one. Now compute the expected results in your head. Does the worksheet give a good answer? Of course, if the first three resistors have very big values, then our missing R will need to be very large, say 1E100. This is another problem we can solve more efficiently with functions as we will see in Chapter 5.

(j) Save your workbook.

COPYING FORMULAS: WHAT HAPPENS TO REFERENCES?

We have seen in the last Exercise that when the formula =1/B3 was copied from B4 to C4, the formula was adjusted to =1/C3. This is very useful, but there are times when we want something else. First, we need to understand how Excel goes about adjusting references when you copy a formula.

	A	B	C	D
1	Adjustment	1.10		
2				
3	Old table			
4	2	3	4	5
5	5	6	7	9
6				
7	New table			
8	2.2	3.3	4.4	5.5
9	5.5	6.6	7.7	9.9

■ **FIG. 2.13**

The formula in B4 was =1/B3; think of this as meaning =1 / (the cell that in the same column, one row above). This is what is meant by a *relative address*. The reference to B3 is interpreted relative to the cell that holds the formula. So when we copy this to C4 it is still =1 / (the cell that in the same column, one row above) and this is, of course, represented by =1/C3. Now let's look at a problem where this automatic adjustment does not work for us. You may wish to make a worksheet of your own to experiment with this.

In Fig. 2.13 we have a range called *Old table* in A4:D5, and we wish to generate a range called *New table* in which the corresponding values in the old one have been adjusted by 10%. In cell B1 we have 1.10. In A8 we type =A4*B1 to get a value that is 10% higher than A4's value. All is well; we get 2.2 when A4 was 2. Now we use the fill handle to drag this down to A9, but we get =A5*B2. That is not what we wanted. We did want A4 to become A5 but wanted B1 to remain as B1. Furthermore, when we copy A8 to B8 we want =B4*B1, but we will get =B4*C1.

We solve this by using an absolute reference for B1. In A8 we type =A4*$B$1. When this is copied to A9 we get =A5*$B$1, and when it is copied to B8 we get =B4*$B$1. You may think of a $ symbol (which has absolutely nothing to do with dollars, United States or otherwise) as an instruction to Excel to make no change to the column or to the row reference that follows it when the formula is copied. We say that a reference such as $B$1 is *absolute* since it is interpreted as the cell in column B, row 1 regardless of what cell it appears in.

Now, look at Fig. 2.14 where we have started a multiplication table for a young person to test her math skills. Row 2 and column G have constant values (that is to say, not formulas). What formula shall we use in H3 such that we may copy it both across the row and down the column? We start with = G3*H2. When this is copied to the right, we want the first term (G3) still to

	G	H	I	J	K	L
1	Mulipication table					
2		*1*	*2*	*3*	*4*	*5*
3	*1*	1	2	3	4	5
4	*2*	2	4	6	8	10
5	*3*	3	6	9	12	15
6	*4*	4	8	12	16	20
7	*5*	5	10	15	20	25

■ FIG. 2.14

point at G3, but when we go down a row we want it to be G4. So we see it is the G that is not to change. On the other hand, the second term (H2) must become I2 as we go across the row and H2 as we go down the column. So it is the 2 that should be immutable. This tells us to use =$G3*H$2. The term $G3 is interpreted as *the cell in column G on the same row as the cell with the formula*. Part of the formula is absolute, part is relative. We call this a mixed reference; both $G3 and H$2 are mixed references. Pressing F4 will toggle through the reference possibilities, that is, A4 to $A$4 to A$4 to $A4 and back to A4.

WHAT'S IN A NAME?

There is an alternative to using an absolute reference. We can give a name to a cell. So in the worksheet shown in Fig. 2.13, the user might have created a name for cell B1. Suppose the name *Adjustment* had been used (we will soon see how to do this), then the formula in A8 could have been =A4*Adjustment. Names are treated generally as absolute references; copying the formula from A8 to A9 would result in =A5*Adjustment. There are ways of making relative names, but we will not investigate that topic. You should also know that names are case insensitive; if the name was created as Adjustment, you can also use ADJUSTMENT or adjustment in a formula. Names may also be given to ranges. It should be obvious that we cannot assign any name that could be confused with a cell reference (such as X1); less obvious is that the names C and R are ruled out since these mean *current column* and *current row* for Excel. If you use a naming method that would result in an illegal name, Excel adds an underscore, as in X1_ and C_.

We will look at three ways of naming cells and ranges.

1. To name a single cell: Select the cell and type a word in the Name Box and press Enter↵ to complete it. So we could select B1 and type Adjustment in the Name Box and press Enter↵. For any given name, this method can be used only once in a workbook—see notes on *scope* later.

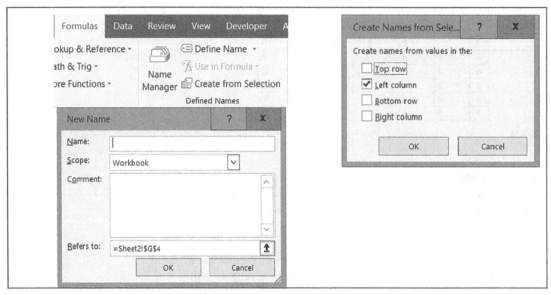

■ FIG. 2.15

2. To name one or more cells using a neighboring cell as the source of the name: Suppose the cells A1:A4 of a worksheet contain text such as *x*, *y*, *z*; and B1:B10 contains numbers. The user can select A1:B1 and issue the command *Formulas / Defined Names / Create Names from Selection* (Fig. 2.15, top left) (*Formulas / Create from Selection* on a Mac) causing a dialog (Fig. 2.15, lower left) to pop up. This permits us to specify which neighboring cell to use for the name. In this example, we would use *Left Column* to have B1 named as *x*, B2 as *y*, and B3 as *z*. With our current worksheet, we may select A1:B1, open the dialog and specify *Left Column*. One could use the shortcut Ctrl + ⇧ Shift + F3 (not available on a Mac) to open the *Create Names from Selection* dialog.

3. To name a cell, a range, or even a formula: The command *Formulas / Defined Names / Define Names* (*Formulas / Define Names* on a Mac in Fig. 2.16) causes a dialog box (Fig. 2.15, right) to pop up. This permits us to

■ FIG. 2.16

specify both the name and the cell(s) to which it refers. The Scope option is explained later. For the current situation, we could select B1 and type Adjustment in the *Name* box leaving =Sheet5!$B$1 in the *Refers to* box.

Note that an Excel name cannot contain a hyphen, but underscores are acceptable. Attempting to name a cell as, for example, mid-x using methods 1 or 3 will result in an error message. If method 2 is used the hyphen in the text gets replaced by an underscore.

Method 1 is quickest for a single name; method 2 is very useful to name a range of cells that have text next to them which can be used for naming purposes, and method 3 is indispensable to give a name to a constant rather than to a cell (see later). We shall see later how to get a list of the names contained by a workbook.

A name always refers to a *formula*. The formula may be: (i) a cell reference as in =Sheet1!$A$1, (ii) a range reference as in =Sheet1!A1:A10, (iii) a text or numeric literal as in =0.82058, or (iv) something more complex such as =SUM(Sheet1!$A$1:$A$8).

What is meant by scope? Suppose you start a new workbook and on Sheet1 you enter the value 10 in A5 which you then name as *alpha* using either method 1 or 2 before. If elsewhere on Sheet1 you use the formula =alpha, the result will be 5. The same formula on Sheet2 will also return the value 5. We say that alpha has a workbook scope: we can refer to Sheet1!$A$5 anywhere in the workbook using =alpha. This also means that if, with Sheet2 active, you type the name *alpha* and press [Enter ↵] you will not cause a name to be generated on Sheet2 but rather you will navigate to that cell on Sheet1 since that is one of the features of the name box. You could use one of the other methods to name a cell (say, B7) as *alpha* on Sheet2. That item will be given a worksheet scope. This means that on Sheet2, and only on that sheet, the formula =alpha will reference Sheet2!$B$7. It is not recommended to have this confusing situation; rather you should use method 3 and specify each *alpha* item to have worksheet scope.

The command *Formulas / Defined Names / Name Manager (Formulas / Define Name / Define Name* on a Mac) may be used to edit or delete a named item. It is, unfortunately, not possible to change a named item's scope. If you have a formula such as =2*A5 and later you name A5 as *alpha* then you can select the cell with the formula and use the command *Formulas / Defined Names / Define Name / Apply Names (Formulas / Define Name / Apply Names* on a Mac) to convert the formula to =2*alpha. Suppose you want to use a name while constructing a formula: at any stage in the formula creation operation you can use the command *Formulas / Defined Names / Use in Formula* (not available on a Mac) to display a list of names from which you can make a selection. The keyboard shortcut for this is the press [F3] while typing the formula. The command *Formulas / Defined Names / Use in Formula / Paste (Insert / Names / Paste* on a Mac) may be used to make a list of your named items.

EXERCISE 6: ANOTHER PRACTICAL EXAMPLE

In this Exercise, we construct a table showing the pressure of a gas at various temperatures and volumes using the van der Waals equation.

Fig. 2.17 shows the completed worksheet. As we proceed we will learn how to spread an entry over a number of cells (Merge Cells) as is the title row 1, fill a range without typing, get superscripts as in CO_2, and use a custom format to display 250 as 250 K.

Before we go too far, here are two reminders: (1) If you start something you do not wish to finish, such as typing in a cell that already has an entry that you would rather not destroy, then hit the Esc key; and (2) if you make a mistake such as deleting a complex formula and realize it in time, use Ctrl+Z to issue an Undo command.

(a) Create a new worksheet in Chap2.xlsx. Start the worksheet by entering the values shown in Fig. 2.18.
(b) We will now use some commands from the *Home / Font* and the *Home / Alignment* groups—Fig. 2.19. Select A1:H1 and click the *Merge & Center* tool to center the title over the first row. With the cells still selected, click the *Bold* tool to add emphasis. Now use the *Merger & Center* tool to get A6 centered over A6:H6.

	A	B	C	D	E	F	G	H
1			The van der Waals Equation of State					
2								
3			Gas	a	b			
4			CO_2	3.59	0.0427			
5								
6			Pressure (atmospheres) at varying T and V					
7								
8		250 K	260 K	270 K	280 K	290 K	300 K	310 K
9	0.05 L	1374.21	1486.61	1599.02	1711.43	1823.84	1936.25	2048.65
10	0.10 L	-0.98	13.34	27.66	41.98	56.30	70.62	84.94
11	0.15 L	31.63	39.28	46.93	54.58	62.22	69.87	77.52
12	0.20 L	40.67	45.88	51.10	56.32	61.53	66.75	71.97
13	0.25 L	41.52	45.48	49.44	53.40	57.35	61.31	65.27
14	0.30 L	39.84	43.03	46.22	49.41	52.60	55.79	58.98
15	0.35 L	37.45	40.12	42.79	45.46	48.13	50.80	53.47
16	0.40 L	34.98	37.27	39.57	41.87	44.16	46.46	48.76
17	0.45 L	32.64	34.65	36.67	38.68	40.70	42.71	44.73
18	0.50 L	30.50	32.29	34.09	35.88	37.68	39.47	41.27

■ FIG. 2.17

	A	B	C	D	E
1	van der Waals Equation of State				
2					
3			Gas	a	b
4			CO2	3.59	0.0427
5					
6	Pressure (atmospheres) at varying T and V				
7					
8		250	260		
9	0.05				
10	0.1				

■ FIG. 2.18

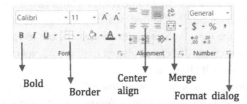

Bold   Border   Center align   Merge   Format dialog

■ FIG. 2.19

(c) Here is a quick way to get the numbers in row 8: Type 250 and 260 in B8 and C8; select the two cells; drag the fill handle to H8. Note the tooltip shows what number will appear in each cell as you move across. This feature is called Auto Fill. Use the same technique to fill A9:A18 with numbers from 0.05 to 0.5 in increments of 0.05.

(d) Select C3:E4, click the downward pointing triangle next to the *Border* tool and click on *All Borders*. With the range still selected, use the *Center Align* tool. Select B8:H18 and click the Border tool; note that the tool icon now shows four cells with borders around each after we used All Borders. Do the same with A9:A18.

(e) To remove gridlines open the *Page Layout* tab and on the *Sheet Options* group remove the checkmark from *Gridlines / View*—see Fig. 2.20. Gridlines may also be hidden from a region of a worksheet by selecting a range and using the Fill Color tool on the *Home / Font* group (its icon, which depicts a jug, is to the right of the Border tool); a fill with white will hide the gridlines in the region without adding color to it.

(f) The numbers in row 8 are integers and need no special numeric formatting, but we wish to emphasize them as headers and show that these are temperature values in Kelvin. Select B8:H8 and click the Italic tool next to the Bold tool. With the cells still selected, launch the Format dialog and select Custom from the Category list on the left-hand

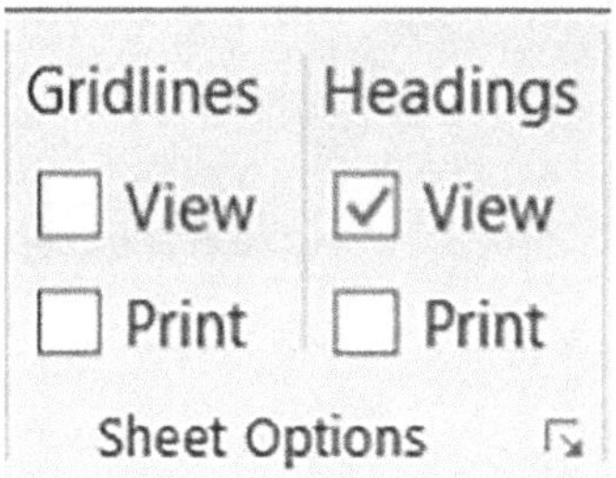

Gridlines	Headings
☐ View	☑ View
☐ Print	☐ Print
Sheet Options	⤡

■ FIG. 2.20

side. Edit the Type box to read General K or 0K with a single space before the K. In like manner italicize the numbers in A8:A18 and with the Increase Decimal tool get them to show two decimal places. Launch the Format Dialog (with the cells selected) and edit the Type box to read 0.00L with a single space before the L. Some may argue that a lowercase l is the symbol for liter, but that is too easily confused for the digit 1.

(g) We wish to have C02 display as CO_2 with a subscript. Select C4 and in the formula bar use the mouse to select the 2 digit. Unfortunately, the common Office shortcuts Ctrl + + and Ctrl + Shift + + do not work in Excel. Use the *Home / Font / Font* launcher (*Format / Cells* on a Mac) and add a checkmark to the subscript box in the Format Dialog—see Fig. 2.21.

The next stage is to define some names. Our formula has three constants: a, b, and R. The first two vary from gas to gas, so we want to be able the change them on the worksheet, but the gas constant R is not something that can be altered (unless we want to work in other units) so we will "hide" it.

(h) Excel allows us to define a name that refers to a constant (numeric or textual). Use *Formulas / Defines Names / Define Name* (*Formulas / Defines Names / Define Name* on a Mac). Complete the New Name

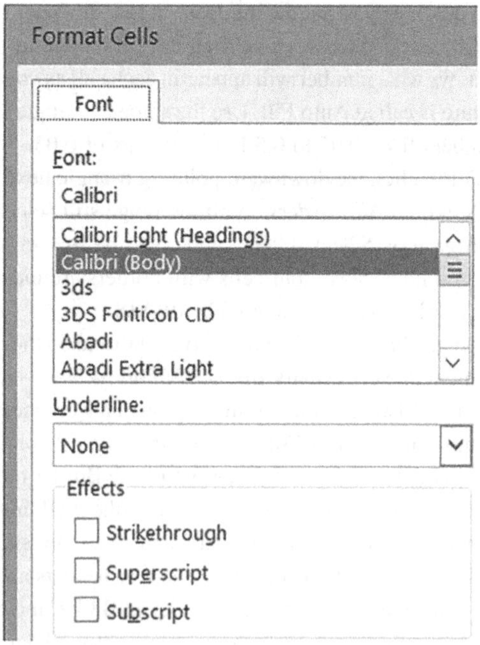

■ FIG. 2.21

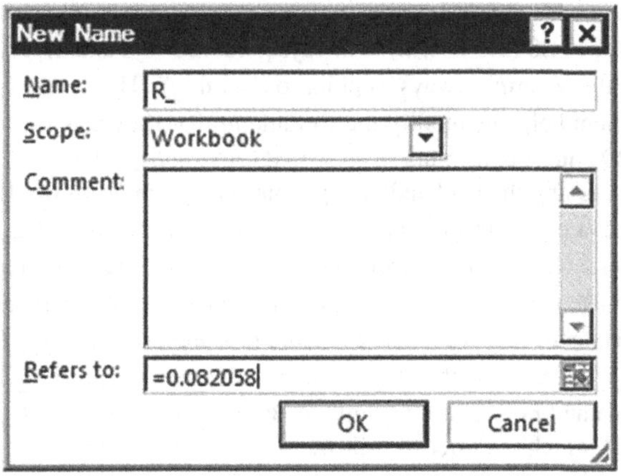

■ FIG. 2.22

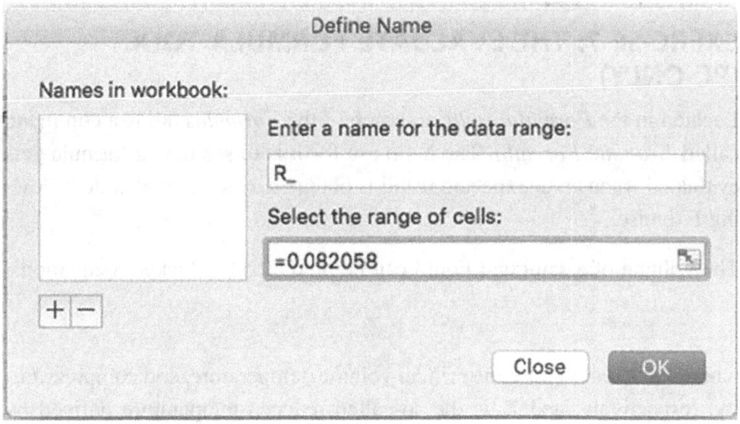

■ FIG. 2.23

dialog as shown in Fig. 2.22 (Fig. 2.23 on a Mac). Note we must use R_ with an underscore since Excel reserves the letters R and C for itself.

(i) To have D4 and E4 named as a and b, respectively: select E3:E4, use *Formulas / Defines Names / Create from Selection* (*Formulas / Create from Selection* on a Mac).

(j) Finally, we need to enter a formula into B9 using relative, mixed, and absolute references in such a way that the formula may be copied across the row and down the column. From the van der Waals equation our initial formula might be = (R_ * B8) / (A9 - b) - (a / (A9 * A9)). This has some redundant parentheses, but these were added to help read the formula. The named cells and the named constant will always be absolute references. We need the reference to B8 (the T term) to always point to row 8 and the reference to A9 (the V term) always to point to column A. This analysis of the problem helps us modify the formula to that shown as follows.

(k) In B9 enter the formula = (R_ * B$8) / ($A9 - b) - a / ($A9 * $A9). Using the fill handle, copy this across and down to fill B9:H18. Select a cell with that range and double-click it. Observing the range finders, check that the formula does point to T and V correctly.

(l) To make your formula more readable, you can also label ranges. Select B8:H8 and then type T in the name box and press [Enter←]. Similarly, select A9:A18 and then type V in the name box and press [Enter←]. With the named ranges, the formula in B9 is = (R_ * T) / (V-b) - a / (V^2) which is easier to follow.

(m) Save the workbook.

EXERCISE 7: THE EVALUATE FORMULA TOOL (PC ONLY)

Located in the *Formula Auditing* group of the *Formulas* tab is a command called *Evaluate Formula*. The main use for it is to see how a formula gets evaluated when an unexpected result is obtained as we demonstrate following Exercise.

The volume of a saturated liquid can be predicted by Rackett's equation:

$$V^{sat} = V_c Z_c^{(1-T_r)^{0.2957}}$$

where V_c, T_c, and Z_c are the critical volume, temperature, and compressibility, respectively, and T_r is the so-called reduced temperature defined by T/T_c. We will use this to compute the expected molar volume of water at 100°C.

(a) Start a new worksheet in the Chap2.xlsx workbook with the text entries shown in Fig. 2.24. Enter the numbers shown in B4:B6.

	A	B	C	D	E	F	G
1	Rackett Equation						
2							
3		Parameters for Water				Calculations	
4	Tc	647.3	K		T	373.15	K
5	Vc	56	cm³/mol		Tr	0.5765	
6	Zc	0.229			Vsat	46.85	cm³/mol

■ FIG. 2.24

56*0.229^(1-*0.576471496987486*)^0.2857

56*0.229^(*0.423528503012514*)^0.2857

56*<u>*0.229^0.423528503012514*</u>^0.2857

56*<u>*0.535638863813508*^0.2857</u>

■ FIG. 2.25

(b) The formulas in F4:F6 are =100+273.15 for the temperature required (this reminds us we must use Kelvin for temperatures), =F4/B4 to get T_r and =B5*B6^(1-F5)^0.2857 to find V^{sat}.

(c) In F6 we get a result of 46.85 but we know that 1 mol of water is approximately 18 g and water's density is more or less 1 at this temperature; so we expect a result close to 18 cm^3/mol. What went wrong and how can we correct it?

(d) Make F6 the active cell and click on the Formula Evaluate tool. Fig. 2.25 shows the results after four evaluations. We see that Excel is raising B6 to the power of (1-F5) and then raises that result to the power of 0.2857. That is not what we want. We require (1-F5) to be raised to the power of 0.2857 and that result be used for the power of B6.

(e) To control the order of mathematical operations we use parentheses. The reader should correct the formula in F6 in this manner and get the correct result of 17.67.

SPECIAL SYMBOLS, SUBSCRIPTS, AND SUPERSCRIPTS

In Exercise 6 we found how to convert CO2 to CO_2 with a subscript. It is obvious from Fig. 2.24 that a similar method works to get superscripts. Superscripted digits 1, 2, 3 may be generated in another way since many fonts contain characters corresponding [1], [2], and [3]. The default font for Excel 2016 is Calibri and this contains both sub- and superscripts for all the digits 0 through 9. There are no simple keys on the keyboard to get these, but in Microsoft Office, we have two ways to produce both these and other symbols such as Δ, Σ, $\pm$, and ½.

The command *Insert / Symbols / Symbol* (*Insert / Symbol* on a Mac) opens the Insert Symbol Dialog shown in Fig. 2.26. The variety of available characters is greatly enhanced if the *Character Code From* box (in the lower right corner) reads *Unicode*, rather than *ASCII*. It is unfortunate that subscripts 1, 2, and 3 are not in a range contiguous with the other superscript

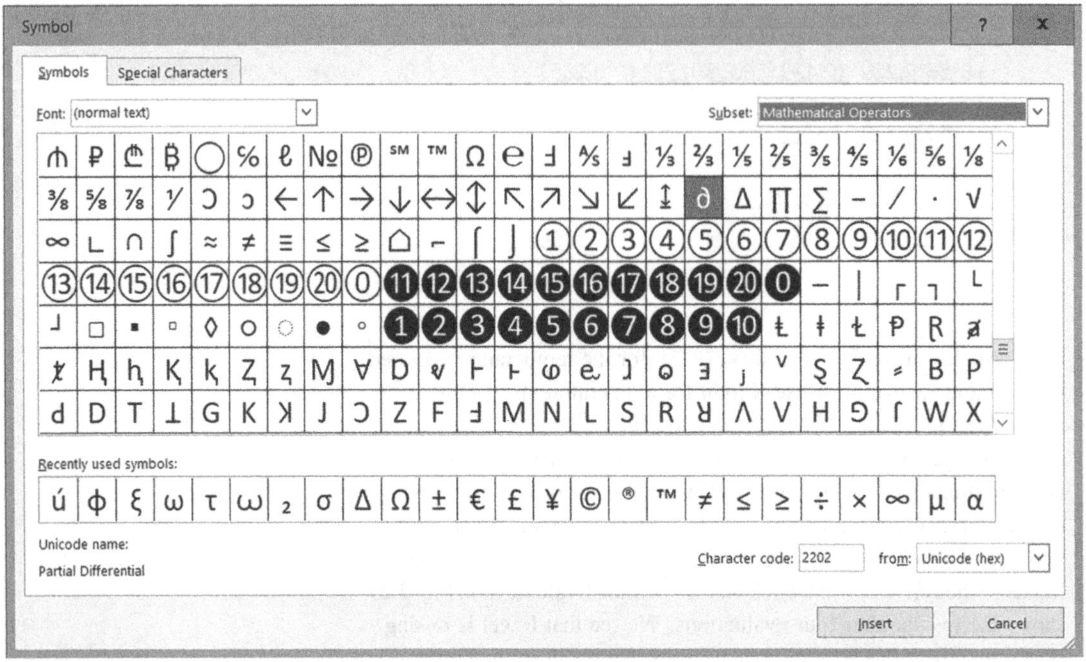

■ FIG. 2.26

digits. Just select a character and click the Insert button to place one of these characters in a cell.

On a PC, a second but more limited method to insert nonkeyboard characters consists of holding down the Alt key while typing on the numeric keyboard a number in the form *0nnn* where *nnn* is a three-digit code. The requirement that the numeric keypad is used makes this method inconvenient for some notebook users. The codes for some commonly used characters are shown in the following table. The symbols available this way and their codes are found from the Insert Symbol dialog with the *Character Code From* box reading *ASCII*.

> The Microsoft Office shortcuts such as Ctrl + +, for subscript, Ctrl + ⇧ Shift + + for superscript, and Ctrl + ⇧ Shift +Q for the Symbol font do not work in Excel; neither does the Alt +X method to change a Unicode to a character.

When our worksheet deals with statistics it can be useful to have symbols such as $\bar{x}$ or $\underline{x}$. In Excel these are made in two stages: (i) type the x (or other letter) and (ii) open the *Insert Symbol* dialog (click on *More Symbols*), in the Subset box locate *Combining Diacritical Marks*, and select with the combining overline symbol (the 13th one on the top row), or the combining macron symbol which is the 12th one on the top line: their Unicode values are 0305 and 0304, respectively.

MATHEMATICAL LIMITATIONS OF EXCEL

Like most computer programs, Excel uses the IEEE 754 standard for storing numbers. A number is converted from digital to a 64-bit binary representation. The fact that a finite number of binary digits are used has two major implications:

1. It limits the range of numbers that can be stored. Excel can store positive numbers from 1.79769313486232E308 to 2.2250738585072E-308. This limitation causes very few problems for users.
2. It is often stated that Excel has 15-digit precision. This is not quite true; what is true is that Excel truncates its stored or computed values to 15 significant digits. We need to understand the ramifications of the way Excel stores and displays numbers.

Integer values: The integer value 123456789012345 with 15 digits is stored and displayed (when the cell has the appropriate format) exactly as typed in. An integer with more than 15 digits (greater than 2^{53}) has trailing digits replaced by zero; it is not rounded. If we type 1234567890123456, Excel displays it as 1234567890123450; the trailing 6 becomes 0. There is no way to recover the lost precision. There is a simple solution when the "number" is actually just a string of digits as in a bank account number: precede the digits with a single quote (a.k.a., *apostrophe*), as in '1234567890123456. The single quote is not visible in the cell, nor will it appear in a printout but can be seen in the formula bar. Its purpose is to format the cell as text and a cell may contain up to 32,767 characters of text, but they may not all be displayed. Note that if 1234567890123450 is stored in A1 and '1234567890123456 is in A2, then =A1/2=A2/2 returns TRUE because when Excel does math on the text-formatted number, it must first lose the trailing 6 and put in a 0. However, we never actually do mathematical operations on bank account numbers or credit card numbers.

Real numbers: It often comes as a big surprise to many users that Excel (and most other computer applications) cannot represent some real (i.e., noninteger) digital numbers with total accuracy. But think about the fraction ⅓; we cannot display its value with complete accuracy in the decimal notation because it is 0.33333333… and the threes go on forever. So it should not be too surprising that Excel cannot store with total accuracy the simple real number 0.1. In binary format, this is 00011001100110011001100… Note that the 1100 goes on repeating forever just like $1/3 = 0.\overline{3}$. Excel uses 53 bits for the fractional part of a number. Using either Excel or a calculator find the

Nnn	137	149	150	176	177	178	179	181	185	186	188	189	190	215	247
Character	‰	·	–	°	±	2	3	µ	1	°	¼	½	¾	×	÷

value of the series $1*2^{-3} + 1*2^{-4} + 0*2^{-3}$ ... ($=1/16 + 1/32 + 0/64 + 0/128 + 1/256 + 1/512$ ....).

The formula $=(67.1 - 67.2)+1$ is computed not as 0.9 but 0.899999999999991 because the intermediate calculation is 0.1, which gets stored with some inaccuracy. We have used 0.1 as an example; other decimal values can cause the same problem.

Make a worksheet like that shown here with six numbers in A1:A6. In A7 enter a formula to find the sum of the six. Everything looks fine. Now make A7 the active cell and use *Home / Number / Increase Decimals* until you have 15 decimals (the maximum displayed precision of Excel) or change the format from General to Scientific. Oh dear! A round-off error gives a slightly incorrect answer.

	A
1	3.99
2	-25.00
3	6.71
4	6.59
5	6.54
6	1.17
7	0.00

TRUE	FALSE
=1.333-1.225=0.108	=1.333-1.225-0.108=0

When the conversion from digital to binary leads to a result that is not exactly correct, we speak of *round-off errors*. Round-off errors are a fact of life in the computer world.

Moral: Unless you are working with integers, never test to see if one number is exactly equal to some other value. Never use $=A20=B20$ to see if the values computed by the two methods give the same result. We can allow for round-off error by using formulas such as $=ROUND(ABS(A20-B20), 10)=0$ or $=ABS(A20-B20)<1E-10$ to see if the absolute difference in the two results expressed to 10 digits is 0 or not. The ROUND and ABS functions are explored in Chapter 4.

Excel sometimes tries to compensate for the IEEE round-off error but the results are not consistent. Compare the results of two similar formulas.

For more information on round-off errors, see any of the following sites or search the Internet with the term Excel IEEE.

Floating-point arithmetic may give inaccurate results in Excel http://support.microsoft.com/kb/78113/en-us.
Understanding Floating Point Precision, aka "Why does Excel Give Me Seemingly Wrong Answers?" https://www.microsoft.com/en-us/microsoft-365/blog/2008/04/10/understanding-floating-point-precision-aka-why-does-excel-give-me-seemingly-wrong-answers/
What Every Computer Scientist Should Know about Floating Point http://docs.sun.com/source/806-3568/ncg_goldberg.html.
Rounding Errors in Microsoft® Excel97 http://www.cpearson.com/excel/rounding.htm.

PLAY IT AGAIN, SAM

A very useful, but not well-known, Excel trick is the repeat shortcut. Here's how it works. Select a range of cells and add a border. Now select another range and press F4. The second range gets the border. This trick works with many formatting features and can be a time saver.

PROBLEMS

1. **(a)** I typed 22.90 but excel displayed 22.9. Name or describe the tool
that I will use to see the trailing zero.
(b) In cells A1:A2 I have typed 43.1, 43.2, and 1, respectively.
In A4, I used the formula =A1-A2+A3 and it displays 0.9 as expected
but if I display 15 digits I see 0.899999999999999. What do we call
this type of error, and why does it occur?
(c) I typed 1/1/2009 in a cell; find how to make the date display as
1-Jan-09. Experiment with Custom format to get 1-Jan-2009.
(d) I typed 2/12 expecting to get a fraction, but Excel displayed 12-Feb
(it might have shown 2-Dec had I been in Europe). What did I forget
to do?
(e) I wish to have column headers that read °F and ft^3. How do I get this
without formatting some characters as superscripts?

2. Referring to Fig. 2.27, make a worksheet to compute the values
in D2:D5 using only cell references (no named cells).

	A	B	C	D	E
1	a	b	c		
2	5			2.236068	D2: $\sqrt{a}$ D3: $\dfrac{c}{a^2-b^2}$
3	5	3	8	0.5	
4	729			9	D4: $\sqrt[3]{a}$ D5: $\dfrac{a-b}{b+c}$
5	16	12	4	0.25	

■ FIG. 2.27

3. The first table in Fig. 2.28 shows the coordinates and masses of three
objects. Using the information in the displayed formulas find the
position of the center of mass of these objects. What formulas are in G3
and G4?

	A	B	C	D	E	F	G
1	Centre of Mass						
2		x	y	mass		centre of mass	
3	Object 1	0.5	2.5	3		$\bar{x}$	$\bar{y}$
4	Object 2	3.5	2.5	3		2	1.7
5	Object 3	2	0.5	4			
6							
7		$\bar{x} = \dfrac{m_1 x_1 + m_2 x_2 + m_3 x_3}{m_1 + m_2 + m_2}$, $\bar{y} = \dfrac{m_1 y_1 + m_2 y_2 + m_3 y_3}{m_1 + m_2 + m_2}$					
8							

■ FIG. 2.28

4. Use a worksheet to answer these questions. (i) A basketball was found to have a volume of 440 in.^{3}. Does it conform to the NBA regulation that the circumference is to be between 29.5 and 29.75 in. for male adults? (ii) A golf ball must not exceed 1.680 in. in diameter nor have a weight over 1.620 oz. What is the maximum density (oz/in.^{3}) of a golf ball? (iii) A soccer ball has a circumference of 28 in.; what is the area of the material required to make one? (iv) In SI units, water has a density of 1 g/cm^3. Given that 1 in. = 2.54 cm (this is the definition of the inch) and 1 oz = 28.3495231 g, what is the density of water in lb/ft^3?

5. *A contractor needs a worksheet to compute the number of packages of shingles to purchase for roofing jobs. Fig. 2.29 shows a draft, but we need at least five rows for each roof shape. The diagrams are optional. Draw up a list of possible improvements to this worksheet. As you learn more Excel, you might wish to return to this worksheet and make improvements.

6. *Columns A and B of Fig. 2.30 show data collected by a team of students working on a solar car. The objective is to compute the average speed between each pair of data points. Hint: after typing the A5 value, use =A5 in A20, then format this as General to display the value 0.04167. How does this relate to the 1:00 in A5? Excel stores dates and times in day units! Column I gets the results in an alternative way not using the data in columns D and E. What formulas are used in D5, E5, G5, and I5?

7. The possibility of a typo increases as the formula to be entered gets more complex. Sometimes it is advisable to break the problem into steps and obtain answers in which you have confidence. Then you can attempt to code the problem in a simple formula. We will look at an example.

The shapes are made with *Insert / Illustrations / Shapes*. The formulas could be typed into cells and the shapes given a transparent fill. Alternatively, the formulas could be added with *Edit Text* (on a PC right-click (on a Mac hold the [Ctrl] key and click) and the shapes given a pale fill).

	A	B	C	D	E	F	G	H	I
1	Roofing Worksheet								
2		Rectangle	Length	Height		Area		Coverage/package	
3	$A = l \times h$	1	20	12		240		100	
4		2	32	12		384			
5		Trapezoid	Bottom	Top	Height	Area		Total Area	
6	$A = (l_1 + l_2)/2 \times h$	1	24	12	8	144		936	
7		2	24	12	8	144			
8		Triangle	Base	Height		Area		Packages Needed	
9	$A = \frac{1}{2}b \times h$	1	6	4		12		9.4	
10		2	6	4		12			

■ **FIG. 2.29**

	A	B	C	D	E	F	G	H	I
1	Average Speed								
2									
3	Clock time	Odometer Reading		Time interval	Distance		Average Speed		MPH
4	0	100.0							
5	1:00:00	158.7		1:00:00	58.7		58.70		58.70
6	2:04:23	218.4		1:04:23	59.7		55.64		55.64
7	2:56:24	267.5		0:52:01	49.1		56.64		56.64
8	3:45:23	315.8		0:48:59	48.3		59.16		59.16
9	4:12:00	340.3		0:26:37	24.5		55.23		55.23
10	5:34:03	422.4		1:22:03	82.1		60.04		60.04

■ FIG. 2.30

An engineer working with a gas-sparge system needs to compute P_m from the equation:

$$\log_{10}\left(\frac{P_m}{P_{mo}}\right) = 192\left(\frac{D_l}{D_T}\right)^{4.38}\left(\frac{D_l^2 N}{v}\right)^{0.115}\left(\frac{D_l N^2}{g}\right)^{1.96\left(\frac{D_l}{D_T}\right)}\left(\frac{Q}{N D_l^3}\right)$$

Make a worksheet similar to Fig. 2.31 computing the four variable terms on the right-hand side of the equation separately in E2:E5. They are combined with the 192 constant in E6 and P_m is computed in E7. You should check each term with a hand calculator. Then in B9 enter a single formula to compute P_m. In a real-life situation, once an agreement has been reached between B9 and E7, the range D2:E7 could be deleted.

8. *When a person stands in the wind the air feels colder than when the air is still. The apparent temperature is called the wind chill (T_{wc}) and is computed from the formula shown in Fig. 2.32 where T_a is the air temperature. The values shown for parameters a through d are for degrees Celsius computations. T_a is the air temperature and V the wind

	A	B	C	D	E
1	Gas-Sparge System				
2	Pmo	794		(Dl/DT)^4.38	0.004768
3	Dl	0.36		(Dl²N/v)^0.115	4.415957
4	DT	1.22		(DlN²/g)^1.96(Dl/Dt)	0.486494
5	N	2.8		(Q/NDl³)	0.031844
6	v	8.93E-07		Right side	0.06263
7	g	9.81		Pm	917
8	Q	0.00416			
9	Computed Pm	917			

■ FIG. 2.31

	A	B	C	D	E	F	G	H	I
1	Wind Chill			$T_{wc} = a + bT_a + cV^{0.16} + dT_aV^{0.16}$					
2									
3	Parameters					Wind speed km/h			
4	a	13.12		Air Temp °C	10	20	30	40	60
5	b	0.6215		10	8.6	7.4	6.6	6.0	5.1
6	c	-11.37		0	-3.3	-5.2	-6.5	-7.4	-8.8
7	d	0.3965		-10	-15.3	-17.9	-19.5	-20.8	-22.6
8				-20	-27.2	-30.5	-32.6	-34.1	-36.5
9				-30	-39.2	-43.1	-45.6	-47.5	-50.3
10				-40	-51.1	-55.7	-58.7	-60.9	-64.2

■ FIG. 2.32

velocity. For Fahrenheit computations use 35.74, 0.6215, −35.75, and 0.4275 for the parameters and express V in mph. The cells B4:B7 have been named by the letters to the right. What formula can be used in E5 such that it can be filled down and across to make the table? Be careful with the name for the cell B6. Can you easily modify your worksheet for Fahrenheit work?

9. The Antoine equation is

$$\log_{10}(p^*) = A - \frac{B}{T+C}$$

where p^* is in mmHg and T is the temperature in degrees Celsius. On a worksheet (Fig. 2.33), the values of A, B, and C for benzene are stored in cells A3:C3, while E4 has a temperature value in Celsius. What Excel formula would you use in F3 to compute $\log_{10}(p^*)$? How would you modify this if E3:E20 had temperature values and you wished to copy the formula down to F20? What formula in G3 computes p^*? Think about the definition of Log_{10}. Test your formula given the fact that benzene boils at 80.1°C.

10. If a volume V_1 of water at temperature T_1 is mixed with another volume V_2 of water at temperature T_2, the resulting temperature T_f can be found using $V_1(T_f - T_1) - V_2(T_f - T_2) = 0$. Construct a worksheet similar to that in Fig. 2.34 to give T_f.

	A	B	C	D	E	F	G
1	Antoine Equation						
2	A	B	C		T (°C)	log10(p*)	p*
3	6.90565	1211.033	220.790		25	1.98	95.18

■ FIG. 2.33

	A	B	C	D	E	F
1	Final Temperature					
2	First volume		Second volume			T_f
3	V_1 (gals)	T_1 (°F)	V_2 (gals)	T_2 (°F)		
4	25	50	60	180		141.76

■ FIG. 2.34

11. The thin lens equation shown in Fig. 2.35 gives the relationship between u, the distance of the object from the lens; v, the distance of the image from the lens; and f, the focal length of the lens. In the form shown we use the Cartesian convention: the incident light shines left to right, distances to the left of the lens are considered negative, and convex lenses have positive focal lengths, while concave lenses have negative f values. An object is placed 12 cm from a convex lens with a focal length of 18 cm. As the light comes from the left, the object must be placed to the left. Show that the image is at distance −36 cm. Since the object is on the same side as the image, it is imaginary. Construct a worksheet similar to Fig. 2.35. Can you get the same result without using the intermediate formulas in F3, F4, and H3?

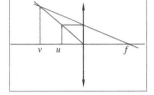

12. In the solvent extraction process, some solvent is mixed with an aqueous solution, shaken, and drained off taking some of the solute for later recovery. The process can be repeated with fresh aliquots of solvent to recover more solute. Let the volumes in each step be V_w and V_s for the water and the solvent, respectively; and let m_0 and m_1 be the mass of solute in the water before and after an extraction step. In the next step, the former m_1 becomes the new m_0. The distribution coefficient may be written as follows. This will be constant for a given solvent-solute pair.

$$K_d = \frac{m_1/V_w}{(m_0 - m_1)/V_s}$$

This can be solved for m_1 to get

$$m_1 = \frac{V_s K_d m_0}{K_d V_s + V_w}$$

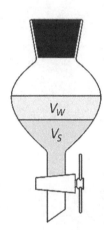

	A	B	C	D	E	F	G	H	I	J
1	Lens Equation									
2	Given data			Intermediate calculations			Final calculations		$\dfrac{1}{v}$	$-\dfrac{1}{u}=\dfrac{1}{f}$
3	Focal length of lens (f)	18		1/f	0.055556		1/v	-0.027778		
4	Object to lens (u)	-12		1/u	-0.083333		v	-36		

■ FIG. 2.35

	A	B	C	D
1	Kd	0.43		
2				
3	m_0	V_s	V_w	m_1
4	5	100	75	1.822034
5	1.822034			
6				
7				

■ FIG. 2.36

A 100 mL aqueous solution (V_s) containing 5 g of solute is extracted four times with 75 mL of solvent (V_w). The value of K_d for this solvent-solute pair is 0.43. Construct a worksheet similar to Fig. 2.36 to: (i) find the mass of solute remaining in the water after each of four consecutive exactions. (After the first extraction, $m_1 = 1.8$), and (ii) show that four extractions of 75 mL are more efficient than two at 150 mL.

13. Recursion may be used to solve certain mathematical problems. The Babylonian (also called Heron's) method to find square roots is summarized by the recursive equation

$$x_{i+1} = \frac{1}{2}\left(x_i + \frac{N}{x_i}\right)$$

or in words: (1) Make a guess, (2) Divide your original number by your guess, (3) Find the average of these two numbers, (4) Use this average as your next guess. Make a worksheet similar to that in Fig. 2.37 and test it for various values of N. Later we shall meet the SQRT function.

14. You wish to restrict the values that may be typed into cell B2 to integers in the range 10–100, inclusive. Experiment with *Data / Data Tools / Validation* (*Data / Validation...* on a Mac) to get this condition. Write a paragraph telling others how to do this.

	A	B	C	D	E
1	Babylonian Square Roots			aka Heron's method	
2	N	169		Sqrt is	13
3					
4	Guess	N/Guess	Average	Test	Error
5	10	16.9	13.45	180.9025	-11.9025
6	13.45	12.56506	13.00753	169.1958	-0.19578
7	13.00753	12.99248	13	169.0001	-5.7E-05
8	13	13	13	169	-4.7E-12
9	13	13	13	169	0

■ FIG. 2.37

15. You wish to select all the cells in a worksheet so that you can change the font. Use Help to find two ways to do this. Write a short paragraph telling others how to do this. Remember to explain what is meant by *current range*.

16. In cells A1:B100 you have some numbers. In C1 you have the formula =A1/B1, and this is copied down to C100. Some B values are zero giving #DIV0! errors. Experiment with *Home / Editing / Find (Edit / Find / Find* on a Mac) such that you are able to select these cells and delete them. Write a paragraph telling others how to do this.

17. (a) In A1 type the number 1. Select the cell and drag the fill handle down to A5. You get a series of 1's. Return to A1 and experiment with *Home / Edit / Fill (Edit / Fill / Series* on a Mac) to make the series 1..0.10. (b) In B1 type a date such as 1/1/2013. Select the cell and drag the fill handle down to B5. You get a series of dates: Jan 1 to Jan 5. Excel is being helpful, but this may be not what you want. Delete B2: B10. Hold (Ctrl) as you drag the fill handle of B1. Now the same date is repeated. (c) What happens if you have a simple number in C1 and you hold (Ctrl) as you drag its fill handle?

18. (a) Excel has many shortcuts. For example, type a number in D1 and move to D2. Now use (Ctrl)+(′) (the key next to the (Enter ↵) key) and the same number appears. Move to an empty cell and use (Ctrl)+(;) and you get today's date. Search the Internet with the term *Excel shortcuts* to learn more about shortcuts. Don't try to learn them all—just the ones you might need in the near future.
(b) If you have a date such as 1/1/2013 in F1 and use the copy-cell-above shortcut (Ctrl)+(′) in F2 what happens? You may get the same date, but you are more likely to get a five-digit number such as 41275. To find what that number means, search the Internet with *Excel dates –function*; using the *–function* term excludes results that talk about date functions.

19. *In 1929 Beattie and Blackman proposed a new equation of state. Compared to the van der Waals equation, it is somewhat better at predicting the behavior of gases.
One form of the B-B equation is

$$P = \frac{RT}{V} + \frac{\beta}{V^2} + \frac{\gamma}{V^3} + \frac{\delta}{V^4}$$

where

$$\beta = RTB_0 - A_0 - Rc/T^2; \gamma = -RTB_0 b + A_0 a - RcB_0/T^2; \delta = RTB_0 bc/T^2$$

and $A_0, B_0, a, b,$ and c are empirically determined values specific to each gas; these parameters are available in various publications. Compare the

Ideal Gas Law and the B-B equation estimations of the pressure of 0.2 L of methane at 200°C; given $A_0 = 2.2769$, $B_0 = 0.05587$, $a = 0.01855$, $b = -0.0187$, and $c = 1.283 \times 105$. Hint: using named (*beta, gamma, delta*) cells to hold the values of β, γ, and δ might help.

20. In mechanics of materials, the extreme axial stress values (principal stresses) can be calculated from the normal stresses in the x and y-direction (σ_x and σ_y) and the shear stress in the XY plane (τ_{xy}). The maximum shear stress is calculated using

$$\tau_{max} = \sqrt{\left(\frac{\sigma_x - \sigma_y}{2}\right)^2 + \tau_{xy}^2} = \left(\left(\frac{\sigma_x - \sigma_y}{2}\right)^2 + \tau_{xy}^2\right)^{0.5}$$

and the extreme axial stresses are

$$\sigma_{p1}, \sigma_{p2} = \left(\frac{\sigma_x + \sigma_y}{2}\right) \pm \tau_{max}.$$

Create a spreadsheet that will calculate σ_{p1}, σ_{p2}, and τ_{xy} from σ_x, σ_y, and τ_{xy}. The units are psi.

21. In statics and dynamics, frequently you need to calculate the cross product of two three-dimensional vectors $\vec{u} = u_x\mathbf{i} + u_y\mathbf{j} + u_z\mathbf{k}$ and $\vec{v} = v_x\mathbf{i} + v_y\mathbf{j} + v_z\mathbf{k}$. The principal directions $\mathbf{i}$, $\mathbf{j}$, and $\mathbf{k}$ point in the x, y, and z directions, respectively. The cross product is calculated using

$$\left(u_y v_z - u_z v_y\right)\mathbf{i} + \left(u_z v_x - u_x v_z\right)\mathbf{j} + \left(u_x v_y - u_y v_x\right)\mathbf{k}$$

Create a spreadsheet that will calculate the calculate the cross product from the components of $\vec{u}$ and $\vec{v}$. A possible spreadsheet is shown in Fig. 2.38.

	A	B	C	D	E	F	G
1	u	3	i	0	j	0	k
2	v	0	i	2	j	0	k
3	cross product	0	i	0	j	6	k

■ **FIG. 2.38**

APPENDIX: SUPPLEMENTARY MATERIAL

Supplementary material related to this chapter can be found on the accompanying CD or online at https://doi.org/10.1016/B978-0-12-818249-9.00002-9.

Chapter **3**

Printing in Excel

Even in this so-called paperless office world we still need to print our worksheets from time to time. Although we have hardly started on our study of Excel 2016, this is as good a place as any to discuss the printing process. Topics we shall learn about this chapter include:

- The Print dialog and its options
- The Print Preview feature that can save paper wastage
- The Print-Area and how to set it
- Setting the margins and orientation
- Setting options such as printing gridlines and row/column headers
- Getting Excel to print the same rows and or columns on every page
- Printing a selection
- Inserting page breaks
- Changing the paper orientation

Excel will not respond to any print command if there is no printer installed on the computer. That does not mean an actual printer must be attached but rather a printer driver must be installed on the PC or Mac. You can perform most of the exercises in this chapter using Microsoft Print to PDF driver if you do not have a printer. Furthermore, Excel will not print or show a print preview if the worksheet is empty. This is true even if you have added a header or footer.

The things we can do while working in Excel may be divided into two categories: operations performed on the workbook (e.g., formatting a cell to display 2 decimal places) and those done with the workbook (e.g., printing it, saving it, etc.). The last set of operations are collected together in the File tab and this area is sometimes referred to as the backstage.

Liengme's Guide to Excel 2016 for Scientists and Engineers. https://doi.org/10.1016/B978-0-12-818249-9.00003-0

EXERCISE 1: QUICK PRINT AND PRINT PREVIEW

In this exercise, we look at a way to quickly print a worksheet. Normally, this method is used only when we know that the printing parameters for the worksheet have previously been set, but since we will work with a small worksheet the paper wasted will be minimal.

(a) Open Sheet6 of Chap2.xlsx where we did the van der Waals calculations.

(b) Print the worksheet by clicking on the Quick Print command on the QAT (see Exercise 1 of Chapter 1 to see how to add it if it is not an option).

(c) Retrieve the printout from the printer. Note that on this worksheet the last cell that has an entry is H18. That means that by default Excel will print the range A1:H18 on as many sheets of paper (pages) as necessary. With the default font size of 11, our worksheet will occupy less than one page.

(d) If you have added the Print Preview command to the QAT, click on it. Otherwise, click on the File tab to open the "backstage" dialog and select Print from the menu on the left-hand panel. In the future, we will condense this to use the command *File / Print*. The keyboard shortcut for the print dialog is [Ctrl]+P, like most windows programs. Fig. 3.1 shows the backstage dialog with Print selected. (Fig. 3.2 shows the Mac Print dialog. Fig. 3.3 shows the full dialog after show details is pressed.) To the left we have the File menu, in the center is the Print dialog which we explore following, and to the right is the print preview. Compare your printed page with the screen—they should be identical. Return to the worksheet by clicking the arrow in the top left corner.

THE PRINT PREVIEW DIALOG

We now take a brief look at the controls on the Print Preview dialog.

1. In the top left corner we see a large button labeled *Print*; clearly, this is the command to send the print preview to the printer.

2. Next to that is the control for selecting how many copies are to be printed. You can use the scroll arrows or type a number in the box to set this value.

3. The Printer control allows us to select which printer we wish to use—assuming we have more than one printer installed.

4. Under this is text reading *Printer Properties*. Clicking on this will bring up a dialog specific to the selected printer. For example, depending on the printer, a user might be able to specify that a multiple page

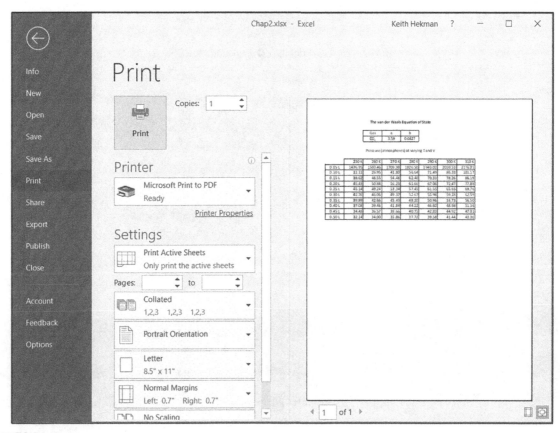

■ FIG. 3.1

output should be printed on both sides of the paper or that everything should be in grayscale rather than color.

5. The first control under the *Setting* heading allows the user to select from the following: print just the active sheets, print all sheets in the workbook, or print a range that has been selected prior to opening the Print Preview dialog.

6. If the print job has many potential pages, the next control lets the user select which pages to print.

7. The Collated control has meaning only when one is printing multiple copies of a multipage job.

8. Next, we have a control from which we may select either portrait or landscape orientation.

9. Then we have a paper size selection tool.

10. The penultimate control allows us to set the margins on the paper.

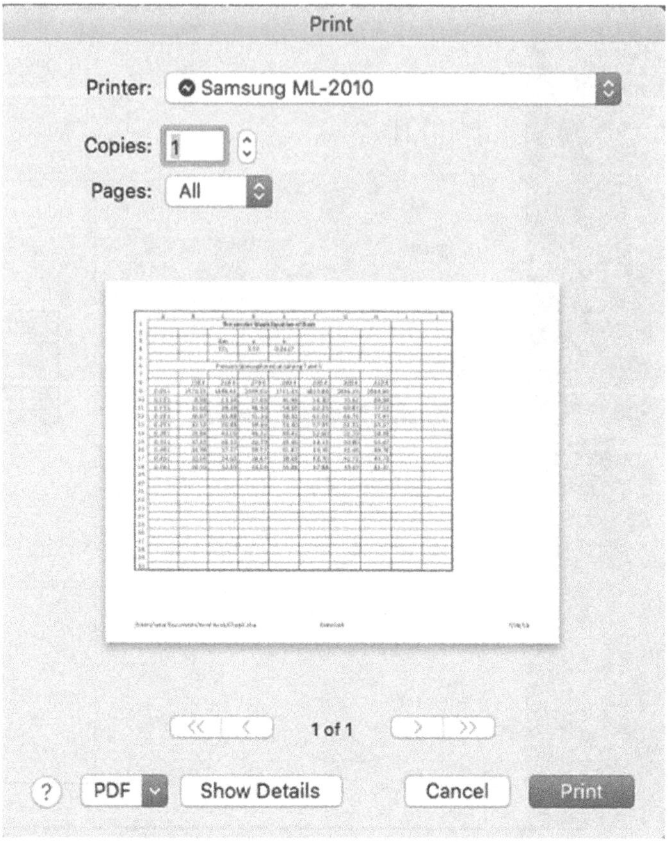

■ FIG. 3.2

11. The final control (scaling) is used to make the printout fit onto a specified number of pages. Suppose you have a print job that needs two pages but the second page would contain perhaps three rows from the worksheet. The *Fit Sheet* control lets us squeeze the print job onto one page. It does this by adjusting the font size on the paper but leaves the worksheet unaltered.

Some of the options (including 8, 9, and 10) on the Print Preview dialog are also available on the Page Layout tab which we look at in the next exercise. On the Print Preview dialog, there is a control (not visible in Fig. 3.1) which gives the user access to the Page Setup dialog.

If your print job takes more than one page then after you have used Print Preview, Excel adds dotted lines to your worksheet showing the page

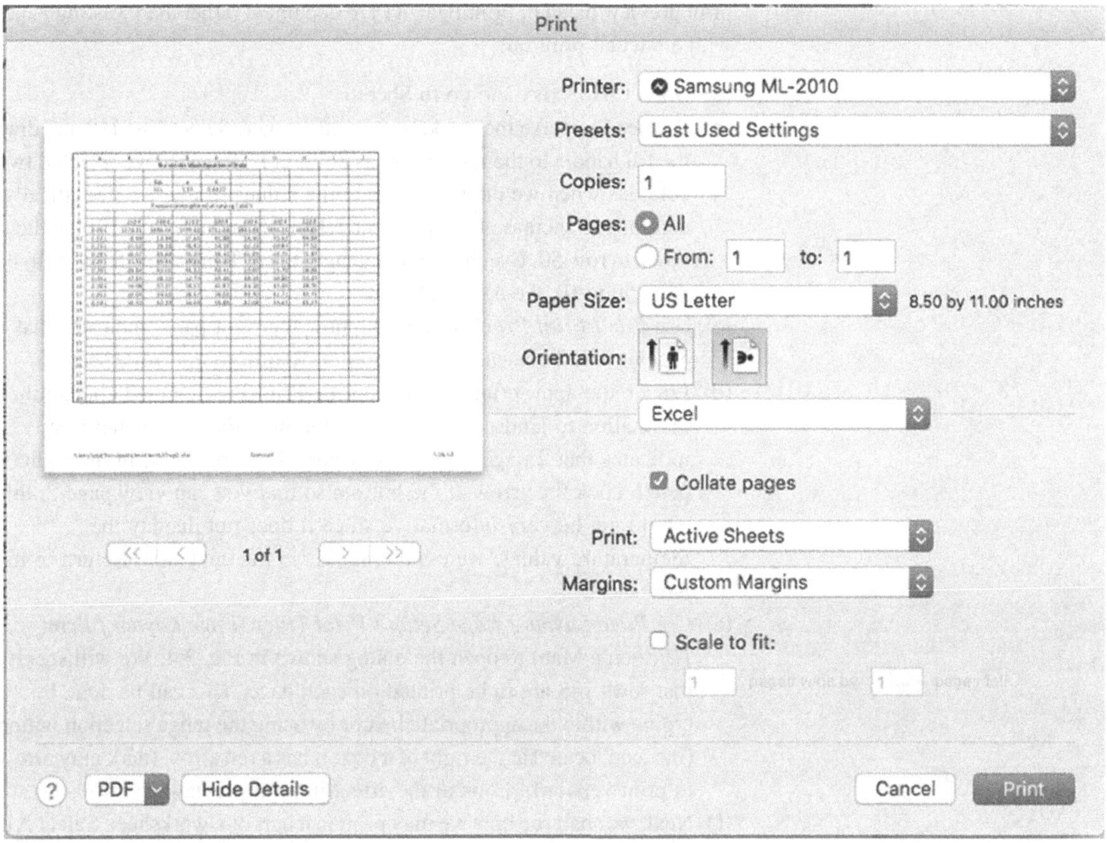

■ FIG. 3.3

breaks. The exact position is printer dependent. On the author's worksheet, the first vertical dotted line was between columns J and K and the first horizontal one, between rows 46 and 47. These automatic page breaks, unlike manually entered page breaks, cannot be removed. If they clutter your worksheet, just close the workbook and reopen it.

EXERCISE 2: THE PAGE LAYOUT TAB

In this exercise, we review some of the commands available on the Page Layout tab. Some are duplicates of those on the Print preview dialog while others are available only on this tab. Please be "green" and just observe the print preview panel or printing to a PDF document rather than wasting paper with an actual print out.

(a) Open Chap2.xlsx and go to Sheet6.

(b) We need to make the worksheet a little larger. Select G8:H18 and drag the fill handle to the right as far as column K. Because we selected two columns when we drag to the right the values in row 8 automatically continue to increase by 10. Select A17:K18 and drag the fill handle down to row 50. Use the name manager to set T to be row 8 and V to be A9:A50 to fix the new formulae.

(c) Use *File / Print Preview* and note how the right panel indicates that 4 pages would be needed for a printout. Return to the worksheet.

(d) Locate the appropriate control on the Print Preview and change the orientation to landscape printing. Note how the right panel now indicates that 2 pages would be needed for a printout. In the right panel, click the arrow at the bottom so that you can view page 2; this would not be very informative since it does not display the temperature values; we overcome this in the next step. Return to the worksheet.

(e) Use *Page Layout / Page Setup / Print Titles (Page Layout / Print Titles* on a Mac) to open the dialog shown in Fig. 3.4. We will specify that rows 1–8 are to be printed on each page. This can be done by typing within the appropriate box or by using the range selection button (the icon located at the right of a box; it has a red arrow (dark gray arrow in print version)). Look at the effect in print preview.

(f) Next, we shall see how we may print just part of a worksheet. Select A1: J40 and use the command *Page Layout / Page Setup / Print Area (Page Layout / Print Area* on a Mac) to set the print area to that selection. Observe the effect in Print Preview with Orientation portrait. Page 2 of the printout has only a few rows of data (in addition to the titles in rows 1–8). Open the Scaling control in *Print Preview* and specify *Fit Sheet on One Page* or *Fit all Rows on One Page* and observe the result. The group *Scale to Fit* on the *Page Layout* tab given even more control. For example, if we had a large worksheet we could specify, for example, that we want it printed 3 pages wide and 5 pages long. Since this scaling is performed by altering the font size one must be careful to maintain legibility! In addition, the *Scale to Fit* group has a control to alter the scale. One possible use would be to use a scale greater than 100% so that your printer work is enlarged. You may wish to experiment with this.

(g) For the next experiment, we need the worksheet reset to print all the rows down to 50. Use one of these methods: (i) with the command *Page Layout / Page Setup / Print Area (Page Layout / Print Area* on a Mac) specify *Clear Print Area* (remember that without a specified Print

Note that if you access the Page Setup dialog using the link on the Print preview dialog you will not be able to set the titles since the worksheet is hidden from view.

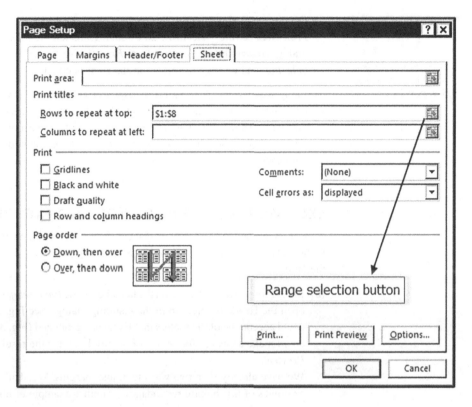

■ FIG. 3.4

Area everything on the worksheet gets printed), or (ii) select A1:K50 and set this as the print area using the method of step (f). What would happen if we set the print area as A9:K50? Since we have set rows 1 to 8 to be printed on every page, we would still get everything printed.

(h) Again, in the *Print Preview* dialog, observe how our worksheet, when in landscape mode, occupies the left-hand part of the paper. In the bottom right corner, there are two dialog controls we have yet to mention. The first of these is *Show Margins*. Experiment with this to expand the left margin such that the data is more or less centered on the paper. The second control magnifies the previewed data and is of limited use. Margins can also be adjusted using *Page Layout / Page Setup / Margins* (*Page Layout / Margins* on a Mac) while the worksheet is being viewed. There is a simpler way to center the material on the printed page. Return to the worksheet, open the Page Setup dialog using the launcher and click on the Margins tab. Not only can you adjust the margins but there are also controls to center both horizontally and vertically.

If you click on any cell (say C15) within the range A9:K50 and use Ctrl + Alt the range A9:K50 will be selected since it is the current range—it is a range surrounded by either a worksheet border or blank cells.

Warning: Do not leave the setting at Print Selection: it will cause confusion when next you go to print: you may see just one cell in the print preview panel.

(i) Finally, we see how to print just part of a worksheet without altering the Print Area setting. Select row 20 through 25. Open the *Print Preview* dialog; change the first control from *Print Active Sheets* to *Print Selection* and observe the effect on the right-hand panel.

You may wonder about the plural in print Active Sheets. It is possible to have more than one sheet active and when this has been accomplished one may wish to print all of the active sheets at once. Also, note there is a setting to print the whole workbook—something one should do only with care.

EXERCISE 3: HEADER & FOOTERS, AND PAGE BREAKS

Now we shall pretend that this worksheet (Sheet6) is a homework assignment[1]; it is to be printed and handed in. So we need some way of identifying the author.

(a) In the *Page Layout* tab click on the launcher of the *Page Setup* group and open the Header/Footer tab in the resulting dialog—see Fig. 3.5. We could accept one of the footers that Excel has generated (Fig. 3.5) or we could make a custom footer as in Fig. 3.6. Examine the result in *Print Preview*.

We have already seen two ways to adjust margins. We shall see other examples of this helpful redundancy—another example of how Excel often provides more than one way to perform a given task.

(b) Close the backstage view to return to the worksheet. To the right of the status bar, just before the zoom slider, there are three controls which alter how the worksheet is viewed: Normal, Print Preview, and Page Break. The same three controls are also available under *View / Workbook Views* (*View* on a Mac.) Click the middle control and the worksheet is displayed more or less as it will print—there is no indication of how many pages will print. You will see that this view also allows you to add customized headers and footers.

(c) Our worksheet has 42 rows of data (rows 9 through 50) and eight rows (1 through 8) that are used as titles on every page. Let us see how we can get the table printed with equal numbers of rows on each of the two pages when the orientation is landscape. Click the third view control on the status bar to display the worksheet in Page Break mode. Under row 33 (this could be elsewhere depending on the user's font size and margin

---

[1]Off topic: A spelling error in a worksheet that only you work with is one thing but a spelling error in a printed page is something to be avoided. The quick way to check spelling is to press F7 .

■ FIG. 3.5

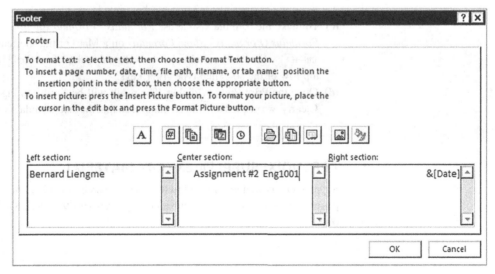

■ FIG. 3.6

	A	B	C	D	E
1	**Resistors in Parallel**				
2					
3	Resistors	1240	1800	2000	4700
4	1/R	=1/B3	=1/C3	=1/D3	=1/E3
5					
6	1/R_e	=B4+C4+D4+E4		R_e	=1/B6
7					
8	Alternative method				
9	R_e	=1/(1/B3+1/C3+1/D			

■ FIG. 3.7

settings) is a blue line indicating a page break. Drag the line until it is under row 29. Use the first view control to return the worksheet to normal view. Now in *File / Print* observe the result of changing the position of the page break.

EXERCISE 4: DOCUMENTATION AND PRINTING FORMULAS

Occasionally one needs to print a worksheet to serve as documentation. This may or may not involve displaying the cell formulas rather than their values.

(a) Open Chap2.xlsx and go to Sheet5 where we computed the equivalent resistance for three resistors in parallel.

(b) In the Sheet Options group of the Page Layout tab click the box for Headings Print. This will cause the column heading (A, B, C, etc.) and the row headings (1, 2, 3, etc.) to be printed.

(c) To make the formula visible use *Formulas / Formula Auditing / Show Formulas (Formulas / Show Formulas* on a Mac) or the shortcut Ctrl+`, where the second key is to be found to the left of the 1 on the top row of the so-called typewriter keys.

(d) Open the Print Preview dialog. The result will be similar to Fig. 3.7. Clearly, we needed to widen column B so that the last formula is clearer.

APPENDIX: SUPPLEMENTARY MATERIAL

Supplementary material related to this chapter can be found on the accompanying CD or online at https://doi.org/10.1016/B978-0-12-818249-9.00003-0.

Chapter 4

Using Functions

CHAPTER OUTLINE

Microsoft Excel 2016 provides over 460 worksheet functions; six of which are new to Excel 2016. Functions are divided into 13 categories: Compatibility, Cube, Database, Date and Time, Engineering, Financial, Information, Logical, Lookup and Reference, Math and Trigonometry, Statistical, Text, and Web. On a PC, to see a full list, open Help by clicking the question mark (?) at the right of the title bar, type functions in the search box and select *Excel functions (by category)* (*Help/Search* ... on a Mac). To learn more about a specific function, use the search box: type either the name of a function (such as SIN) or the text All, or select *Excel functions (alphabetically)*. In Chapter 8 we shall see that the user may construct user-defined (custom) functions.

Liengme's Guide to Excel 2016 for Scientists and Engineers. https://doi.org/10.1016/B978-0-12-818249-9.00004-2

In addition to using Help, you should also consider using the internet to get advice on Excel. For example, using the search term **Excel 2016 trig** in Google or Bing will locate many thousands of sites with helpful information. One of the best sites is Charley Kyd's http://www.exceluser.com/excel_help/functions/.

Functions are always used as part of a formula as in =SIN(A1) or =8+LOG (B1,2). When a cell contains a formula, the formula bar displays that formula, but the cell generally displays the value produced by the formula. We often use the phrase "the value returned by the formula" for this quantity. In the two examples at the start of this paragraph, the terms SIN and LOG are the names of the functions and the quantities A1, B1, and 2 are called arguments.

Arguments: Arguments are contained within parentheses, and in the English-language version of Excel they are separated by commas; semicolons are used in other language versions. The number and types of arguments (cell reference, range reference, text, number, Boolean term, etc.) depend on the syntax (the rules governing its use) of the function.

Depending on the function, the number of arguments may be fixed, variable, or even zero. For example:

zero arguments	=PI( )
one argument	=SQRT(A2)
two arguments	=ROUND(A2, 2)
variable number	=SUM(A1:A20) or =SUM(A1:A10,B3,B4)

The syntax for SUM is SUM(*number1*, [*number2*], ...). The square brackets around the second argument indicate that it is optional, while the ellipsis (three dots) tells us that we may add more arguments if needed. The number arguments can again be a cell reference, a literal, or an expression but can also be a range reference as in =SUM(A1:A100). There is nothing in the syntax to tell you this; one needs some basic knowledge of each function to use it correctly. While SQRT(range) would be senseless, SUM(range) is meaningful. At other times, Help is more detailed, and the text of the Help entry gives details on the arguments.

In some circumstances, one may wish to specify an entire row or column as an argument, as in =SUM(A:A), which will sum all of the numbers in column A. Of course, one would not want to put this formula itself in column A that gives a circular error.

Limit on arguments: When the number of permitted arguments is variable, the maximum number is 256. Note that a range such as A1:A100 counts as one argument, not 100.

Nesting: In the example =ROUND(SQRT(A2)/2, 2), the first argument contains another function: we refer to this as nesting. All versions after Excel 2007 permit nesting to 64 levels.

Error values: If a syntax rule is not followed, the formula will return an error value. If A10 holds a nonnumerical value, then =SQRT(A10) cannot give a valid result, so it returns an error value (in this case #VALUE!).

The error values are as follows:

#DIV/0!	Division by zero. This would be the result, for example, of =A1/B1 if B1 had a zero value. Note that a blank cell is treated as having a zero value when used in a numeric context like this
#NAME?	This results when a formula contains an undefined variable or function name
#N/A	No value is available
#NULL!	A result has no value
#NUM!	Numeric overflow; for example, a cell with =SQRT (Z1) when Z1 has a negative value
#REF!	Invalid cell reference. This can be caused by deleting a row or column that is referred to in a formula or copying a formula inappropriately. For example, trying to drag the formula =A1 in F2 to F1
#VALUE!	Invalid argument type. For example, a cell with =LN (Z1) when Z1 contains text would return this error

While not a true error value, we should also mention:

######	Column is set too narrow for the value/format used in a cell

When a cell having an error value is referenced in the formula of a second cell, that cell will also have an error value. A worksheet with an error value needs attention. The only exception is #N/A (note: it alone has no exclamation or question mark), which can be taken to mean not applicable or not available. This "error" even has its own function; enter =NA() in a cell and it will display #N/A.

An error you are sure to meet once or twice is the circular reference error. A formula cannot contain a reference to the cell address of its own location. For example, it would be meaningless to place in A10 the formula =SUM (A1:A10). If you try this, Excel displays an error dialog box. If you click OK and do not correct the error, the status bar will display CIRCULAR REFERENCE. There are some specialized cases when circular references are used purposefully.

EXERCISE 1: THE AutoSum TOOL

We are about to introduce the SUM function whose purpose is to compute the sum (i.e., add) of the values in a range of cells. Some novice Excel users treat it like the word *sum* as used in elementary school to mean an arithmetic operation and will use =SUM(A1, A2) when =A1+A2 will suffice.

Perhaps the most basic operation done with a spreadsheet is to add a column of numbers. For this reason, Excel has always had an AutoSum tool that can be used to very quickly construct a formula such as =SUM(A1:A6). We shall look at this tool and explore its other features (it can also generate Average, Count, Max, and Min formulas).

(a) We will be making the worksheet shown in Fig. 4.1. Open a new Excel workbook, and on Sheet1 enter the text shown in A1, C1:C6. Select C1:D1 and use the *Merge and Center* command on the *Home / Alignment* group (*Home* on a Mac). To enter the sequence of numbers, type 1 and 2 in A2 and A3, respectively, select the two cells and drag the *fill handle* (the small solid square in the lower right corner of the last selected cell) down to row 7.

The final step in completing a formula is called *committing*. We do this using one of the [Enter↵] key; the [Tab⇥] key; any of the navigation keys [↓], [←], [→], [↑] or the check mark ✓ in the formula bar. Avoid the temptation to click on another cell to perform the commit; it will work with literals but will ruin a formula!

(b) Make A8 the active cell and click once on the AutoSum tool found in the top left-hand corner of the *Home / Editing* (*Home* on a Mac) group; look for the Greek uppercase sigma Σ symbol.

(c) Cell A8 will now contain the formula =SUM(A2:A7) and there will be a mobile dotted line (the *ant track*) around the range A2:A7—see Fig. 4.2. Commit the formula by clicking the check mark on the Formula Bar. Note that had we double-clicked the AutoSum tool Excel would have inserted the formula and committed it in one operation.

To quickly sum a column (or row) of numbers: move to the cell below (to the right of) the last number and while holding down the [Alt] key press the [=] key. Use any method to commit the formula.

Note how the AutoSum tool was successful in finding the correct range of addends (numbers to be added). Frequently, the sum value is needed at the bottom of the column, but for this exercise, we shall move it.

	A	B	C	D
1	Numbers		Results	
2	1		Sum	21
3	2		Average	3.5
4	3		Count	6
5	4		Maximum	6
6	5		Minimum	1
7	6			

■ FIG. 4.1

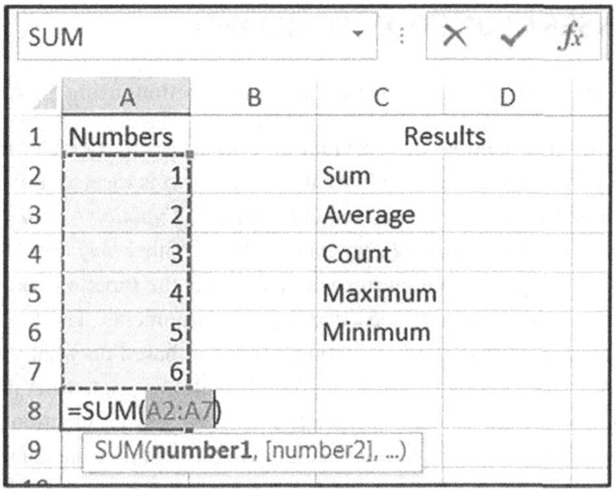

■ FIG. 4.2

(d) Click on A8 and use the shortcut [Enter ←]+X to cut the formula. Move to D2 and use [Ctrl]+V to paste the formula.

(e) Move to A8 and click the disclosure triangle (▼) on the AutoSum command. Select the Average option to get =AVERAGE(A2:A7). Commit the formula and move it to D3.

   The active cell need not be directly below (or to the right) of the values to use the AutoSum tool. We will get the Count formula into D4 in one step.

(f) With D4 as the active cell, open the AutoSum dialog and select Count Numbers. The ant track will be around D2:D3 as these are the closest numbers to the active cell. Use the mouse to select A2:A7 and note how the ant track moves and the formula changes to =COUNT (A2:A7). Now you can commit the formula. The COUNT function returns the number of cells in the range that have a numeric value. The function COUNTA enumerates the cells having nonblank values; it counts text, numbers, dates, and Boolean entries.

(g) Repeat step (f) to include the Maximum and Minimum formulas.

(h) Use the mouse to select A2:A7. Look at the status bar and you should see Sum: 21. If you right-click the status bar you will see how to add Count, Average, Max, and Min. This can be useful to quickly find the sum (or average, etc.) of a range without using a formula.

(i) Save the workbook as Chap4.xlsx. The formula for F10 will be entered in Exercise 3.

A little experiment: Replace the 1 in A2 with the word CAT. What happens to the formulas? They just ignore that value and work with the remaining numbers. But =A2 +A3+A4 will give a #VALUE! error.

THE INSERT FUNCTION COMMAND

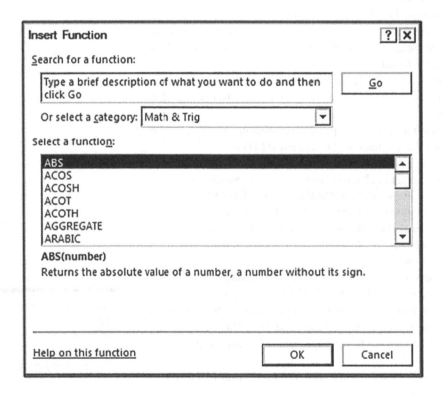

fx
Insert
Function
So far we have composed formulas with functions using the AutoSum too. The Insert Function tool provides us with access to a greater range of functions. We may access it either with the icon (*fx*) is located to the left of the Formula Bar or with the command *Formulas / Function Library / Insert Function* (*Formulas / Insert Function* on a Mac). Either way opens the dialog shown in Fig. 4.3. Within this you may select the function you wish to use, learn a little of what it does, and insert its arguments. The Search box provides some limited help in locating a function based on what you type, but it is not very intelligent. The Categories box allows you to filter the list of functions to just one category or to a list of recently used functions. In the next Exercise, we see how to use the dialog to create a formula containing a function.

Insert Function [?] [X]

<u>S</u>earch for a function:

| Type a brief description cf what you want to do and then click Go | <u>G</u>o |

Or select a <u>c</u>ategory: Math & Trig ▼

Select a functio<u>n</u>:

| ABS |
| ACOS |
| ACOSH |
| ACOT |
| ACOTH |
| AGGREGATE |
| ARABIC |

ABS(number)
Returns the absolute value of a number, a number without its sign.

<u>Help on this function</u>   OK   Cancel

■ FIG. 4.3

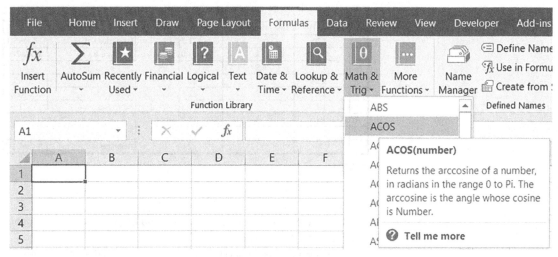

■ FIG. 4.4

When you become more familiar with the functions you can use one of the other tools in the *Formulas / Function Library* (*Formulas* on a Mac) group to locate a function based on its category—Fig. 4.4 (on a Mac, the description is not provided if you hover over a function).

EXERCISE 2: COMPUTING A WEIGHTED AVERAGE

For the purpose of this Exercise, let us imagine a student has measured the voltage of a battery many times and recorded his results in a table such as that shown in columns A and B of Fig. 4.5. His objective is to compute the weighted average of the results so he needs to calculate

$$\overline{V} = \frac{\sum V_i \times n_i}{\sum n_i}$$

In the course of this exercise and the next one (cells E10:F10) we shall compute this value in several ways.

(a) On Sheet2 of Chap4.xlsx, type in the text shown in A1:C1 and in column E; note that the entry in E10 is in preparation for Exercise 3. Use [Alt]+0215 to make the multiplication sign in C1. Type in the values shown in A2:B10. Use the *Center* command on the *Home / Alignment* group with A1:B10 selected. Use the Border command on the *Home / Fonts* group to add the borders as shown. Gridlines are removed using the tool found in the *Page Layout / Sheet Options* group.

	A	B	C	D	E	F
1	Voltage (V)	Observations (n)	V × n		Sum of Observations	35
2	1.2	1	1.2		Sum of V × n	55.1
3	1.3	3	3.9		Average V	1.574286
4	1.4	5	7.0			
5	1.5	7	10.5		Sumproduct	55.1
6	1.6	8	12.8		Average V	1.574286
7	1.7	5	8.5		Rounded	1.57
8	1.8	3	5.4			
9	1.9	2	3.8			
10	2	1	2.0		Average V	1.57

■ **FIG. 4.5**

(b) Enter the formula =A2*B2 in cell C2 and fill down to C10 by double-clicking the fill handle.

(c) The formulas in F1 and F2 are =SUM(B1:B10) and =SUM(C1:C10), respectively. Compute these using the AutoSum tools as in step (f) of Exercise 1 but using the Sum option. The shortcut to =SUM(is [Alt]+[=].

(d) Compute the average in F3 with =F2/F1.

We will complete the exercise by demonstrating Excel's SUMPRODUCT function. This powerful function makes column C unnecessary. The function computes the sum of the products of elements in two or more ranges. We will compose a formula with this function using the *Insert Function* command.

(e) With F5 as the active cell, click the *Insert Function* icon to open the dialog shown in Fig. 4.3.

(f) Use the *Math & Trig* option in the *Category* box and then select SUMPRODUCT from the lower window. This will open the *Function Arguments* dialog shown in Fig. 4.6. Fill this in as shown by clicking first in the *Array1* box and then dragging the mouse over the range to be used. If the dialog gets in the way use the *Collapse* and *Expand* dialog icons to the right of the text box. Note that the Function Arguments dialog gives information on the purpose of the function and displays the final value when all required arguments have been inserted. Click OK and note that the formula in F5 is =SUMPRODUCT (A2:A10, B2:B10).

(g) To round the result to two decimal places in F7, we will use =ROUND (F6,2). Use the *Insert Function* command to compose this. ROUND is in the *Math & Trig* category.

(h) Save the workbook.

There is a more direct way in Excel 2016 to get to the Function Argument dialog (Fig. 4.6) when you already know the function's category and do not

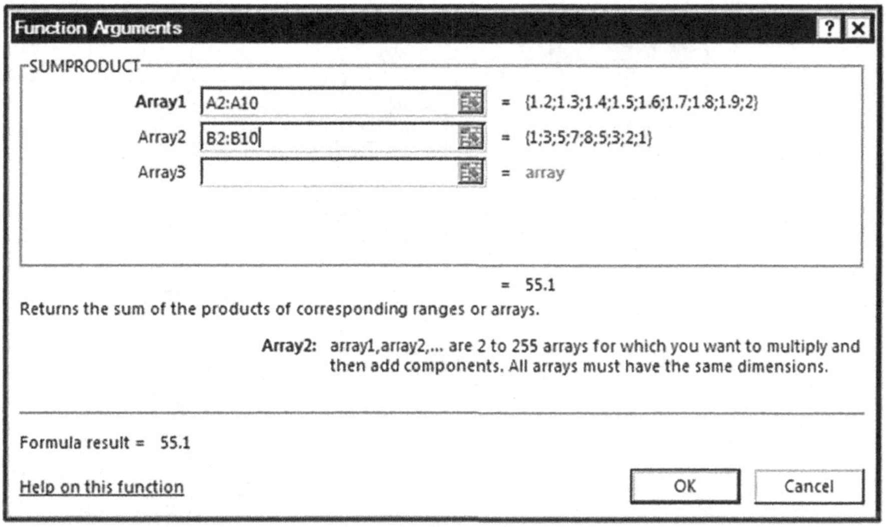

■ FIG. 4.6

need to use the Search tool in the *Insert Function* dialog. All you need to do is open the *Formulas* tab on the ribbon, select the correct category from the *Function Library* group (see Fig. 4.5), and click on the required function name. The reader is encouraged to experiment with this method in G5 to compute the SUMPRODUCT result.

EXERCISE 3: ENTERING FORMULAS BY TYPING

We will complete the worksheet shown in Fig. 4.5 by typing a formula in F10. This Exercise shows: (i) how to type formulas with functions and (ii) a two-level nesting formula. The formula we will use computes the average voltage from the experimental data in columns A and B and rounds the answer to two decimal places. We shall nest SUMPRODUCT and SUM within ROUND: `=ROUND (SUMPRODUCT(A2:A10,B2:B10) / SUM(B2:B10), 2)`.

(a) In F10 start the formula by typing `=R` (or `=r`, since Excel will convert formula names to upper case as it proceeds). A popup menu opens, showing a list of all functions beginning with R (see Fig. 4.7). As you continue to type, the list is amended until with `=ROU` there are only four functions showing. This feature is known as *Intellisense*. You can continue typing or you may click on the ROUND function name to get `=ROUND(` in the cell.

Average V	=r

RADIANS	Converts degrees to radians
RAND	
RANDBETWEEN	=rou
RANK.AVG	
RANK.EQ	ROUND	Rounds a number to a specified number of digits
RATE	ROUNDDOWN
RatkaisinLataa	ROUNDUP
RatkaisinLisää	
RatkaisinLopeta	Average V	=ROUND(
RatkaisinLopetaVIkkuna		ROUND(**number**, num_digits)
RatkaisinMuuta	
RatkaisinNouda	

Average V	=ROUND(SUMPRODUCT(
	SUMPRODUCT(**array1**, [array2], [array3], ...)

■ FIG. 4.7

(b) Continue typing until you have: =ROUND(SUMPRODUCT(in the formula bar.

(c) The arguments may be entered by typing or by dragging the mouse over the appropriate range until you have =ROUND(SUMPRODUCT(A2:A10,B2:B10)/SUM(B2:B10),2).

(d) Save the workbook.

This is a reasonably complicated formula. It is instructive to use the Evaluate Formula command from *Formulas / Formula Auditing* group to see how the result is obtained (not available on a Mac). You may also wish to experiment with the Trace Precedents and Remove Arrows commands in the same group. That group also contains the Show Formulas command.

> If you plan to share your workbook with others, it is important to remember that formulas using functions that exist only in Excel 2016 can result in #NAME? errors when opened in earlier versions of Excel.

EXERCISE 4: TRIGONOMETRY FUNCTIONS

Until the 2013 version, Excel's trigonometry functions were (excluding the hyperbolic functions) COS, SIN, TAN together with the inverse functions ACOS, ASIN, ATAN, and ATANT2. But in starting in Excel 2013 we have a much fuller range of functions: use *Formulas / Function Library / Math & Trig (Formulas / Math & Trig* on a Mac) to see what is available.

It is essential that the reader remembers that Excel's trigonometric functions expect the arguments to be in radians, not degrees, and that the inverse functions return values in radians. While the relationship $\pi\ (radians) \equiv 180\ (degrees)$ may be used to convert between the two measurements, it is wiser to use the conversion functions: RADIANS and DEGREES; their use is demonstrated as follows.

We will construct the worksheet shown to the left of Fig. 4.8 then we shall modify it as shown to the right.

(a) On Sheet3 of Chap4.xlsx enter the text in rows 1 and 2. In A3 enter =PI ( )/4. In A4 enter =A3+PI( )/4 and use the fill handle to drag this down to row 10. Do not format the cells yet.

(b) The formulas in B3:E3 are =SIN(A3), =ASIN(B3), =COS(A3), and =ACOS(D3), respectively. These are filled down to row 10. Note the inexact values in some cells. Where zero is expected, we sometimes get values such as −2.45E − 16 (meaning -2.45×10^{-16}). This is a result of the IEEE rounding error mentioned in Chapter 2. The displayed values revert to zero when the cells are formatted as Numbers with a limited number of decimals. We will see later how to avoid this with the ROUND function.

(c) Go to *Home / Number* (*Home* on a Mac) and change General in the top box to Number and then use the Increase Decimal tool to have four decimal places showing.

(d) In column F test to see if your data correctly demonstrates that $\sin^2(\theta) + \cos^2(\theta) = 1$ for each radian value in column A. In G3 use =TAN (A3)=B3/D3 to test if $\tan(\theta) = \sin(\theta)/\cos(\theta)$. Copy the formula down the column. Hopefully, each cell in G3:G10 displays the Boolean TRUE value.

(e) Change A2 to read Degrees. Replace the formulas in A3:A10 by the values 45, 90… 360. Edit the formulas in B3:E3 to read =SIN(RADIANS (A3)), =DEGREES(ASIN(B3)), =COS(RADIANS(A3)), and =DEGREES (ACOS(D3)), respectively. Select B3:E3 and double-click E3's fill handle to copy the formulas down to row 10.

	A	B	C	D	E
1	Trigonometry Functions				
2	Radians	SIN	ASIN	COS	ACOS
3	0.7854	0.7071	0.7854	0.7071	0.7854
4	1.5708	1.0000	1.5708	0.0000	1.5708
5	2.3562	0.7071	0.7854	-0.7071	2.3562
6	3.1416	0.0000	0.0000	-1.0000	3.1416
7	3.9270	-0.7071	-0.7854	-0.7071	2.3562
8	4.7124	-1.0000	-1.5708	0.0000	1.5708
9	5.4978	-0.7071	-0.7854	0.7071	0.7854
10	6.2832	0.0000	0.0000	1.0000	0.0000

	A	B	C	D	E
1	Trigonometry Functions				
2	Degrees	SIN	ASIN	COS	ACOS
3	45	0.7071	45.0000	0.7071	45.0000
4	90	1.0000	90.0000	0.0000	90.0000
5	135	0.7071	45.0000	-0.7071	135.0000
6	180	0.0000	0.0000	-1.0000	180.0000
7	225	-0.7071	-45.0000	-0.7071	135.0000
8	270	-1.0000	-90.0000	0.0000	90.0000
9	315	-0.7071	-45.0000	0.7071	45.0000
10	360	0.0000	0.0000	1.0000	0.0000

■ FIG. 4.8

(f) Do you still get the expected results in columns F and G? If not, what corrections are needed?

(g) Save the workbook.

We have looked at the trigonometry functions SIN, COS, and TAN, and the inverse functions ASIN and ACOS. We have also seen the use of the functions RADIANS and DEGREES as well as PI(). In column G we have seen a formula that returns a Boolean value.

EXERCISE 5: EXPONENTIAL FUNCTIONS

In this Exercise, the reader is asked to construct a worksheet on Sheet4 of Chap4.xlsx similar to that in Fig. 4.9 using the text in row 1 as hints as to the formulas to use. Use Help to discover why log(n), log10(n), and log(n,10) all return the same values. Could you have known that log(8,2) has the value 3 without a computer?

EXERCISE 6: THE ROUND, ROUNDUP, AND ROUNDDOWN FUNCTIONS

We frequently need to round numbers in calculations. We may be attempting to follow the rules of significant numbers, or we may wish to avoid troubles resulting from binary round-off. Whatever the reason, Excel provides a variety of functions to round or truncate numbers. In this Exercise, we look at the most common ones: ROUND, ROUNDUP, and ROUNDDOWN. Remember that formatting changes how a value is displayed but does not affect the stored value—see Exercise 3 in Chapter 2.

(a) On Sheet5 of Chap4.xlsx enter the text shown in Fig. 4.10.
(b) Enter the values in column A.
(c) In B3 enter =ROUND($A4,3) and fill this down to row 6.
(d) Using row 3 as your guide enter formulas in C4:I4 and fill them down the column. Note the use of a mixed reference ($A4) allows you to drag the formula from B4 to I4 and make some minor changes.
(e) Save the workbook.

> Significant digits: There is a very useful formula to round a number to n significant digits. You may wish to experiment with =ROUND(A1, B1-1-INT(LOG10(ABS(A1)))) where A1 holds the value to be rounded and B1 the number of significant digits required. A literal may replace B1 in the formula. Note that the number may be displayed with extra trailing zeroes that are not to be counted as significant. Credit for this formula goes to John Walkenbach. An alternative formula is =--TEXT(A1,"0."&REPT("0",B1-1&"E+0")) but this fails when A1 is <1.

◢	A	B	C	D	E	F	G
1	n	exp(n)	ln(n)	log(n)	log10(n)	log(n,10)	log(n,2)
2	2	7.38906	0.69315	0.30103	0.30103	0.30103	1
3	5	148.413	1.60944	0.69897	0.69897	0.69897	2.32193
4	8	2980.96	2.07944	0.90309	0.90309	0.90309	3

■ FIG. 4.9

	A	B	C	D	E	F	G	H	I
1	ROUND, ROUNDUP, ROUNDDOWN								
2									
3	n	ROUND(n,3)	ROUND(n,2)	ROUND(n,0)	ROUND(n,-1)	ROUND(n,-3)		ROUNDUP(n,2)	ROUNDDOWN(n,2)
4	1234.5678	1234.5680	1234.5700	1235.0000	1230.0000	1000.0000		1234.5700	1234.5600
5	456.9838	456.9840	456.9800	457.0000	460.0000	0.0000		456.9900	456.9800
6	-3456.8970	-3456.8970	-3456.9000	-3457.0000	-3460.0000	-3000.0000		-3456.9000	-3456.8900

■ **FIG. 4.10**

The syntax for ROUND is =ROUND(number, num_digits). Clearly, when num_digits is 2 we get two decimal places. Note that all cells are formatted with four decimal places to show the effect of rounding. When num_digits is −1 rounding occurs to the nearest multiple of 10; when −2, to the nearest multiple of 100, and so on.

Round uses the common rule: if the last digit to be rounded is <5 then round down (so 4.43 rounded to one decimal becomes 4.4) otherwise round up (making 4.46 become 4.5). To force the rounding to always go down (toward zero) we can use ROUNDDOWN and conversely ROUNDUP always rounds up. Note that with negative values the rounding rule is applied to the absolute value and the result is then made negative.

Most of us round 4.3 to 4 and 4.6 to 5. But what about 4.5? While many would reply 5, others use the round-to-even rule. Thus 4.5 rounds to 4 as does 3.5. Unfortunately, Excel does not provide a function that follows this rule but one can construct a user-defined function (see Chapter 8) that does.

OTHER ROUNDING FUNCTIONS

Excel provides an almost bewildering variety of functions for rounding in addition to ROUND, ROUNDUP, and ROUNDDOWN as can be seen in the following table.

ODD	Returns the nearest odd integer: =ODD(12.2) returns 13
EVEN	Returns the nearest even integer: =EVEN(12.4) returns 12
CEILING.MATH	This is a new function introduced in Excel 2013; in Excel 2010 it was called CEILING.PRECISE. This function is used to round a number up to the nearest integer or to the nearest multiple of significance. The syntax of this function is CEILING.MATH(number,[significance],[mode]) where *significance* and *mode* are optional. The *significance* indicates the multiple of the number to which we want to round off the number. For example = CEILING.MATH (56.3,3) will round the number 56.3 up to the nearest integer that is a multiple of 3 (57) so it will return 57. When the mode argument is present and nonzero rounding will be away from zero. So =CEILING.MATH (−45.6,0.5) or =CEILING.MATH(−45.6,0.5,0) will each return −45.5 but =CEILING.MATH(−45.6,0.5,1) returns −46

	The old CEILING function is still available for compatibility with earlier versions of Excel. However, if backward compatibility is not required, you should consider using the new functions from now on, because they more accurately describe their functionality There is also the new ISO.CEILING function but we will not explore it
FLOOR.MATH	This is very similar to CEILING.MATH except that rounding occurs down rather than up
INT	Rounds a number down to the nearest integer (cf. TRUNC): =INT(−5.6) returns −6
TRUNC	Truncates a number to an integer (cf. INT):=TRUNC(1.55) returns 1; =TRUNC(−5.6) returns −5. INT(x) and TRUNC(x) differ only when the argument is negative. TRUNC(x,n) truncates to n decimals places: TRUNC(66.79,1) returned 66.7
MROUND	Returns a number rounded to the required multiple: =MROUND(6.89,4) returns 8

SOME OTHER MATHEMATICAL FUNCTIONS

As we progress in this book we will meet other mathematical functions such as the matrix functions (MINVERSE, MMULT, MDETERM); functions to generate random numbers (RAND and RANDBETWEEN); various summation functions (SUMXMY2, SUMX2MY2, SUMX2PY2); and so on. The following table lists a few other commonly used functions.

PI()	The formula =PI() returns 3.14159265358979. Note there is no argument
SUMSQ	Returns the sum of the squares of a range of numbers. =SUMSQ(A1:A10)
SUMPRODUCT	Returns the sum of the products of the elements of two ranges—see Exercise 2. Very useful for conditional summations—see Chapter 5
SQRT	Returns the square root of a positive number
SQRTPI	Returns the square root of a multiple of π. Thus SQRTPI(2) is equivalent to =SQRT(2*PI())
FACT	=FACT(4) returns the value 4! or $4 \times 3 \times 2 \times 1 = 24$ See also FACTDOUBLE in Help
GCD	Returns the greatest common divisor. =GCD(9, 18, 24) returns 3
LCM	Returns the largest common multiple. =LCM(9, 18, 24) returns 72
PRODUCT (limited use)	May be used in place of the multiplication operator; =PRODUCT(A1:A2) and =A1*A2 are equivalent. Can be useful with larger range involved as in =PRODUCT(A1:A10)
QUOTIENT (limited use)	Returns the integer portion of a division. Use this function when you want to discard the remainder of a division. QUOTIENT(A1,B1) is equivalent to =INT(A1/B1)
POWER (limited use)	May be used in place of the exponentiation operator: =POWER(A1,2) is equivalent to =A1^2

ARRAY FORMULAS

All the functions we have looked at so far produce a single result in one cell. There are a number of Excel functions that produce results in a range of cells. When you have finished typing the formula containing one of these functions you must commit the function with Ctrl + ⇧ Shift + Enter↵. None of the other methods of committing a formula will work. A formula that requires this is called *an array formula* in that they return an array of values. We shall be using array formulas throughout the book. In the next chapter, we shall see examples of array formulas that produce a single result from an input array.

EXERCISE 7: MATRIX FUNCTIONS

Excel includes these functions for working with matrices:

MMULT(A, B)	for matrix multiplication AB
MINVERSE(A)	for finding an inverse A^{-1}
MUNIT(n)	to generate an $n \times n$ unit matrix
MDETERM	for finding the determinant

We will examine these functions in this exercise and show a more practical use of them in the next.

For our example of an array formula, we shall look at MMULT, which is the function used to find the matrix product of two matrices. In mathematical terms, MMULT computes **AB** where *necessarily* the number of columns in **A** equals the number of rows in **B.** We will also experiment with MINVERSE and MDETERM.

(a) On Sheet6 of Chap4.xlsx, copy all the entries seen in Fig. 4.11 other than G4:H5, A9:B10, and D9.

(b) Select G4:H5 and use the Insert Function tool to create the formula =MMULT(A4:B5,D4:E5) and then holding down Ctrl + ⇧ Shift, tap the Enter↵ key.

The range G4:H5 now holds the matrix C defined by C = AB. If you did not get the expected result, try again. If you are familiar with matrix algebra, you may wish to do the calculation manually.

The MMULT function returns a #VALUE! error if the number of columns of A is not equal to the number of rows of B, or when any cells contain nonnumerical values.

If you happen to forget to use Ctrl + ⇧ Shift + Enter↵ then you just get a value in the top left corner if you use only Enter↵, and #VALUE! error if you forget the ⇧ Shift.

	A	B	C	D	E	F	G	H
1	Matrix Functions							
2	Multiplication							
3	Matrix A			Matrix B			Matrix C	
4	2	3		1	2		11	16
5	4	5		3	4		19	28
6								
7	Inverse			Determinant				
8	Matrix A			Determinant of A				
9	-2.5	1.5		-2				
10	2	-1						

■ FIG. 4.11

(c) Click on G4 and note what the formula bar displays: {=MMULT(A4:B5,D4:E5)}. Your formula has been enclosed within braces { } by Excel. This is a "trademark" of array functions.

(d) With G4 still selected, try to delete it with the D key. You get a message stating that you cannot change just one cell in an array formula.

(e) To find the inverse of matrix A: Select A9:B10, enter =MINVERSE (A4:B5), and again use [Ctrl]+[⇧ Shift]+[Enter ↵] to commit the formula. The MINVERSE function returns a #VALUE! error if the number of columns and rows of A are not equal, or when any cells contain nonnumerical values. Some square matrices cannot be inverted and will return the #NUM! error value with MINVERSE. The determinant for a noninvertible matrix is 0.

(f) The determinant of matrix A is found in D9 with the nonarray formula =MDETERM(A4:B5).

(g) A reader who is familiar with matrix algebra might wish to experiment with (i) =MMULT(A3:B5, A9:B10) in G9:H10, and (ii) =MUNIT(2) in G12:H13.

(h) Save the workbook.

VOLATILITY: CALCULATE MODE

Whenever a value in a cell is changed, Excel normally recalculates every cell that is dependent on the changed value. In a more complex worksheet with many thousands of interrelated formulas, you may see a message such as *Calculating 10%* in the status bar. Usually, the work gets done too quickly for this to be visible.

Some functions get recalculated whenever there is any change made to a worksheet regardless of whether or not the changed cell has an effect on them. Such functions are said to be *volatile*. Some obvious examples are NOW (the current time and date), TODAY (the current date), and RAND (random number). But there are some less obvious ones such as INDIRECT, OFFSET, CELL, and INFO.

If you open a workbook with formulas containing volatile functions and later close it, you will be asked if you wish to save the changes. This message gets displayed even when the user has done nothing to the workbook. The presence of a chart will cause the same message in Excel 2016. Saving is always the best option.

Large workbooks, especially those with many volatile functions, can have long recalculation times. When the workbook gets very complex, the recalculation time can cause a loss of productivity since the user must pause between cell entries. In such cases, it is common to set the calculation mode to manual in *File / Options / Formulas* (*Formulas / Calculation Options / Manual* on a Mac). With this setting, the status bar displays *Calculate* whenever a recalculation is needed. Generally, the user presses F9 every so often to have Excel recalculate.

EXERCISE 8: SOLVING SYSTEMS OF EQUATIONS

A system of linear equations may be represented in the matrix form. Consider the system of two equations:

$$2x + 3y = -1$$
$$4x + 5y = 5$$

This may be represented by the matrix equation.

$$\begin{bmatrix} 2 & 3 \\ 4 & 5 \end{bmatrix} \begin{bmatrix} x \\ y \end{bmatrix} = \begin{bmatrix} -1 \\ 5 \end{bmatrix}$$

The proof of this is fairly simple. Firstly, we perform the matrix multiplication on the left side to get

$$\begin{bmatrix} 2x + 3y \\ 4x + 5y \end{bmatrix} = \begin{bmatrix} -1 \\ 5 \end{bmatrix}$$

Secondly, we note that when matrix A equals matrix B then the corresponding elements are equal. So it follows that $2x + 3y = -1$ and $4x + 5y = 5$; these are the equations with which we started, thereby justifying the statement that we may represent a system of linear equations in the matrix form.

Let A represents the matrix of the coefficients, X the matrix of the variables, and C the matrix of the constants.

Let us write the system of equations in the form:	$AX = C$
Multiply both sides by A^{-1} (the inverse of A) giving	$A^{-1}AX = A^{-1}C$
But since $A^{-1}A = I$, this becomes	$IX = A^{-1}C$
We know that $IX = X$, therefore	$X = A^{-1}C$

From this, we see that the value of the X matrix may be obtained by computing $A^{-1}C$.

In Exercise 7 we found for

$$A = \begin{bmatrix} 2 & 3 \\ 4 & 5 \end{bmatrix} \quad A^{-1} = \begin{bmatrix} -2.5 & 1.5 \\ 2 & -1 \end{bmatrix}$$

So we may write

$$\begin{bmatrix} x \\ y \end{bmatrix} = \begin{bmatrix} -2.5 & 1.5 \\ 2 & -1 \end{bmatrix} \begin{bmatrix} -1 \\ 5 \end{bmatrix}$$

Performing the matrix multiplcation we get

$$\begin{bmatrix} x \\ y \end{bmatrix} = \begin{bmatrix} 10 \\ -7 \end{bmatrix}$$

From which we see that $x = 10$ and $y = -7$.

This may have left you less than impressed; you could have solved the two simultaneous equations in your head. But the method may be applied to more challenging problems. In this exercise we solve:

$$2x + 3y - 2z = 15$$
$$3x - 2y + 2z = -2$$
$$4x - y + 3z = 2$$

The completed worksheet will resemble Fig. 4.12.

(a) In Sheet7 of Chap4.xlsx, enter all the all text values. Enter the equation coefficients and constants in A4:C7 and D4:D7, respectively.
(b) Next, we compute the inverse (A^{-1}) of the matrix of coefficients. Select the range A10:C12, enter the formula =MINVERSE(A5:C7), and press Ctrl+⇧Shift+Enter↵. Format the cells to display five places.
(c) The final step to find the solutions is to compute $A^{-1}C$. Select D10:D12, enter the formula =MMULT(A10:C12, D5:D7), and press Ctrl+⇧Shift+Enter↵.

◢	A	B	C	D	E	F
1		Using Matrix Functions to Solve				
2		a System of Linear Equations				
3						
4	Matrix of Coefficients (A)			Matrix of Constants		
5	2	3	-2	15		
6	3	-2	2	-2		
7	4	-1	3	2		
8						
9		Inverse A^{-1}		A^{-1}C = X		
10	0.19048	0.33333	-0.09524	2	x	
11	0.04762	-0.66667	0.47619	3	y	
12	-0.23810	-0.66667	0.61905	-1	z	
13						
14		Reconstructed equations				
15	4	9	2	15		
16	6	-6	-2	-2		
17	8	-3	-3	2		

■ **FIG. 4.12**

The solutions have now been found. We may wish to check that these agree with the system of equations.

(d) Name the cells D10:D12 as *x*, *y*, and *z*, respectively.

(e) The formulas in row 15 from left to right are: =A5*x, =B5*y, =C5*z, and =SUM(A15:C15).

(f) These formulas are filled down to row 17.

(g) Save the worksheet.

The values in D15:D17 agree with those in D5:D7, thus confirming that we have solved the system of equations. In Chapter 11 we show the use of this method to solve some practical problems.

FINANCIAL FUNCTIONS

As one would expect, Excel offers a very wide range of financial functions. This is not a book on finance, but we shall briefly look at the ones relating to loans and savings.

Financial analysts use a sign convention for the flow of money. Consider the situation where you take out a loan from a bank and make monthly repayments to amortize (pay off) the loan. From your perspective, the initial money coming from the bank to you is considered a positive quantity. The payments you make flow from you to the bank and are considered negative quantities from your viewpoint. The bank, of course, views things the other way around. If you use Excel to evaluate a potential loan, do not be surprised if an Excel calculation differs slightly from the bank's data: there could be banking fees and rounding adjustments.

Let's say you deposit $100 in a savings account and the bank offers an interest rate of 3%. Generally, the advertised rate is a nominal annual rate, but your interest is accumulated monthly. The rate that is used each month is the nominal rate divided by 12. We will assume you leave the original money (the principal) and the interest in the bank for a set period of time.

The $100 deposited today is called the present value or *pv*. After a certain number of interest periods (*nper*), the savings may be worth say $125. This is called the future value or *fv*. The quantity payment or *pmt* is what you pay the bank to amortize the loan or what the bank pays you (deposits into the saving account) as interest earned. The rate is the compounding rate. There is just one more quantity: if payments are made at the end of the month then the type is 0 (this is the default value if you do not enter the type argument) and is 1 if payments are made at the start of the month. All these quantities come into play when performing calculations on loans and savings. Excel uses the following equation:

$$pv \times (1 + rate)^{nper} + pmt \times (1 + rate \times type) \times \left(\frac{(1 + rate)^{nper} - 1}{rate} \right) + fv = 0$$

There are Excel functions to compute various quantities. For example, the FV function computes the future value. Its syntax is FV(rate, nper, pmt, [pv], [type]); arguments within square brackets are optional. We use uppercase for functions and lowercase for arguments.

You plan to deposit $100 a month into a savings plan for 5 years at a nominal rate of 4%. How much will you have at that time? The answer is found with =FV(4%/12, 5*12, −100). Why is *pv* not used? Because you did not start out with a lump-sum deposit. What would happen if you forgot the negative sign for *pmt*? The result would be negative—you would have taken $100 every month and would be left with a debt!

You win a lottery prize and have to choose between (i) getting $300 a month for 6 years, or (ii) a lump sum of $6000. Which will you take? We need to look at the present value of each. The present value of a $6000 check is exactly $6000. We can compute the present value for the first option with =PV(5%, 6*12, 300), which gives a result of −$5821. You would need to deposit (hence the negative sign) that amount of money to generate the

$300 monthly payment if the bank rate is 5%. Option (ii) is more than this and could generate more than $300 pm if deposited. So option (ii) wins unless there are tax implications.

The function NPER can be used to compute how many periods are needed for a certain scenario, and PMT gives the size of each deposit for another scenario. The function to compute how much is being applied to pay interest is IPMT, while PPMT tells how much is used to pay off the principal. The syntax for RATE (to compute the needed rate for a certain financial scenario) is RATE(nper, pmt, pv, [fv], [type], [guess]). The last argument may look strange. Look at the financial equation before and you will see it cannot be solved explicitly for the rate. Excel needs to perform an iterative routine to get an answer. Generally, we can omit the guess argument. If the successive results of RATE do not converge to within 0.0000001 after 20 iterations, RATE returns the #NUM! error value. Under these circumstances, we can try to assist Excel by giving a guess at the answer.

EXERCISE 9: BORROWER BEWARE

John and Anne each set out to purchase a new computer system. John is rather compulsive; he rushes off to PCs-R-US, selects an expensive gaming unit and arranges credit with the store. Anne does some research; she finds a unit that will satisfy all her needs for classwork and social media (she is not a great games player) and she visits the bank to arrange a line of credit. We will compare the payments made by each—see Fig. 4.13.

	A	B	C
1	Borrower Beware		
2			
3		John	Anne
4	Principal	$3,000.00	$1,000.00
5	APR	26.0%	6.50%
6	Term (years)	3	2
7			
8	PMT	$120.87	$44.55
9			
10	Total paid	$4,351.39	$1,069.11
11	Interest	$1,351.39	$69.11

■ FIG. 4.13

(a) On Sheet8 of Chap4.xlsx, enter the text shown in A3:C4 and A4:A11 of Fig. 4.13.
(b) Enter the numerical data in B4:C6. John's systems costs 3 times as much as Anne's and his interest rate is 4 times Anne's so he takes an extra year to pay off the loan.
To compute the monthly payments enter this formula in B8 =PMT (B5/12,B6*12,B4). In keeping with how Excel handles financial functions, this will display as a negative number: the amount in B4 is positive since the money flowed to John but in B8 we have money flowing away so the sign is negative. In complex worksheets, it is better to stay with the convention chosen by Excel but for our simple case, we will add a unitary negation operator (minus sign) as in =−PMT(B5 / 12, B6 * 12, B4) to get a positive value.
(c) The reader should enter appropriate formulas in B10 and B11 to find the total of all payments and the total amount of interested included in this.
(d) Format the cells as shown.
(e) Select B4:B11 and drag the fill handle to the right to compute Anne's payments and interest data. Note how the formatting gets carried along with the drag-to-copy.

Note how much more John is paying; he eventually pays $4500 for his $3000 system. Experiment with the IPMT function to compute how much interest they each pay in the first month; John's payment is about 50% interest! Even if he had purchased the $1000 system and settled his loan in 2 years he would still be paying a great deal more because of the high interest rate—experiment with your worksheet to show this.

PROBLEMS

1. *Numerical differentiation on tabulated data may be done using one of the formulas as follows

Forward	Central	Backward
$\frac{dy}{dx} = \frac{y_1 - y_0}{h}$	$\frac{dy}{dx} = \frac{y_1 - y_{-1}}{2h}$	$\frac{dy}{dx} = \frac{y_0 - y_{-1}}{h}$

	A	B	C	D	E	F	G
1	Numerical Differentiation						
2							
3	L	20					
4							
5	t	0	0.05	0.1	0.15	0.2	0.25
6	i	4.9550	4.89936	4.73369	4.46172	4.08954	3.62552
7	V	-22.2560	-44.2620	-87.5280	-128.8300	-167.2400	-185.6080

■ FIG. 4.14

The voltage drop (V) across an inductance (L) is given by $V = L\dfrac{di}{dt}$. In Fig. 4.14 the central difference formula is used for interior points and the forward and backward formulas for end-points. What are the Excel formulas in B7:G7?

2. *To measure the index of refraction μ of a liquid with an Abbé refractometer, a drop is placed between two prisms and a mirror is rotated until the boundary of the light and dark zones align with the crosshairs in a microscope. The index of refraction of the liquid is given by the following equation in which the angle of rotation is θ and μ_g is the refractive index of glass (1.51). For liquid a, the value of θ was found to be 150° and for B it was 75°. Using a worksheet, find μ for each liquid

$$\mu = \frac{1}{\sqrt{2}}\left[\left(\mu_g^2 - \sin^2\theta\right)^{1/2} + \sin\theta\right]$$

3. Many engineering applications require *normalizing* an n-element vector. If the original vector is V and the normalized one is W, then the elements of W are given by

$$w_i = \frac{v_i}{\sqrt{\sum_1^n v_1^2}}.$$

	A	B	C	D	E	F
1	Normalized Vector					
2						
3	V	2.3	3.6	5.7	6.8	8.9
4	W	0.173276	0.271214	0.429422	0.512293	0.670501

■ FIG. 4.15

What formula would you use in B4 of Fig. 4.15 such that it can be copied by dragging to column F?

4. *To fit n ordered pairs of data to the equation $y = mx + c$, we can use the formulas

$$m = \frac{n \sum x_i y_i - \sum x_i \sum y_i}{n \sum x_i^2 - \left(\sum x_i \right)^2}$$

$$c = \frac{\sum y_i - m \sum x_i}{n}$$

To fit the data in A3:G4 of Fig. 4.16, what formulas would you use in B6:B10 and E7:E8? In Chapter 8 we will use the SLOPE and INTERCEPT functions to get the same result in a simpler manner.

◢	A	B	C	D	E	F	G
1	Least Squares Fit						
2							
3	x	1	1.5	2	2.5	3	3.5
4	y	5.52	7.67	8.06	9.57	10.55	12.14
5							
6	n	6					
7	Σ(xy)	131.21		slope	2.471429		
8	Σ(x)	13.5		intercept	3.357619		
9	Σ(y)	53.51					
10	Σ(x²)	34.75					

■ FIG. 4.16

5. *A trough[1] of length L has a semicircular cross section with radius r. When filled with water to within a distance h of the rim, the volume V is given by

$$V = L \left[0.5 \pi r^2 - r^2 \arcsin \left(\frac{h}{r} \right) - h \left(r^2 - h^2 \right)^{1/2} \right]$$

Our task is to make the worksheet shown in Fig. 4.17. Typing a long formula is error-prone, so we do it stepwise. What are the formulas in E3:E6? Excel has a CONVERT function, but this does not help here; therefore Google to find a conversion factor for cubic feet to U.S. or Imperial gallons.

---

[1]J. D. Faires and R. Burden. Numerical Methods, Brooks/Cole, Pacific Grove, CA, 1998 (page 41).

	A	B	C	D	E	F	G	H
1	Trough gauge							
2								
3	Length	10.0 ft		Term 1	3.534292			
4	Radius	1.5 ft		Term 2	0.633330			
5	Height	5/12 ft		Term 3	0.600403			
6				Volume	23.0 ft³			
7				Volume	172.1 gallons			

■ FIG. 4.17

6. Electrical engineers use a color-coded system to identify the value of a
resistor. Thus a resistor with the banding colors green, blue, yellow, and
gold has a value of $560\,k\Omega \pm 5\%$. Construct a worksheet similar to
Fig. 4.18. The user enters the digit 1 next to the appropriate color for
each band and the resistance is computed in C19 with the help of table
G3:I13. The multiplier is 10 raised to the power in column A. Check
your results on the Internet. Use a custom format of ##0.0E+0 in cell

	A	B	C	D	E	F	G	H	I	J
1	Resistor Four-color Code									
2										
3	R	Color	Bar 1	Bar 2	Bar 3		digit 1	digit 2	Multiplier	
4	0	Black					0	0	0	
5	1	Brown					0	0	0	
6	2	Red					0	0	0	
7	3	Orange			1		0	0	1000	
8	4	Yellow					0	0	0	
9	5	Green	1				5	0	0	
10	6	Blue		1			0	6	0	
11	7	Violet					0	0	0	
12	8	Grey					0	0	0	
13	9	White					0	0	0	
14										
15	Tolerance			Bar 4						
16	5%	Gold		1						
17	10%	Silver								
18	20%	None								
19										
20	Resistance		56E+3			Tolerance		5%		

■ FIG. 4.18

C19 as which gives exponents in multiples of three.[2] We shall expand on this problem in later chapters. Clearly, we really should arrange things so that there is only a single 1 in each of the *Bar* columns.

7. Create a worksheet to solve the following system of equations using matrix algebra.

$$3x_1 - 4x_2 + 5x_3 + 6x_4 + 2x_5 = 62.5$$
$$x_1 + 2x_2 + 3x_3 + 4x_4 + 5x_5 = 19.5$$
$$6x_1 + 7x_2 - 4x_3 + 2x_4 - x_5 = 15$$
$$5x_1 - 5x_2 + 2x_3 + 5x_4 + 7x_5 = 32$$
$$-3x_1 + 5x_2 + 6x_3 + 2x_4 + x_5 = 16$$

8. You decide to deposit $100 in a savings account on the last day of each month. The bank's nominal rate is 5% per year, but interest is paid monthly. (i) What will be the value of the savings after 2 years? (ii) Draw up an amortization table as shown in Fig. 4.19. Do you get the same answer? (iii) The Excel financial functions do not account for the fact that banks will round interest to the nearest penny. Change your amortization table to round the interest. Change the monthly payment to $100,000. How does this affect the difference between the Excel function result and your amortization table?

	A	B	C	D	E
1	Finance				
2					Future Value
3	pmt	100		FV function	$2,518.59
4	rate	5%		table	$2,518.59
5	nper	2			
6					
7	month	principal start of month	interest	principal end of month	
8	0			100	
9	1	100	0.42	200.42	
10	2	200.42	0.84	301.25	
11	3	301.25	1.26	402.51	
30	22	2298.98	9.58	2408.56	
31	23	2408.56	10.04	2518.59	
32	24	2518.59			

■ FIG. 4.19

---

[2]The custom format [<0.001]##0E+0;[<1000]#0.00;##0E+0 will give exponents in multiples of three, except for values between 0.001 and 1000 which will not have the exponent.

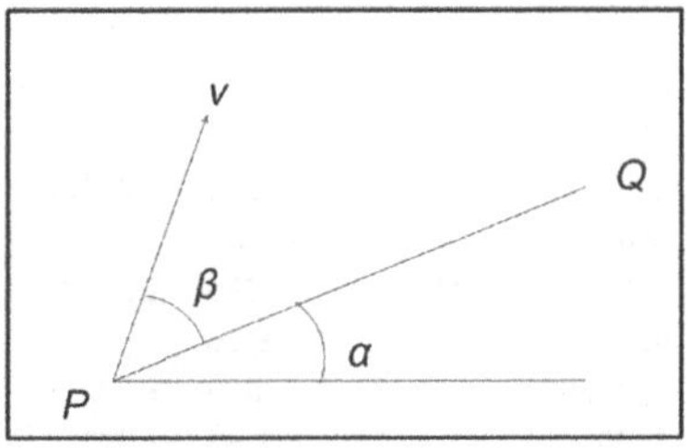

■ FIG. 4.20

9. Referring to Fig. 4.20, PQ is a plane inclined at angle α to the horizontal. A particle is projected from P at a velocity v ft./s at an angle of β to the plane. Construct a worksheet to compute the range on the inclined plane and the time of flight. The time of flight is calculated using

$$t = \frac{2v\sin\beta}{g\cos\alpha}$$

and the range is

$$R = (v\cos\beta)t - \frac{g\sin\alpha}{2}t^2$$

What values do you get when $\alpha = \beta = 30°$ and $v = 900$ ft./s using $g = 32$ ft./s^2?

10. In Fig. 4.21 the mantissa and the exponents of the numbers in column A are computed in columns B and C. What formulas will give these results?

	A	B	C
1	N	Mantissa	Exponent
2	1.234	1.234	0
3	24.567	2.4567	1
4	-25.6789	-2.56789	1
5	0.000146	1.456	-4
6	-0.0123	-1.23	-2

■ FIG. 4.21

11. The Fourier series coefficients for the given data can be calculated for each value of N that can be calculated using

$$A_N = \frac{2}{k}\sum_1^k F(t)\cos\left(\frac{2\pi N}{T}t\right)$$

	A	B	C	D	E	F	G	H	I	J	K	L
1		N	1	2	3	4	5	1	2	3	4	5
2	t	F(t)	Cos					Sin				
3	0.25	5.11	4.72102	3.61332	1.95551	3.1E-16	-1.95551	1.95551	3.61332	4.72102	5.11	4.72102
4	0.5	2.83	2.00111	1.7E-16	-2.00111		-2.83	-2.00111	2.00111	2.83	2.00111	3.5E-16	-2.00111
5	0.75	-2.94	-1.12509	2.07889	2.71621	5.4E-16	-2.71621	-2.71621	-2.07889	1.12509	2.94	1.12509
6	1	-6	-3.7E-16	6	1.1E-15	-6	-1.8E-15	-6	-7.4E-16	6	1.5E-15	-6
7	1.25	-2.94	1.12509	2.07889	-2.71621	-9E-16	2.71621	-2.71621	2.07889	1.12509	-2.94	1.12509
8	1.5	2.83	-2.00111	-5.2E-16	2.00111	-2.83	2.00111	2.00111	-2.83	2.00111	1E-15	-2.00111
9	1.75	5.11	-4.72102	3.61332	-1.95551	-2.2E-15	1.95551	1.95551	-3.61332	4.72102	-5.11	4.72102
10	2	2	-2	2	-2	2	-2	2.5E-16	-4.9E-16	7.4E-16	-9.8E-16	1.2E-15
11	2.25	-2.28	2.10645	-1.6122	0.87252	-1.3E-15	-0.87252	0.87252	-1.6122	2.10645	-2.28	2.10645
12	2.5	-2.83	2.00111	-8.7E-16	-2.00111	2.83	-2.00111	2.00111	-2.83	2.00111	-1.7E-15	-2.00111
13	2.75	0.12	-0.04592	-0.08485	0.11087	-2.9E-16	-0.11087	-0.11087	0.08485	0.04592	-0.12	0.04592
14	3	2	-3.7E-16	-2	1.1E-15	2	-5.4E-15	-2	7.4E-16	2	-1.5E-15	-2
15	3.25	0.12	0.04592	-0.08485	-0.11087	-1.2E-16	0.11087	-0.11087	-0.08485	0.04592	0.12	0.04592
16	3.5	-2.83	-2.00111	1.2E-15	2.00111	2.83	2.00111	2.00111	2.83	2.00111	-2.4E-15	-2.00111
17	3.75	-2.28	-2.10645	-1.6122	-0.87252	6.1E-15	0.87252	0.87252	1.6122	2.10645	2.28	2.10645
18	4	2	2	2	2	2	2	-4.9E-16	-9.8E-16	-1.5E-15	-2E-15	-2.5E-15
19		Coefficients	2.2E-16	1.99879	9.4E-16	3.3E-16	-5.6E-16	0.0008	-6.7E-17	4.00018	-5.8E-16	-0.00094
20			A_1	A_2	A_3	A_4	A_5	B_1	B_2	B_3	B_4	B_5

■ FIG. 4.22

and

$$B_N = \frac{2}{k}\sum_{1}^{k} F(t)\sin\left(\frac{2\pi N}{T}t\right)$$

Where there are k samples in the period, that is, T seconds. These calculations can be done using excel by calculating the product of F(t)*cos(2*pi()*N/T*t) or F(t)*sin(2*pi()*N/T*t) for each value of t, and then summing up each of the products as seen before. Use the proper cell referencing to efficiently do the products, and then use the sum function to calculate the totals. Label the sums as given as follows. The numerical solution is given in Fig. 4.22. Enter the values for t, $F(t)$, and N and calculate the rest. In this example $k = 16$ and $T = 4$.

APPENDIX: SUPPLEMENTARY MATERIAL

Supplementary material related to this chapter can be found on the accompanying CD or online at https://doi.org/10.1016/B978-0-12-818249-9.00004-2.

Chapter 5

Conditional Functions

This chapter deals with making decisions or having a cell display a value that is conditional upon what is in one or more other cells. We begin with the logical comparison operators and Boolean functions such as AND, OR, NOT, and NOR. We examine the IF and IFERROR functions and show how to use SUMIF, SUMIFS, COUNTIF, and so on, together with SUMPRODUCT. Then we examine the table lookup functions LOOKUP, VLOOKUP, and HLOOKUP. The chapter concludes with some notes on conditional formatting.

Liengme's Guide to Excel 2016 for Scientists and Engineers. https://doi.org/10.1016/B978-0-12-818249-9.00005-4

LOGICAL COMPARISON OPERATORS

The logical comparison operators (you may know these operators as relational operators if you have studied computer programming) are used to make tests or to establish a criterion. We will concentrate on numerical comparisons such as *is the value in A1 greater than 4?*

The comparison operators are as follows:

=	Equal to
>	Greater than
>=	Greater than or equal to
<	Less than
<=	Less than or equal to
<>	Not equal to

Let A1 hold the value 10, and B1 the formula =A1>10. Since this is untrue, the formula returns the Boolean value FALSE. If we make the formula =A1>=10 then the result will be TRUE.

In the formula =A1>=10, the A1>=10 part is called a *logical expression*. Logical expressions evaluate to either TRUE or FALSE in Excel. A logical expression has the form:

Expression-1 Logical-operator Expression-2

It can be useful to have logical expressions evaluated to 1 or 0. Excel treats the Boolean values as 1 and 0 when combined with mathematical operations. Following the previous example, the formula =(A1>10)*1 will return the value 0 while =(A1>=10)*1 returns the value 1. Using two negation operators is a very efficient method to coerce Boolean values to numeric values; we may use a formula such as =--(A1>=10). When a Boolean value is expected, Excel will accept any nonzero numeric value as TRUE, and a zero value as FALSE.

EXERCISE 1: BOOLEAN FUNCTIONS

The functions AND and OR may be used to test two or more logical expressions, while the NOT function is used to reverse the truth value of a logical expression. The XOR function returns TRUE when one, and only one, of the logical expressions is true.

(a) On Sheet1 of a new workbook, enter the values in A1:B5 of Fig. 5.1. Use the information in columns D and F to enter formulas in columns C and E.

	A	B	C	D	E	F
1	Boolean Functions					
2	a	4	FALSE	=AND(B2>=5,B3>=5)	TRUE	=OR(B2>=5,B3>=5)
3	b	6	TRUE	=AND(B3>=5,B4>=5)	TRUE	=OR(B3>=5,B4>=5)
4	c	10	FALSE	=NOT(B4=10)	TRUE	=AND(B2+B3=10,B4=10)
5	d	7	FALSE	=XOR(B5=7,B3=6)	TRUE	=XOR(B5=7,B3=12)

■ **FIG. 5.1**

(b) Save the workbooks as Chap5.xlsx.

By default, Excel aligns Boolean values (TRUE and FALSE) centered horizontally in their cells.

Note that if in this worksheet you entered the formula =A2>5, the result would be TRUE. Excel would compare the letter a (a text data type) with the literal 5 (also a text data type): the ASCII value for the letter a is 97, and that for the digit 5 is 53.

There are some common combinations that are useful to know. In the following table, A and B may be expressions or references to cells containing the values TRUE or FALSE. You may wish to experiment with these nested formulas on Sheet1.

Logic	Formula	TRUE returned if
NAND	=NOT(AND(A,B))	Not both true
NOR	=NOT(OR(A,B))	Neither is true

EXERCISE 2: PRACTICAL EXAMPLE

The scenario for this Exercise is: In a manufacturing plant 10 items are tested every hour. For each item, two quantities (P and Q) are measured; the P value must be at least a certain value (*pmin*), while the Q value must not exceed a certain value (*qmax*). Fig. 5.2 shows the worksheet we need to find what percentage of our product is up to specification.

(a) On Sheet2 of Chap5.xlsx, enter the text and numbers shown in A1:C3, A5:E5, and A6:B15 of Fig. 5.2.

(b) Name C2 and C3 as *pmin* and *qmax*, respectively. Select A5:B15. Use *Create from Selection* to label cells in the P and Q columns P and Q, respectively.

	A	B	C	D	E
1	Quality control				
2		pmin	1.25		
3		qmax	0.5		
4					
5	P	Q	P test	Q test	Two test
6	1.70	0.65	1	0	0
7	1.36	0.50	1	1	1
8	1.44	0.40	1	1	1
9	1.57	0.45	1	1	1
10	1.90	0.82	1	0	0
11	1.52	0.32	1	1	1
12	1.23	0.75	0	0	0
13	1.65	0.50	1	1	1
14	1.29	0.36	1	1	1
15	1.15	0.45	0	1	0
16		Percentage passing	80.0%	70.0%	60.0%

■ **FIG. 5.2**

(c) The formula in C6 is `=--(P>=pmin)` (carefully observe the double negation to make the answer 1 instead of TRUE) and this is copied down the column to row 15 using the shortcut method of double-clicking the fill handle.

(d) In D6 we need the formula `=--(Q<=qmax)`; and in E6 `=--AND(P>pmin, Q<=qmax)`. These are to be copied down to row 15. Note that an alternative formula for E6 would be `=C6*D6`.

(e) Row 16 summarizes the results giving the percentage that passed the tests. We might be tempted to use `=SUM(C6:C15)/COUNT(C6:C15)` in C16 to get the fraction that passed the *P*-test. But this is just an average, so why not use `=AVERAGE(C6:C15)` here and corresponding formulas in D16 and E16? NOTE: the text in B16 is made into two lines using [Alt]+[Enter⏎] between the two words, and C16:E16 have been formatted to display percentage values.

(f) Save the workbook.

THE IF FUNCTION

The IF function returns one of two values depending on the truth value of a logical expression called the *test*. The syntax is `=IF(logical_test, true_value, false_value)`. If the test is true then whatever is entered as true_value gets returned, otherwise false_value gets returned.

The logical test is generally a logical expression of the type we looked at before; for example, A1 > 10. A logical test must return either TRUE or FALSE. An example of a simple IF is =IF(A1>=10, "OK", "Too small"). However, Excel also allows for simple arithmetic expressions in an IF test. Consider the arithmetic expression A1–10, if this evaluates to a nonzero value, Excel treats it as TRUE; only zero is taken as FALSE. The formula =IF(A1=0, "Zero", "Not zero") could be coded as =IF(A1, "Not Zero", "Zero") but do not try to be too clever with this approach since others may not follow the logic.

Here are some examples of simple IF formulas.

  (i) =IF(A2<0, "Negative", "Positive")
     Returns the text "Negative" if A2 has a value <0; otherwise, it returns "Positive."

 (ii) =IF(ABS(A10-B10)<=EPSILON, "Equal", "Unequal")
     A cell called EPSILON contains a value such as $1.0E-6$. Rather than doing a direct comparison of two values, we test if they differ by more than this value.

(iii) =IF(ABS(A10-B10)<= 0.001,1,0)
     Returns 1 if ABS(A10 − B10) is less than or equal to 0.001, otherwise, it returns 0. We could also use the simpler formula =--(A10-B10<=0.001).

 (iv) =IF(SUM(A12:A20)>0, SUM(A12:A20), "Error")
     If the sum of the range is >0, that value is returned, otherwise, the text "Error" is displayed.

  (v) =IF(A1<>0, TRUE, FALSE) or =IF(A1, TRUE, FALSE)
     These will return the value TRUE if A1 contains a nonzero value, a formula giving a nonzero value, or the TRUE value. If A1 is empty or has the value 0, the FALSE is returned. A simpler formula would be =A1<>0, and this would be easier to understand.

 (vi) =IF(B5<>0, A5/B5, "") or =IF(B5<>0, A5/B5, NA())
     Here is a way of preventing the #DIV0! error. These formulas will return a blank or #N/A when the divisor is zero, otherwise, the division result is returned.

NESTED IFs

Nested formulas in Excel refer to combining the operations of two or more functions in one cell. For example, we may use the formula =ROUND(SUM (A1:A10),2) to compute the sum of a range and round the answer to two decimals. The syntax for ROUND is =ROUND(number, digits) but for the first argument (number) we may use an expression that returns a number such as SUM(A1:A10).

We have seen how an IF formula can return one of two values; what if we wish to return more? For example, the formula =IF(A1>10, "A1>10","A1 not >10") tells us either A1 is >10 or it is not—of course, it is more likely we would want some sort of calculation rather than a text but this example is easy to follow. How can we get the formula to tell us whether A1 is >10, A1 is equal to 10, or A1 is <10? The answer is with nesting; we replace one of the arguments in the simple IF formula with a second IF expression. The following flow diagram may clarify this concept. A lozenge shape indicates a test; the path to the left is what happens when the test is true and the path to the right is what happened when it is false.

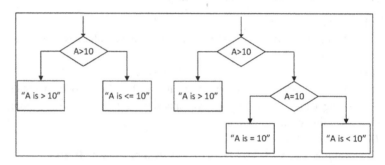

Here are some examples of nested IF formulas.

(i) =IF(A1>10, IF(A1>50, "Big","Medium"), "Small")
It is clear that if the condition A1 > 10 is false then the outer IF returns "Small." What happens if the condition is true? The inner IF comes into play. When A1 > 50, the inner IF returns 'Big,' otherwise it returns "Medium."

(ii) =IF(A1>10, IF(A1>50, "Big","Medium"), IF(A1<0, "Negative","Small"))
Here both the true-value and the false-value of the outer IF are themselves IF functions. What results when A1 is 0?

Nesting is permitted to 64 levels in Excel 2016, but getting the logic correct with anything this complex would be a major achievement! Generally, it is better to look for a solution using one of the lookup functions—see later.

IF FORMULAS WITH BOOLEAN FUNCTIONS

The logical functions AND(), OR(), NOT(), and XOR() may be used within an IF formula.

(i) =IF(AND(A2>0, A2<11), A2, NA())
The value A2 is returned if A2 is >0 and <11. Otherwise, the function NA() returns the error value #N/A.

(ii) =IF(OR(A2>0, B2>A2/2), 3 , 6)

Returns the value of 3 if either A2 > 0 or B2 > A2/2. If neither condition is true, the value 6 is returned.

(iii) `=IF(NOT(A2=0), TRUE, FALSE)`

This is the same as `=IF(A2=0, FALSE, TRUE)`.

(iv) `=IF(NOT(OR(A1=1, A2=1)), 1, 0)`

This is a somewhat contrived example. It returns 1 only when both A1 and A2 have a value that is not 1.

THE IFERROR FUNCTION

An earlier example used `=IF(B5<>0, A5/B5, "")` to compute A5/B5 if B5 was nonzero or to return a blank otherwise. An alternative formula for this is `=IFERROR(A5/B5, "")`. This will return the value of A5/B5 unless that arithmetic results in an error; such an error would result when B5 was zero (or blank) or when either cell had a nonnumeric value. As we proceed we will see other examples of the use of IFERROR.

EXERCISE 3: RESISTORS REVISITED

In Exercise 5 of Chapter 2, we developed a worksheet that computed the effective resistance of four resistors in parallel. We had to invent a workaround to allow us to use the worksheet with fewer than four resistors. With the information in this chapter, we can improve our work.

(a) Open Chap2.xlsx, select A1:E6, and click the Copy command on the Clipboard group of the Home tab (or use the Ctrl+C shortcut). Open Chap5.xlsx and move to Sheet3. With A1 as the active cell, use Ctrl+V to paste the copied material.

(b) Select D6:E6 and then click the eraser icon (clear) on the Editing group of the Home tab; select *Erase All* to remove both cell entries and formats.

Now we need to change some of the formulas to give us a worksheet as shown in Fig. 5.3.

	A	B	C	D	E	F
1	Resistors in Parallel					
2						
3	R	1240	1800	2000	0	
4	1/R	0.000806	0.00056	0.0005		
5						
6	R_e	540 ohms				

■ FIG. 5.3

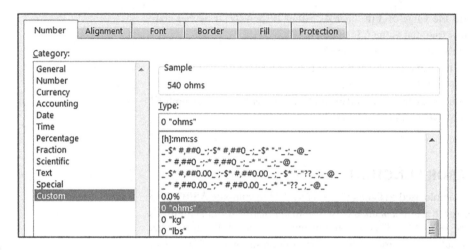

(c) In B4 replace `=1/B3` by `=IFERROR(1/B3,"")`. Copy this across to E4. Place a zero value in E4 to see that the new formula no longer gives #DIV0! but an apparently empty cell when the divisor is zero.

(d) We will find the reciprocal of all sums of the four 1/R values with one formula. At the same time, we will round the result to the nearest 10. In the ROUND function, the first argument gets rounded to multiples of 10, 100, 100 when the second argument is negative. What we need in B6 is `=ROUND(1/SUM(B4:E4),-1)`.

(e) Select B6:C6 and use the Merge and Center tool. Open the Formatting dialog and give this cell the custom format *0 "ohms"*—see Fig. 5.4.

(f) Test your worksheet to see that it gives correct results with other data.

(g) If this were a worksheet in the workplace, the user would have no need to see row 4. Furthermore, we would want to ensure that the user could not inadvertently change any formulas in that row. On a PC, right-click the row header 4 and select *Hide*. (On a Mac, hold down the ⌃Ctrl key and click on the row header 4 and select *Hide*.) Later we will learn how to protect cell B6. Save the workbook.

THE IFNA FUNCTION

A new function named IFNA was introduced with Excel 2013. It performs in a similar manner to IFERROR except that it is specify to the #N/A error. Referring to Fig. 5.5, a user has named the range as *Mydata* and had intended to sue this name in a VLOOKUP formula. He is surprised that

B2   ▾   ⋮   ✕   ✓   *fx*   =IFERROR(VLOOKUP(A2,mydate,2,FALSE),"not found")

	A	B	C	D	E	F	G	H	I
1		IFERROR	IFNA			Mydata			
2	b	not found	#NAME?		a	3			
3	c	not found	#NAME?		b	4			
4	z	not found	#NAME?		c	5			
5					d	6			

■ FIG. 5.5

his formula (in column B) returns *not found* for values that are clearly in the table. The formula =IFNA(VLOOKUP(A2, mydate,2, FALSE),"not found") in column C shows him that his VLOOKUP is not returning an #N/A error. The #NAME? error alerts the user to the true problem—the misspelling.

EXERCISE 4: QUADRATIC EQUATION SOLVER

In this exercise we will design a worksheet to solve a quadratic formula in the form $ax^2 + bx + c = 0$ using the quadratic formula:

$$x = \frac{-b \pm \sqrt{b^2 - 4ac}}{2a}$$

The quantity $b^2 - 4ac$ is called the discriminator. Clearly when the discriminator is negative there can be no real roots when it is zero only one real root is possible (or, if you prefer, two identical roots), otherwise, there will be two different real roots. For information on imaginary roots, see the file *Imagroots.xlsx* on the companion website.

Our completed worksheet will resemble that in Fig. 5.6.

	A	B	C	D	E	F
1	**Quadratic Equation Solver**					
2						
3	a	b	c		disc	
4	1	0	-9		36	
5						
6	Number of real roots		2			
7	Root 1	3	Root 2	-3		
8						

■ FIG. 5.6

(a) Open Chap5.xlsx. On Sheet4 enter the values and text shown in A1:
C4. Select A3:C4 and horizontally center the entries with the button in
the *Home / Alignment* (Home on a Mac) group.

(b) With A3:C4 still selected, use the command *Formulas / Defined
Names / Create from Selection (Formulas / Create from Selection* on a
Mac) to give the cells A4:C4 the names in the cells above them.

(c) Type `disc` (short for discriminant) in E3. Enter the formula `=b*b -
4*a*c_` in E4. Center E3:E4 and create the name *disc* for the cell E4.

(d) Temporarily ignore the entries in row 6.

(e) Type the text shown in A7 and C7. Enter these formulas
    B7: `=(-b+SQRT(disc))/(2*a)`
    D7: `=(-b-SQRT(disc))/(2*a)`

(f) Save the workbook Chap5.xlsx.

You now have an operational worksheet. Test it with quadratic
equations whose roots you know. What happens if the value of the
discriminant is negative? Try the values 1, 3, and 6 for a, b, and c,
respectively. Cells B7 and D7 show the error value #NUM! since it is
impossible to evaluate the square root of a negative number without
entering the realm of imaginary numbers.

The next steps will improve the behavior of the worksheet when the
discriminant is negative and present some additional information.

(g) Enter the text in A6 and Merge & Center this over A6:B6. In C6 enter
the formula `=IF(disc<0, 0, IF(disc=0, 1, 2))`. This returns 0
when the discriminant is negative, 1 when it is zero, and 2 in all other
cases. Starting in Excel 2016 IFS can be used instead of nested IF
statements. In this case the formula is `=IFS(disc<0,0,disc=0,1,
TRUE,2)`. The final TRUE is the default case. Using IFS would make it
so the formula would have an error if the worksheet was opened in an
earlier version of Excel.

(h) Replace the text in A7 with `=IF(C6=0,"",IF(C6=1,"Double
Root", "Root 1"))` or `=IFS(C6=0,"",C6=1,"Double Root",
TRUE,"Root 1")`. If there is one root, this returns the text "Double
Root," if there are two identical roots, it returns "Root 1." When there
are no real roots, it returns an empty text string.

(i) We require the formula in B7 to return a root when the discriminant
has a zero or positive value, and an empty text string otherwise. We
can achieve this by modifying it to read: `=IF(disc>=0,(-b +
SQRT(disc))/(2*a)`, ""`).

(j) Replace the text in C7 by `=IF(C6=2`, "Root 2", ""`)` to return the
text "Root 2" only when the discriminant has a positive nonzero value.

(k) Modify D7 to `=IF(C6=2,(-b - SQRT(disc))/(2*a)`, ""`)` to
return the value of the second root only when there are two
unique roots. Note that in B7 we tested to see if *disc* $>=0$ while

Do not rely on this type of
protection for supersensitive
information as many websites offer
password breakers.

in D7 we tested if there are two roots. This prevents a double root appearing twice.

(l) Make up some simple quadratics whose roots you know; for example, $(x-4)(x+3)=0$ gives $x^2 - x - 12 = 0$ and the roots are clearly 4 and -3. Test that $x^2 - 9 = 0$ reports a "double root." Make any required adjustments. Save the workbook.

EXERCISE 5: PROTECTING A WORKSHEET

Imagine that you have developed a worksheet for use by yourself or others to solve real-world problems. It would be wise to guard against accidental changes being made to cells, especially those with formulas. We will use the quadratic solver as an example. We will arrange things such that the user can visit only the cells needed to define the problem: A4:C4.

There are two steps to the process: (i) specify which cells the user may change by unlocking those cells (by default all cells on a new worksheet are locked) and (ii) switch on worksheet protection.

(a) Open Sheet4 of Chapt5.xlsx. Select A4:C4. Use the command *Home / Cells / Format (Format / Cells* on a Mac) to open the menu shown in Fig. 5.7. We could just click the Lock Cells item (last but one from the bottom), which acts as a toggle to lock and unlock cells. Alternatively, we may use the last item Format Cell... and open the Protection tab to reveal the dialog shown in Fig. 5.8. Here we will uncheck the Locked box and click the OK button.

(b) Use the command *Home / Cells / Format / Protect Sheet (Tools / Protection / Protect sheet* on a Mac) to open the dialog shown in Fig. 5.9. Our objective is just to prevent accidental (rather than malicious) changes so we will not use a password. Note that we have the option of allowing the user to visit both locked and unlocked cells.

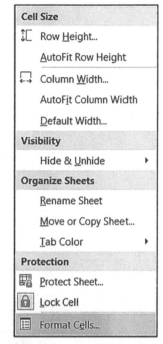

■ **FIG. 5.7**

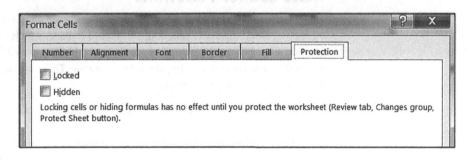

■ **FIG. 5.8**

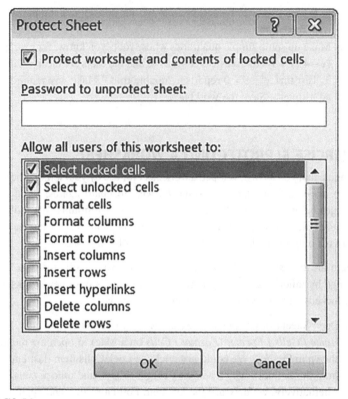

■ FIG. 5.9

If we did not wish the user to see our formulas, we could deselect Locked in this dialog. We shall leave all the other boxes unchecked.

(c) Test the worksheet to see that only the unlocked cells (A4:C4) can be changed. Save the workbook.

TABLE LOOKUP FUNCTIONS

Table lookup functions have a range of uses. Whenever you find yourself composing a multinested IF function, you should consider whether a lookup function would be more appropriate. A vertical table has its headings in a row, while a horizontal one has them in a column. There are no inherent advantages to one over the other.

The functions VLOOKUP and HLOOKUP have similar syntax:

VLOOKUP(lookup_value, table_array, column_index_num, range_lookup)
HLOOKUP(lookup_value, table_array, row_index_num, range_lookup)

Lookup_value	Is the value to be located in the first column of a vertical table (or the first row of a horizontal table). Lookup_value may be either a numeric or text value or a cell reference
Table_array	Is the range reference or name of the table
Column_index_num (row_index_num)	Is the column (or row) of the table from which the value is to be returned
Range_lookup	Is a logical value (TRUE or FALSE) specifying whether you want an approximate or an exact match. If range_lookup is TRUE or omitted, and there is no exact match, then the function returns the next largest value that is less than the lookup value. If FALSE and no exact match is found, the function will return the error value #N/A. If lookup_value is less than the lowest value in the first column (first row with HLOOKUP), the function returns the #N/A error value

There is also the LOOKUP function—see Exercise 8.

The MATCH function returns the relative position of a lookup_value in an array. Its syntax is: MATCH(***lookup_value, lookup_array,*** *match_type*). The first two arguments have the same meaning as before. Use 1 for match_-type when the table is sorted in ascending order, and you wish to find the largest value that is less than or equal to lookup_value. Use 0 when you needed an exact match; the table need not be sorted. Use −1 when the table is sorted in descending order, and you wish to find the smallest value that is greater than or equal to lookup_value. When *lookup_value* is nonnumeric, MATCH, VLOOKUP, and HLOOKUP are not case sensitive. MATCH may also be used with wildcards.

The INDEX function returns an element from an array and has two forms. The syntax of the first form is INDEX(***array***, *row_num, column_num*). Thus =INDEX(A1:C10, 2, 3) returns the value at the intersection of row 2 and column 3 of the Table A1:C10. Here it returns the value from cell C2.

EXERCISE 6: A SIMPLE LOOKUP

For the purpose of this exercise, a geologist wishes to grade some ore samples based on their rare metal content. Ore with 50 to 59 ppm is to be given a low grade: from 60 to 79 merits a medium ranking, from 80 to 99 is

◢	A	B	C	D	E	F
1	Vertical table lookup example					
2						
3	Site	ppm metal	Grade		Lookup Table	
4	A	75	medium		ppm	Grade
5	A	56	low		50	low
6	B	86	high		60	medium
7	B	60	medium		80	high
8	C	34	#N/A		100	very high
9	C	120	very high			

■ FIG. 5.10

considered high, and anything above that is very high. Our completed worksheet will resemble Fig. 5.10.

(a) Open Chap5.xlsx and start on Sheet5. For convenience, we will enter the lookup on the same worksheet as the ore data. Type the entries shown in E3:F8.

(b) Enter the text and numbers shown in A1:B10 and C3.

(c) The formula in C4 is =VLOOKUP(B4,$E$5:$F$8,2,TRUE). Since we do not want an exact match, we could omit the fourth argument and entered =VLOOKUP(B4,$E$5:$F$8,2). The $ symbols within the references are, of course, needed to keep the reference to the table unchanged as we copy the formula.

(d) Copy the formula down the column by double-clicking C6's fill handle.

(e) Save the workbook.

Cell C8, which displays #N/A, has a small green triangle in its upper left corner. When the cell is selected, a warning tip appears, which, if opened, gives information on the error value. One option is Ignore Error; use this to hide the triangle. Later in the chapter, we see how conditional formatting could be used to hide the NA() entry. Alternatively, we could make the formula in C4 to be =IFERROR(VLOOKUP(B4,$E$5:$F$8,2,TRUE),"") and fill this down the column. This will result in a blank cell when the lookup fails. We could, of course, have an informational message as in =IFERROR(VLOOKUP(B4,$E$5:$F$8,2,TRUE),"Value not located").

Here are some "experiments" you may wish to try. Between each one, close the file without saving and then reopen it.

(f) Modify the formulas in column C with an IF function such that when the B value is <50, the cell appears empty.

(g) Get the same effect by adding a new row to the table. Select E5:F5, on a PC right-click (on a Mac hold the [Ctrl] key and click) and use the command sequence *Insert / Shift Cells Down / OK*. Now add new data in E5 and F5. You will need to modify the formulas in column C.

(h) Cut the table and paste it on a new sheet. Note how the formulas in column C automatically adjust.

(i) Since the table is (i) sorted and (ii) only two columns wide, the LOOKUP function could be used in place of VLOOKUP. Read Help and make the change.

A very simple lookup. Suppose a teacher wishes to convert number grades to letter grades. Here is the start of a nested IF: =IF(B2<50,"F",IF(B2<60, "C", IF(B2<70,"B-")... But here is a much simple LOOKUP formula: =LOOKUP(B2,{0,50,60,70,75,80,85,90},{"F","C","B-","B","B +","A-","A","A+"}). It is much easy to type without making errors and much easier to change if the need arises.

EXERCISE 7: A TWO-VALUED LOOKUP

In this example our tables have more than two columns, so we need some way of indicating in the VLOOKUP formula which one to use. For this Exercise, we shall use MATCH.

Scenario: A nutritionist enters a client's height, frame type, and weight, and the worksheet gives the person's optimal weight and a comment on his actual weight. To keep the Exercise to a reasonable size, we limit ourselves to just male clients. Our final product will resemble Fig. 5.11.

(a) Begin by entering the table in E1:H16 within a new worksheet in Chap5.xlsx.

(b) Enter the text shown in A1:A12; and the values in B3:B5. Use custom format # ??/12 in B3 so we can use feet and inches. In B5 use custom format 0 "lbs" and apply this to B11 with the Format Painter. We will treat B12 differently.

(c) Create the following names: frame =F1:H1; height =E2:E16, and optimal = E2:H16.

(d) The formula in B8 is =MATCH(B4,frame,0)+1. Observe how this works: The L in B4 corresponds to the third column in the frame; we add 1 since we are working with the table *optimal*.

(e) In B11 we have =VLOOKUP(B3*12,optimal,B8). The value in B3 times 12 gives 68; this value is found in the first column of the

◢	A	B	C	D	E	F	G	H
1	Optimal male weight				height (ins)	S	M	L
2					62	130	134	143
3	Height	5 8/12			63	132	137	145
4	Frame	L			64	134	139	148
5	Weight	154 lbs			65	137	141	152
6					66	139	143	154
7	Match functions				67	141	148	159
8	frame	4			68	143	150	161
9					69	145	154	165
10					70	148	156	170
11	Optimal	161 lbs			71	150	159	172
12	Comment	7 lbs	under		72	154	163	174
13					73	156	168	179
14	Putting it all together				74	161	170	183
15	Optimal	161 lbs	under by 7 lbs		75	165	174	190
16					76	168	179	194

■ FIG. 5.11

optimal, and the function returns the corresponding value in the fourth column since B8 evaluates to 4.

(f) The formulas in B12 and C12 are, respectively

```
=IF(B5=B11,"OK",ABS(B11-B5)&" lbs")
=IF(B5=B11,"",IF(B11<B5,"over", "under"))
```

The range A14:C15 shows an alternative approach. The formula in B15 uses INDEX and MATCH without the need for a separate cell while C15 combines what is in B2:C12.

(g) Enter the test text in A14; select A14:C14 and merge the cells. The formulas in B15 and C15 are, respectively:

```
=INDEX(F2:H16,MATCH(B3*12,height,1),MATCH(B4,frame,0))
=IFS(B11=B5, "OK", B5>B11, "over by "&B5-B11& " lbs",
TRUE, "under by "&B11-B5& " lbs")
```

(h) Save the workbook.

Change the values in B3:B5, observe the results, and ensure you understand how the formulas work.

EXERCISE 8: CONDITIONAL ARITHMETIC

The Excel functions SUMIF, COUNTIF, and AVERAGEIF may be used to conditional sum, count, or AVERAGE a range of values subject to specified criterion being satisfied. In the case of SUMIF and AVERAGEIF, the range to be summed may differ from the range to be tested. If more than one criterion is to be tested we use SUMIFS, COUNTIFS, or AVERAGEIFS.

	A	B	C
1	**Unit**	**Problem**	**Downtime**
2	A	P2	9
3	B	P4	14
4	A	P1	20
5	B	P4	26
6	C	P3	14
7	C	P6	12

■ FIG. 5.12

To appreciate these functions, one needs a moderately large dataset. To save the reader having to construct one, the file ConditionalSums.xlsx is available on the companion website. Fig. 5.12 shows the first few rows of Sheet1 in this workbook. Note that each column is named by the label in the top row.

(a) Download and open the file ConditionalSum.xlsx. Type Unit into the name box to see that the data in column A is named Unit.

(b) Open Chap5.xlsx and insert a new worksheet. Use the command *Formulas / Defined Names / Name Manager (Formula / Defined Names / Defined Names* on a Mac). You will see that named ranges from ConditionalSum.xlsx are not available. You will have to recreate the names *Unit, Problem,* and *Downtime* in Chap5.xls as well. Click the New... button to bring up the dialog box is in Fig. 5.14 by selecting the range in ConditionalSum.xlsx. You will have close out of the name manager each and return to the workbook each time you add a name in the workbook. (On a Mac, you need to have the Unit, Problem, and Duration columns visible in the background to select them.)

	A	B	C	D	E	F	G	H	I
1	How many problems for each Unit?				Average downtime for each Unit				
2	A	#VALUE!	=COUNTIF(Unit,A2)		A	#VALUE!	=AVERAGEIF(Unit,E2,Downtime)		
3	B	#VALUE!	=COUNTIF(Unit,A3)		B	#VALUE!	=AVERAGEIF(Unit,E3,Downtime)		
4	C	#VALUE!	=COUNTIF(Unit,A4)		C	#VALUE!	=AVERAGEIF(Unit,E4,Downtime)		
5									
6	Total downtime for Unit A				Average downtime for each Unit by Problem				
7		#VALUE!	=SUMIF(Unit,"A",Downtime)			P1	P2	P3	P4
8					A	#VALUE!	#VALUE!	#VALUE!	#VALUE!
9	Total stoppages over 10 minutes				B	#VALUE!	#VALUE!	#VALUE!	#VALUE!
10	10	#VALUE!	=COUNTIF(Downtime, ">"&A10)		C	#VALUE!	#VALUE!	#VALUE!	#VALUE!
11	20	#VALUE!	=COUNTIF(Downtime, ">"&A11)						
12	25	#VALUE!	=COUNTIF(Downtime, ">"&A12)		=AVERAGEIFS(Downtime,Unit,$E8,Problem,F$7)				
13	Requires workbook ConditionalSum.xlsx to be open								
14									

■ FIG. 5.13

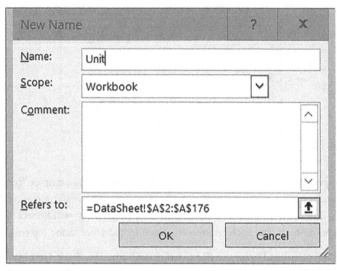

■ FIG. 5.14

(c) In Chap5.xlsx construct the worksheet shown in Fig. 5.13. All numerical values result from formulas that are shown in the figure.

(d) Save the workbook.

Compare the formulas in B2 and B10. In the former, we have used a cell reference (A2) for our criterion. In the later we have used a literal (text)—note the quotation marks around >*10*. Had we wished to count stoppages of exactly 10 we could have used either =COUNTIF(Downtime,"=10") or more simply =COUNTIF(Downtime,10). Had we wished to make a table for various stoppage values we could have used something like that shown in Fig. 5.15.

What happens if you close both files and then reopen Chap5.xlsx on the newly constructed worksheet?

Very often rather than having to develop formulas to summarize a dataset it is easier to let Excel do the work by way of a Pivot Table. If you are feeling adventurous, do a Bing/Google with the term "Excel 2016 pivot table tutorial." Then see if you can make a report similar to that in Fig. 5.16.

	A	B	C
10	10	121	=COUNTIF(Downtime, ">"&A10)
11	20	64	=COUNTIF(Downtime, ">"&A11)
12	25	24	=COUNTIF(Downtime, ">"&A12)

■ FIG. 5.15

Average of Downtime	Column Lab ▾						
Row Labels ▾	P1	P2	P3	P4	P5	P6	Grand Total
A	16.5	15.5	14.8	14.7	14.3	24.6	16.1
B	16.8	16.0	19.4	13.0	18.9	15.3	16.6
C	16.2	18.5	11.4	15.3	17.6	15.7	16.1
Grand Total	16.5	17.0	15.4	14.4	17.0	17.5	16.2

■ FIG. 5.16

An Excel user has a worksheet with numbers (integers for simplicity) in A1: A100. She finds the maximum with =MAX(A1:A100); this turns out to be 55, but then she wonders if this is unique. In cell D4 the formula =COUNTIF(A1:A100,55) would tell her how many cells hold this value. However, in case the numbers get changed at a later date, this formula would be safer =COUNTIF(A1:A100, "="&MAX(A1:A100)). To sum all the numbers that are equal to, or exceed, the maximum value one could use =SUMIF(A1:A1000,">="&MAX(A1:A1000)). Likewise, the formula =SUMIF(A1:A20, ">="&E2) will sum the numbers in A1:A20 that are equal to, or exceed, the value in E2.

EXERCISE 9: ARRAY FORMULAS

An array formula is a formula that can perform multiple calculations on one or more of the items in an array. Some array formulas return a single result while others can return results to a range of cells. We will look at the first type here and at multicell array formulas in later chapters.

We will develop a worksheet similar to that in Fig. 5.17.

◢	A	B	C	D	E	F	G	H	I	J	K
1	Array Formulas										
2	95	33	99	30	99	24	42	73	75	64	
3											
4		Sum of N$_i$ -5		584		=SUM(A2:J2)-COUNT(A2:J2)*5					
5				584		{=SUM(A2:J2-5)}					
6		Sum of top 3		293		=LARGE(A2:J2,1)+LARGE(A2:J2,2)+LARGE(A2:J2,3)					
7				293		=SUM(LARGE(A2:J2,{1,2,3}))					
8				293		{=SUM(LARGE(A2:J2,ROW(1:3)))}					
9		Sum of first 100		5050		=(100*(100+1))/2					
10				5050		{=SUM(ROW(1:100))}					

■ FIG. 5.17

(a) In a new worksheet in Chap5.xlsx, enter the text shown in column A and the numbers in row 2. You may optionally merge the cells A4:C4, A6:C6, and A9:C9.

(b) In D4 we wish to compute $\sum_i(N_i-5)$. Clearly, this is the same as $\sum_i N_i - 5\sum_i 1$. We could count the cells in the range A2:J2 but for safety sake, we will use the COUNT function. So our formula is =SUM(A2:J2)-COUNT(A2:J2)*5.

(c) In D5 we use an array formula. Enter =SUM(A2:J2-5) and pause for a moment. You have learned various methods of committing a formula such as using one of the keys [Enter←], [Tab⇄], [→] or clicking the check mark in the formula bar. None of these will work for an array formula. You must commit it with [Ctrl]+[⇧ Shift]+[Enter←]. Now look in the Formula Bar: the formula appears within curly braces {=SUM(A2:J2-5)}.

Because it is necessary to commit array formulas with Control + Shift + Enter, some writers have started calling them CSE formulas—we shall not adopt this terminology!

> If you somehow open D5 for editing, you must remember to use [Ctrl] + [⇧ Shift] + [Enter←] to complete the edit and return to the Ready state.

(d) It is interesting to see how the formula works. Make D5 the active cell, use the command *Formulas / Formula Auditing / Evaluate Formula* (not available on a Mac) and click the Evaluate button a few times.

(e) In D6:D8 we tackle the problem of summing the three largest values in the data. The simple formula is as shown next to D6; enter it and confirm you get the correct result. In D7 we get a little smarter. We enter an array {1,2,3} within the LARGE function. Type the formula and commit it in the normal way. Next, we go for a real array formula =SUM(LARGE(A2:J2, ROW(1:3))) which we commit with [Ctrl]+[⇧ Shift]+[Enter←]. Again, it is instructive to use *Formulas / Formula Auditing / Evaluate Formula* to observe how it works. We see that ROW(1:3) generates the array {1,2,3} that we used in the formula in the cell above.

(f) Finally, in D9 and D10 we find the sum of the first 100 integers. In D9 we use the formula attributed to Gauss: $\sum_1^N N_i = \frac{N\times(N+1)}{2}$. Then in D10, we use a simple array formula to get the same result. Again, use the command *Formulas / Formula Auditing / Evaluate Formula*.

(g) Save the workbook.

Now we look at a problem that could arise. Imagine you have made a worksheet similar to the one in this Exercise; you have checked the formulas, and all looks fine. The next day you decide to make it look neater by adding a blank row.

(h) On a PC, right-click (on a Mac hold the [Ctrl] key and click) on the header for row 3 and insert a new row. Oh dear! The results where the formulas contain the ROW function have now produced incorrect values.

◢	A	B	C	D	E	F	G	H	I	J
13	a	b	a	c	b	a	a	a	a	b
14	z	x	z	x	x	x	z	z	x	x
15			0			2				
16					{=SUM((A13:J13="a")*(A14:J14="x"))}					
17	{=SUM(--AND(A13:J13="a",A14:J14="x"))}									

■ FIG. 5.18

(i) Let's see if the use of absolute references helps. Delete the inserted row and replace the formula in D8 by =SUM(LARGE(A2:J2,ROW($1:$3)). Do not forget to use [Ctrl]+[⇧ Shift]+[Enter ←]. Insert a new row 3 and again D8 give the wrong answer. The formula becomes =SUM(LARGE(A2:J2, ROW($1:$4)).

(j) The solution for a bulletproof array formula using ROW is the INDIRECT function. Delete the inserted row and replace D8's formula by =SUM(LARGE(A2:J2,ROW(INDIRECT("1:3")))) and commit with [Ctrl]+[⇧ Shift]+[Enter ←]. Now if you insert a row above the formula it still gives the correct answer. The INDIRECT function is, in effect, saying "use the stuff in quotes and never change it."

It is important to know that Boolean operators (AND, OR, NOT, and XOR) cannot be used within an array formula. However, if we wish to use an AND condition we need only multiply the two conditions, and for OR we use addition. Remember that logical expressions evaluate to either FALSE or TRUE but when arithmetical operations are done on these they are treated as 0 and 1. In Fig. 5.18 we have two rows of letters and we wish to know how many times the letter *a* appears above the letter *x*. The formula with AND fails, but the one using multiplication works. Use Evaluate formula to ensure you understand why. This problem is used just for a demonstration; in a real situation, we would use SUMPRODUCT (see later) for this problem.

EXERCISE 10: THE SUMPRODUCT FUNCTION

The primary purpose of the SUMPRODUCT function is to compute the sum of the products of the elements of two or more arrays. Thus SUMPRODUCT (A1:A3,B1:B3) evaluated A1*B1 + A2*B2 + A3*B3. However, Excel users have expanded its use to perform conditional summations.

◢	A	B	C	D	E	F
1	The SUMPRODUCT function					
2						
3	PPM	1	2	3	4	5
4	#	1	11	10	2	1
5						
6	Average	2.64		Stdev	0.86	
7						
8		Result	2.64 ± 0.86			

■ **FIG. 5.19**

We will start with the primary use. Scenario: A process engineer has taken 25 samples from a product stream and analyzed them for an impurity. The results are tabulated in rows 3 and 4 of Fig. 5.19. He needs to compute the average and standard deviation. When computing an average where a measurement x_i occurs n_i times, we speak of a weighted average. When computing averages we frequently also wish to compute the standard deviation of the samples. These two quantities are found using the formulas:

$$avg = \frac{\sum_i x_i n_i}{\sum_i n_i} \quad std^2 = \frac{\sum_i (x_i - \bar{x})^2 n_i}{\sum_i n_i - 1}$$

The numerator for the formula for the average is exactly what SUMPRO-DUCT computes, while the denominator is found with SUM. With some imagination, we can also use SUMPRODUCT for the sample standard deviation.

(a) Open new worksheet in Chapt5.xlsx and enter everything shown in Fig. 5.18 other than cells B6, E6, and C8.

(b) In cell B6 use =SUMPRODUCT(B3:F3,B4:F4)/SUM(B4:F4).

(c) In E6 use =SQRT(SUMPRODUCT((B3:F3-$B$6)^2,B4:F4) / (SUM(B4:F4)-1)) to get the standard deviation. Note how SUMPRO-DUCT accepts the argument (B3:F3-$B$6)^2 without requiring that we make it an array function. This is a major strength of the function. It would be instructive to use *Formulas / Formula Auditing / Evaluate Formula* to see how this works.

(d) To summarize the results in C8 we use =ROUND(B6,2) & " ± " & ROUND(E6,2). Recall from Chapter 2 that ± is produced with [Alt] + 0177 on the numeric keypad. In this formula the ampersand (&) is used as the concatenation operator—it joins text together.

	A	B	C	D	E	F	G	H
1	The SUMPRODUCT function							
2								
3	Test1	a	a	b	a	a	b	a
4	Test2	x	y	y	x	x	x	y
5								
6		How many samples have Test1 = a and Test2 = x?						
7		COUNTIFS		3	SUMPRODUCT		3	

■ FIG. 5.20

SUMPRODUCT is also used in ways that the developers may never have considered. It can be used for single and multiple criteria summations. Indeed, before Excel 2007 introduced SUMIFS and COUNTIFS, this was the only way to handle multiple criteria. But there are still times when SUMPRODUCT outpaces these new functions: SUMPRODUCT allows you to specify criteria that the other functions do not permit.

To see how SUMPRODUCT works with conditional summations we will make a worksheet similar to that in Fig. 5.20.

(e) Open a new worksheet in Chapt5.xlsx and enter everything shown in Fig. 5.20 other than cells D7 and G7.

(f) In D7 enter =COUNTIFS(B3:H3,"a",B4:H4,"x") and visually confirm its accuracy.

(g) In G7 enter =SUMPRODUCT(-(B3:H3="a"),-(B4:H4="x")) and confirm it agrees with the COUNTIFS result.

We have two logical expressions: B3:H3="a" and B4:H4="x." Normally these would return arrays of Boolean values of FALSE or TRUE. But when you perform an arithmetical operation on Boolean values they convert to the numerical values of 0 and 1. With −−(B3:H3="a") and with −−(B4:H4="x") we have performed a double unitary negation to perform the conversion. If you use *Evaluate Formula* on G7, at some point you will see something like Fig. 5.21. Note that we could just as well used =SUMPRODUCT((B3:H3="a")*(B4:H4="x")) for the same purpose; here the multiplication coerces Boolean to numeric.

Now we will see a SUMPRODUCT formula where the other functions are of little or no use. We will examine a range of numbers (albeit, a rather small dataset but this is just a demonstration) and sum only those that are even.

(h) On the same worksheet, enter the text and numbers shown in rows 10 of Fig. 5.22. Enter the text shown in rows 11 and 12.

Evaluation:

= | SUMPRODUCT({1,1,0,1,1,0,1},--({*TRUE,FALSE,FALSE,TRUE,TRUE,TRUE,FALSE*}))

■ FIG. 5.21

◢	A	B	C	D	E	F	G	H	I	J	K	L	M	N	O	P
10	Data	23	7	16	16	20	21	16	21	18	9	8	7	6	9	7
11	Helper	0	0	16	16	20	0	16	0	18	0	8	0	6	0	0
12	Sum of even numbers	Old way			100	SUMPRODUCT		100								

■ FIG. 5.22

(i) We have a Helper Row (at other times we might use a Helper Column). This is data that is generated from the original data for the purpose of selecting just relevant values. In B11 enter =ISEVEN(B10) *B10 and drag the fill handle to P11. The ISEVEN function returns either FALSE or TRUE but when we multiply its result by the corresponding number we get either 0 or the number since False acts like 0 and True like 1.

(j) In E12 we sum the selected data with =SUM(B11:P11).

(k) In C11 enter =SUMPRODUCT(-(MOD(B10:P10,2)=0),B10:P10). Visually confirm it gives the correct result. Why did we not use the ISEVEN function? Because it will not accept an array while MOD does. The formula MOD(N, D) returns the remainder of N/D; MOD stands for Modulus. This is an example of SUMPRODUCT saving us cluttering the worksheet with a helper row. However, helper rows/columns can be useful to check that a complex formula is giving the correct result and then the helper data can be erased.

(l) Use *Evaluate Formula* to see how the formula in C11 works.

(m) Save the workbook.

EXERCISE 11: CONDITIONAL FORMATTING

This chapter has dealt with formulas that return values that depend on one or more conditions (tests, or criteria). In the last exercise, we look at conditional formatting: how to give cells a format that depends on a criterion. We may wish values above (or below) a certain value to be displayed in a different color or to be in cells with an eye-catching background fill. Another use is to hide certain values.

We will start by seeing how to hide specified formula values. In Exercise 6 we had a formula that may return #N/A. If you fixed this using an if statement or an expanded lookup table, adjust the formula so that it only uses the values from 50 to 100 in the lookup table. We will see how we may hide this by changing the font's color or highlight the problem by coloring the cell.

(a) Select C5:C10 in Sheet5. Use the command *Home / Styles / Conditional Formatting* (*Home / Conditional Formatting* on a Mac) and in the resulting drop-down menu select *New Rules*. In the next dialog box, click on *Format only cells that contain....* and select *Errors* (see Fig. 5.23). Click the Format button and in the resulting dialog (Fig. 5.24) open the Font tab to set the color to match the cell background color (most likely it will be white). Return to the worksheet and the #N/A value is invisible.

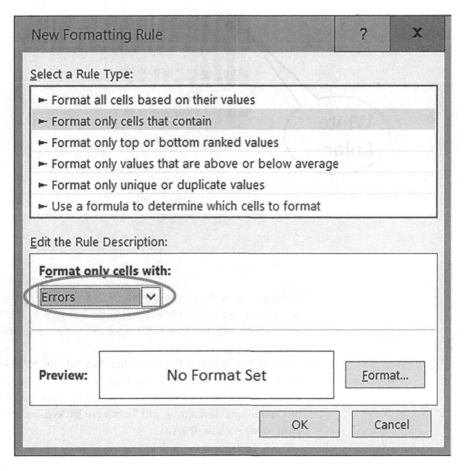

■ FIG. 5.23

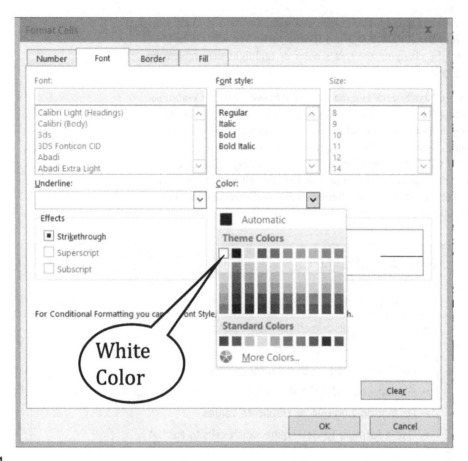

Alternatively, at the start of the process, select *Use a formula to determine which cells to format* and in the box type =ISERROR(C4), refer to Fig. 5.25. We specify C4 since this is the top left cell of the selected range.

(b) If we wish to highlight the cell rather than hide the NA() by altering the font color, we would open the Fill tab (Fig. 5.26) and specify a fill color (say, red) as our formatting. Any cell to which we have applied this conditional formatting will have a red background if it holds an error value such as #N/A.

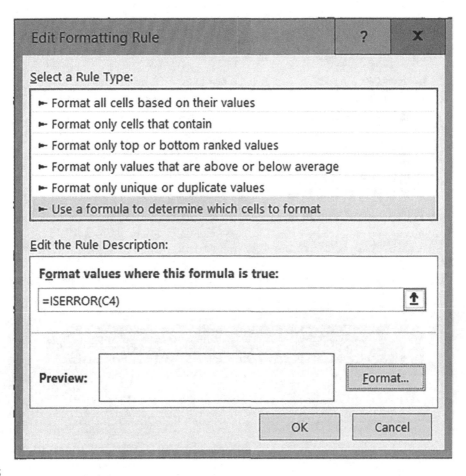

Edit Formatting Rule   ?   X

Select a Rule Type:

- ► Format all cells based on their values
- ► Format only cells that contain
- ► Format only top or bottom ranked values
- ► Format only values that are above or below average
- ► Format only unique or duplicate values
- ► Use a formula to determine which cells to format

Edit the Rule Description:

Format values where this formula is true:

=ISERROR(C4)   ⬆

Preview: Format...

OK   Cancel

■ FIG. 5.25

(c) To demonstrate another Conditional Formatting feature, select the range B4:B9 on Sheet5, open the Conditional Formatting dialog and click on Color Scales and select one of the options in the displayed gallery. Now the cells in the range will have a color fill depending on their relative values.

We have only touched upon the features available in Conditional Formatting. We will see more throughout the book. The file ConditionalFormat. xlsx on the companion website shows other examples. To learn more do an internet search with "Excel 2016 conditional formatting."

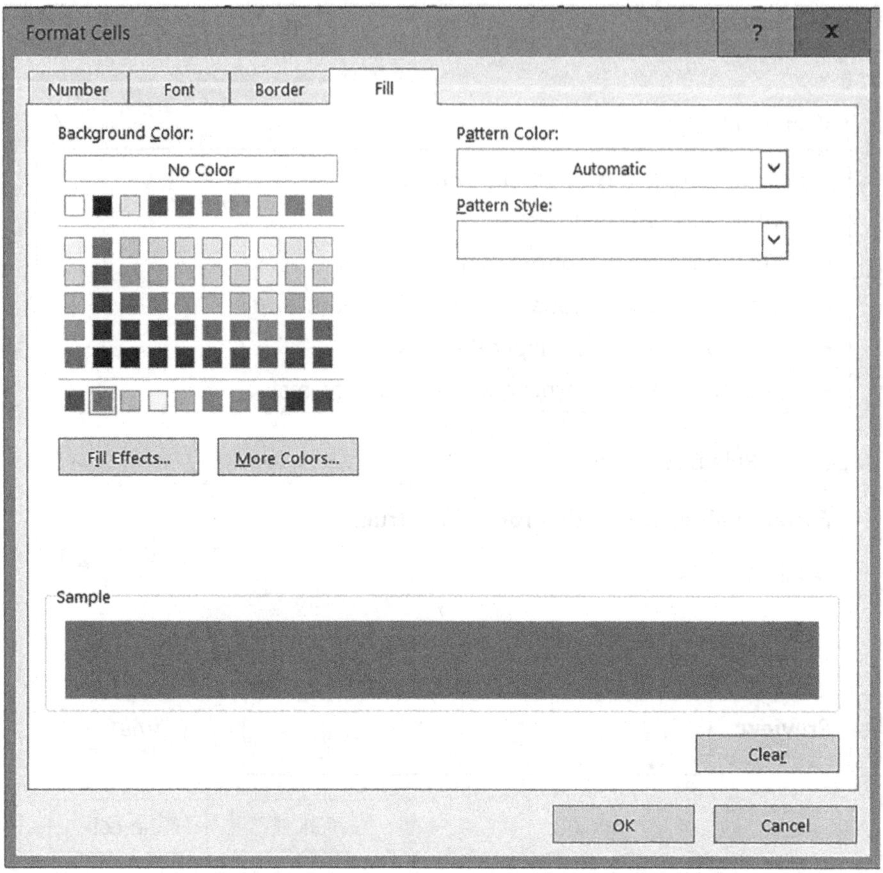

■ FIG. 5.26

PROBLEMS

1. In the hydraulic jump,[1] a liquid stream of depth D_1 flowing at velocity v_1 suddenly increases its depth to D_2. Fig. 5.27 shows the equation that governs this effect. What formula will you use in E5 that can be dragged across to H5?

2. *Refer to Fig. 5.11 of Exercise 7. We saw that =MATCH(B4,frame,0) tells us which column in the range frame matches the frame type entered in B4. Write a formula to find the row position in the range height to match the client's height entered in B3. With the existing data in B3:B4, our client's height and type place him in row 7 and column 3 of the table F2:H16. Write a formula beginning =INDEX

---

[1]Carnahan et al., Applied Numerical Methods, Wiley, New York, 1969 (page 203).

	A	B	C	D	E	F	G	H
1	Hydraulic Jump							
2				$D_2 = \dfrac{D_1}{2}\left[\sqrt{1+\dfrac{8v_1^2}{gD_1}} - 1\right]$ if $v_1 > \sqrt{gD_1}$				
3								
4								
5	D_1	g		v_1	5	10	15	20
6	10	32		D_2	No Jump	No Jump	No Jump	15.8

■ FIG. 5.27

that will locate the optimal weight within this. Finally, combine the INDEX formula and the two MATCH formulas into one.

3. *The range A2:A11 in a worksheet contains both positive and negative values, and you wish to sum only the positive ones. Give a formula that will accomplish this.

4. * With the same numbers as in Problem 3, find the sum of the squares of only the positive ones. Give a formula that will accomplish this. Hint: try either of these:
   (i) SUMPRODUCT, or
   (ii) IF nested inside a SUMSQ as an array formula.

5. *With the same numbers as in Problem 3, find the average of the squares of the positive values.

6. Construct a worksheet similar to that in Fig. 5.28 to make a simple molar mass calculator. Cell C10 uses a SUMPRODUCT formula. Each cell in row 7 uses two IF formulas joined with the concatenation operator &. The first IF gets the symbol and the next gets the number if

	A	B	C	D	E	F	G	H	I	J
1	Molecular Mass Calculator									
2										
3	At. Wt	12.01	1.008	16	14.01	30.97	32.07	35.45	79.9	126.904
4	Element	C	H	O	N	P	S	Cl	Br	I
5	Number	2	6				1			
6										
7	Formula	C2	H6				S			
8										
9	Compound		Molar Mass							
10	C2H6S		62.135							

■ FIG. 5.28

◢	A	B	C	D	E	F	G	H	I
1	Resistor Four-color Code								
2									
3	R	Color	Bar1	Bar2	Bar3		Tolerance		Bar 4
4	0	Black					5%	Gold	x
5	1	Brown					10%	Silver	
6	2	Red					20%	None	
7	3	Orange							
8	4	Yellow			x				
9	5	Green	x						
10	6	Blue		x					
11	7	Violet					Resistance		560E+3
12	8	Grey					Tolerance		5%
13	9	White							

■ FIG. 5.29

it is >1. Then the row 7 cells are themselves concatenated in A10. Hiding rows 3 and 7 would make the worksheet more interesting!

7. Refer back to Problem 6 in Chapter 4. This time we will solve the problem without the helper columns. Construct a worksheet similar to Fig. 5.29. The cells I11 and I12 each contain formulas that use SUMIF. Alternatively, you may wish to use SUMPRODUCT in your formulas. Again we need to protect against having more than one X in a column. Use the same approach as in Chapter 4's problem but with COUNTA rather than COUNT.

8. Redo Problem 7 using formulas in I11 and I12 with MATCH nested inside INDEX.

9. Rev. Rebecca is a recycler; she finds 49 candle stubs and makes exactly seven new candles. These in turn yield seven stubs, allowing her to make candle number eight from which she later gets a stub. This is illustrated in A3:C7 of Fig. 5.30 where we represent the process by $49 \Rightarrow 8R1$.

   (i) What formulas are used in F5, G5, and E6?
   (ii) Show that $59 \Rightarrow 9R5$, $67 \Rightarrow 11R1$, $79 \Rightarrow 13R1$, and $88 \Rightarrow 14R4$.
   (iii) From this data, you might conclude that N stubs always yield INT(N/6) candles. You might reason that this is so because, although it takes seven stubs to make a candle, only six get consumed. Show that for $N=72$ this is incorrect; under what circumstance does it break down? Algorithms must be fully tested!
   (iv) What formulas are used in J5 and K5?

◢	A	B	C	D	E	F	G	H	I	J	K
1	Recycle										
2											
3		49 → 8 R 1				79 → 13 R 1					
4	stubs	candles	remainder		stubs	candles	remainder		stubs	candles	remainder
5	49	7	0		79	11	2		49	8	1
6	7	1	0		13	1	6		79	13	1
7	1	0	1		7	1	0		72	11	6
8					1	0	1		125	20	5

■ FIG. 5.30

10. In the left-hand part of Fig. 5.31, we see the solution of a system of equations, while in the right-hand part we see a system with no solution. Refer to Exercise 8 of Chapter 4 and tell what array formula is used in G4: G6.

11. In B11 Exercise 7 we ended up with:
    `=VLOOKUP (B3*12, optimal, MATCH(B4,frame,0)+1)`
    (i) Modify it to return "?" if B3 is less than E2 or B4 has an invalid value.
    (ii) Further modify it to return "??" if B3 exceed the largest value in E2:E16.
    (iii) Modify B12:C12 to return nothing when B11 is "?" or "??."

12. Refer to Fig. 5.32. Your task here is to explain how formulas work. You may need to do an internet search on the topic of prime numbers and the use of checksums in ISBN numbers. If this is a written

	A	B	C	D	E	F	G		A	B	C	D	E	F	G
1	Matrix							1	Matrix						
2								2							
3		Coefficients		Constants			Soln	3		Coefficients		Constants			Soln
4	1	2	4	4		x	2	4	1	2	4	4		x	?
5	4	6	9	20		y	5	5	3	6	12	20		y	?
6	2	2	8	-2		z	-2	6	2	2	8	-2		z	?

■ FIG. 5.31

◢	A	B	C	D
1	SUMPRODUCT formulas			
2				
3	Is number a prime?	703	Not Prime	
4				
5	10-digit ISBN checksum	075065614x	10	valid
6	13-digit ISBN checksum	978-0-306-40615-7	7	valid

■ FIG. 5.32

assignment, use the Window's Snipping Tool to illustrate your work with screen captures from *Evaluate Formula*.

(a) B3 holds an integer value and the formula in C3 tells if the integer is a prime number or not. The formula is =IF(SUMPRODUCT(-(MOD(B3, (ROW(INDIRECT("2:"& INT(SQRT(B3))))))=0)) "Not Prime", "Prime"). Can you explain how it works?

(b) Formerly books were identified by 10-digit International Standard Book Numbers (ISBN) but more recently 13-digit ISBNs were introduced. Explain either of these formulas beginning by telling the purpose of each function. Either
C5:=IF(MOD(SUMPRODUCT(-(MID(B5,ROW(A1:A9),1)), {10;9;8;7;6;5;4;3;2}),11)=0,0,11-MOD(SUMPRODUCT(-(MID(B5,ROW(A1:A9),1)),{10;9;8;7;6;5;4;3;2}),11))
Or
C6:=MOD(10-MOD(SUMPRODUCT(-(MID(SUBSTITUTE(B6,"-",""), ROW(INDIRECT("1:12")) ,1)), {1;3;1;3;1;3;1;3;1;3;1;3}),10),10)
Can you modify one of these formulas to report either "valid" or "invalid" depending on if the last character in the ISBN matches the computed checksum number?

APPENDIX: SUPPLEMENTARY MATERIAL

Supplementary material related to this chapter can be found on the accompanying CD or online at https://doi.org/10.1016/B978-0-12-818249-9.00005-4.

Chapter

6

Data Mining

CHAPTER OUTLINE

The purpose of this chapter is to demonstrate how to work with lists containing alphanumeric data. The first topic shows how to import a data list from a non-Excel file. We will then see how to count, sum, and average a column of numbers using criteria that relate to values in a column of text. We will also see the powerful Pivot Table feature. Finally, we will do simple sort and filtering.

The original data were generated by this scenario: Acme Manufacturing has a number of machines in which a particular part must be frequently replaced. There are two sources for this part, Alpha and Beta, each of whom supplies the part in brass, nickel, or stainless steel. The maintenance engineer has kept track of how many hours each part lasts. He keeps his data in a Notepad file; a sample line reads "Alpha, Brass, 450." The file Chap6.txt, available from the companion website, contains 100 records.

EXERCISE 1: IMPORTING TXT FILE

We will open the text file and see that Excel provides a tool to parse a record into fields. Before beginning the exercise you may wish to open the file in Notepad to see its contents.

Liengme's Guide to Excel 2016 for Scientists and Engineers. https://doi.org/10.1016/B978-0-12-818249-9.00006-6

(a) Start Excel and use the command *Office / Open* (*File* / Open on a Mac) and navigate to the folder where you store files for this book. Make sure the bottom right box of the file open dialog reads *Text* or *All Files*, and point to the Chap6.txt file. Use *Open* or double-click its name to bring up Step 1, the *Text Import Wizard* dialog shown to the left in Fig. 6.1.

(b) The fields in our text file have a separator between fields (they are delimited) rather than having a fixed width. Having specified *delimited*, click Next to open the Step 2 second dialog box—see right side of Fig. 6.1.

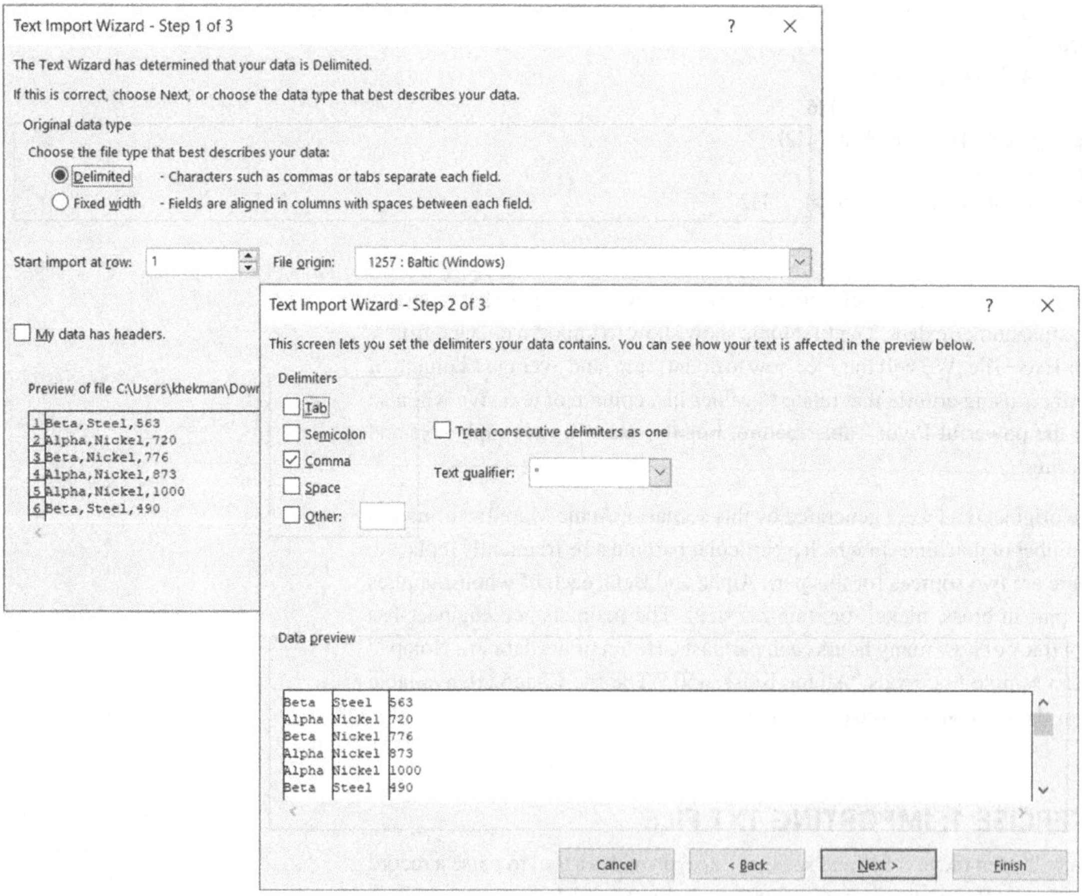

■ **FIG. 6.1**

(c) Our delimiters are commas, so uncheck the Tab box and check Commas. Note in the preview box the vertical lines that show where the fields begin and end. Click the Finish button.

(d) On a newly inserted row 1, add headers to the columns; Brand, Type, and Hours.

(e) The worksheet holding the data will be named Chap6 after the TXT file. On a PC, right-click the worksheet tab and change the name to Sheet1. (On a Mac, hold down the Ctrl key and click on the worksheet tab.) Save your work as an Excel file named Chap6.xlsx.

EXERCISE 2: COUNTING AND SUMMING WITH CRITERIA

In this Exercise, we look at the functions COUNTIF, SUMIF, AVERA-GEIF, COUNTIFS, SUMIFS, and AVERAGEIFS, some of which were briefly touched upon in Chapter 5.

(a) On Sheet1 of Chap6.xlsx enter the text shown in cells E1:I23 of Fig. 6.2.

(b) Use the following formulas with one criterion:

F2: =COUNTIF($A$2:$A$101, E2) to count Alpha items
G2: =SUMIF($A$2:$A$101, E2, $C$2:$C$101) to sum Alpha Hours
H2: =G2/F2 to compute average Hours of Alphas
I2: =AVERAGEIF(($A$2:$A$101, E2, $C$2:$C$101) as above

Copy F2:I2 down to next row to find the corresponding Beta values.
Since it is clear that row 1 would have no effect on our computations, and provided we added nothing to columns A through C (other than new data), we could simplify all the equations before by using "full column" references. An example would be for G2 =SUMIF($A:$A, E2, $C:$C). We will adopt this method for the remaining equations.
The functions whose names end with *IFS* allow us to Count and Sum subject to multiple criteria. We will find, for example, how many brass parts were supplied by Alpha. It is important to note that because these functions allow for a variable number of criteria, their syntax differs from those of the corresponding functions with names ending with *IF*.

(c) Use the following formula to count with two criteria:

F7: =COUNTIFS($A:$A, $E7, $B:$B, F$6)

◢	A	B	C	D	E	F	G	H	I
1	Brand	Type	Hours			Count	Sum	Average	
2	Beta	Steel	563		Alpha	47	28756	611.83	611.83
3	Alpha	Nickel	720		Beta	53	35034	661.02	661.02
4	Beta	Nickel	776						
5	Alpha	Nickel	873			Count by 2 criteria			
6	Alpha	Nickel	1000			Brass	Nickel	Steel	Total
7	Beta	Steel	490		Alpha	13	18	16	47
8	Alpha	Brass	301		Beta	17	12	24	53
9	Alpha	Nickel	709						
10	Alpha	Nickel	758			Sum by 2 criteria			
11	Alpha	Brass	420			Brass	Nickel	Steel	Total
12	Beta	Nickel	555		Alpha	7755	11135	9866	28756
13	Alpha	Steel	614		Beta	12211	7942	14881	35034
14	Alpha	Steel	432						
15	Beta	Brass	765			Average by division			
16	Alpha	Steel	703			Brass	Nickel	Steel	
17	Beta	Brass	930		Alpha	596.54	618.61	616.63	
18	Beta	Steel	590		Beta	718.29	661.83	620.04	
19	Alpha	Steel	922						
20	Alpha	Steel	615			Average by 2 criteria			
21	Alpha	Steel	496			Brass	Nickel	Steel	
22	Alpha	Nickel	565		Alpha	596.54	618.61	616.63	
23	Beta	Steel	318		Beta	718.29	661.83	620.04	

■ **FIG. 6.2**

Use of the absolute and relative references allows us to copy this across to column H and down to row 13. The argument list is first_range, first_criteria, second_range, second_criteria, etc.

(d) Use the following formula to sum the hours for each type of part:

F12:   =SUMIFS($C:$C, $A:$A, $E12, $B:$B, F$11)

Again, this can be copied across and down. Note that the arguments are in the order: range to sum, first criteria range, first criteria, second criteria range, second criteria, etc.

(e) Average hours are calculated in F17 with =F12/F7 and in F22 with =AVERAGEIFS($C:$C, $A:$A, $E22, $B:$B, F$21). The argument list is the same as with SUMIFS.

(f) Save the workbook.

EXERCISE 3: FREQUENCY DISTRIBUTION

The Hours column has data varying from 300 to 1000. Perhaps we would like to know how many fall in the ranges 300–399, 400–499, and so on. We do this with a FREQUENCY function. Many novices have trouble with the Excel Help's use of the term *bin*. A bin is any container—the Brits talk about *garbage bins*. Think about sorting red and blue balls into two boxes or two *bins*. We are going to sort our Hours into various bins as shown in Fig. 6.3.

	K	L
1	Frequency	
2	Bin	Count
3	300	1
4	400	14
5	500	16
6	600	11
7	700	18
8	800	18
9	900	10
10	1000	12
11		0

■ **FIG. 6.3**

(a) On Sheet1 of Chap6.xlsx enter the text shown in rows 1 and 2 of Fig. 6.3 and the bin values shown in K3:K10.

(b) Select L3:L11 and enter =FREQUENCY($C:$C, K3:K10) as an array formula using Ctrl + ⇧ Shift + Enter ↵. Whereas the bins extend over rows 3 through 10, the frequency formula uses rows 3 through 11; the extra cell is a "safety valve" in case there are values in the data that exceed the last value in the bin.

(c) Save the workbook.

In Exercise 2 of Chapter 16, we will make a normal curve from data generated with the FREQUENCY function.

EXERCISE 4: PIVOT TABLES

The Pivot Table feature is extremely powerful, and we will only be able to scratch the surface of it. Pivot tables are used to summarize data: we will summarize our data as shown in Fig. 6.5. If you have problems with this exercise, view the video PivotTable.MP4 in the companion website.

(a) Use the command *Insert / Table / Pivot Table (Insert / Tables / Pivot Table* on a Mac) to open the dialog shown in Fig. 6.4. We need to indicate where the data is (A1:C101) and where the output should go (cell N1 of this sheet). Click OK. If the active cell is within A1:C101 when the pivot table dialog is opened, Excel will automatically find the required dataset.

(b) In the Pivot Table Field List dialog (see right side of Fig. 6.5) we need to specify where we wish each *field* to be. Click on the *Brands* button in the top section of the dialog and drag it to the COLUMNS area in the lower part. You will see the worksheet change as you do this. Drag the Type button to the ROWS area and the Hours button to the VALUES area. The order for doing this is unimportant. Note that each field gets checked as you place it in the lower area. When ready, close the Pivot Table dialog with the X in its title bar.

Create PivotTable
Choose the data that you want to analyze
Select a table or range
Table/Range: Sheet1!$A$1:$C$101
Use an external data source
Choose Connection...
Connection name:
Choose where you want the PivotTable report to be placed
New Worksheet
Existing Worksheet
Location: Sheet1!$N$1
Choose whether you want to analyze multiple tables
Add this data to the Data Model
OK
Cancel

FIG. 6.4

N
O
P
Q
Sum of Hours
Column Labels
Row Labels
Alpha
Beta
Grand Total
Brass 7755 12211 19966
Nickel 11135 7942 19077
Steel 9866 14881 24747
Grand Total 28756 35034 63790
Drag the Type button to the Rows area
PivotTable Fields
Choose fields to add to report:
Search
Brand
Type
Hours
MORE TABLES...
Drag fields between areas below:
FILTERS
COLUMNS
Brand
ROWS
Type
VALUES
Sum of Hours
Defer Layout Update
UPDATE

FIG. 6.5

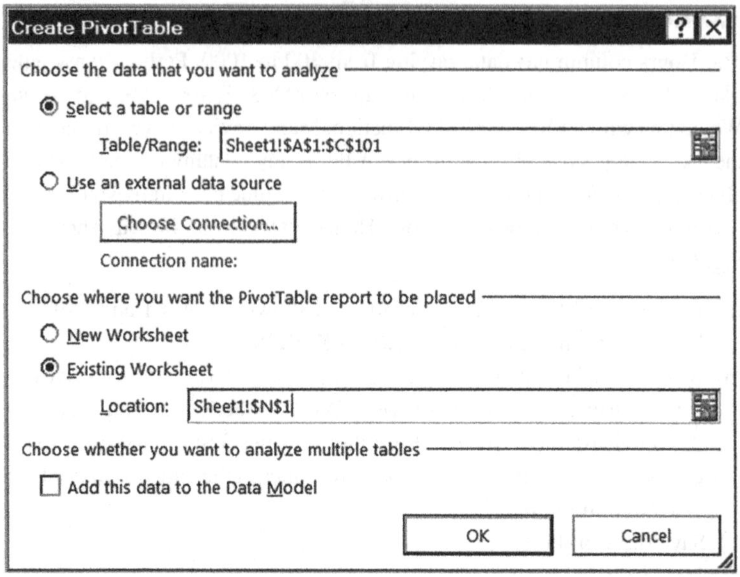

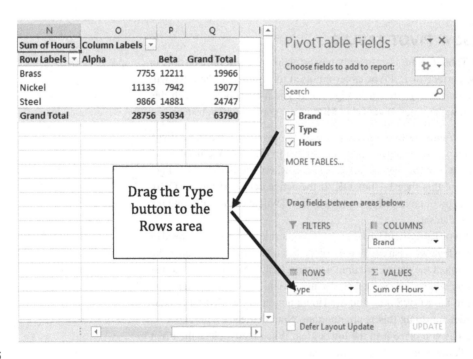

(c) We have generated the results shown on the left side of Fig. 6.5. But what if we wanted average values rather than sums? Method 1: Click anywhere in the Pivot Table to open the *Pivot Table Field List*, click on the *Sum of Hours* button, and choose *Value Field Settings* to bring up the dialog from which average may be selected. Note you cannot only select *Average* but you can have a custom name that will appear in cell N1. Method 2: On a PC, right-click any of the numbers in the pivot table and select *Summarize Values By* and select *Average* from the popup menu. (For a Mac, while holding down the [Ctrl] key, click on a number in the pivot table and select *Summarize Values By …*) You may wish to experiment further.

(d) Save the workbook.

EXERCISE 5: SORTING

In this exercise, we look at some of the tools Excel provides for sorting a data table.

(a) For convenience, we will copy the data to Sheet2. Use the plus sign on the tab list to make a new sheet. On Sheet1 click on any cell within the range A1:C101 and select all the data using [Ctrl]+A. Using the shortcuts [Ctrl]+C and [Ctrl]+V copy and paste to Sheet2.

We shall first look at a quick way to sort by just one field at a time.

(b) Select any cell in column A (Brand) of the data on Sheet2. Click the A→Z icon in the *Data / Sort & Filter* group. The data is now sorted with the Alpha records first followed by Beta records. Note that Excel has sorted the entire dataset not just the first column; rows of data remain intact. How did Excel know not to sort the top row? It recognized that the labels in that row were table headers. But that cannot always be relied upon; in our case Excel's clue was the text *Hours* at the top of a column of numbers.

(c) Select any cell in column B (Type) of the data on Sheet2. Click the A→Z icon *Data / Sort & Filter* (*Data* on a Mac) group. Since the data was previously sorted by brands, we now have data sorted in groups Alpha and Brass, Beta and Brass, Alpha and Nickel, Brass and Nickel, and so on.

Next, we will do a multiple field sort. Excel 2016 allows you to sort on up to 64 fields. However, we have just three fields. We will end up with

> Pivot tables are an excellent tool for summarizing data; furthermore, they avoid all the work of developing formulas. However, they have a minor drawback in that they are not dynamic. If any of the data is edited (perhaps to correct a typing error where 10 was entered but 100 was required), the pivot table, unlike formulas, will not automatically reflect the change. On a PC, one needs to right-click the pivot table and use the *Refresh* command. (For a Mac while holding down the [Ctrl] key click on the pivot table and use the *Refresh* command.)

the data sorted first by brand, within each brand it will be sorted by type, and the hours for each brand-type group will be sorted in descending order.

(d) Select a cell anywhere in the table and use the command Data / *Sort & Filter / Sort* (Data / *Sort* on a Mac) to open the sort dialog—Fig. 6.6.

(e) Using the drop-down arrows, sort first on Brand specifying Values and A to Z (i.e., ascending alphabetical order). Then click on *Add Level* and sort on Type again specifying Values and A to Z and, finally, add another level to sort on Hours specifying Values and Largest to Smallest. This is shown in Fig. 6.9. Experiment as much as you like with variations in the sort; using columns in different orders, sorting with increasing or decreasing values, and so on.

(f) Save the workbook.

As mentioned before, Excel frequently provides more than one path to perform a given operation. So it is with sorting, and with filtering which is covered in the next exercise. We have accessed the sort tools from the Data tab but they are also available in the Editing group of the Home tab.

EXERCISE 6: FILTERING

Filtering is the process by which we cause only part of our data to be displayed on the worksheet. No data is actually lost; it is just temporarily hidden. Again, this is a powerful feature, so we have space for only a cursory look.

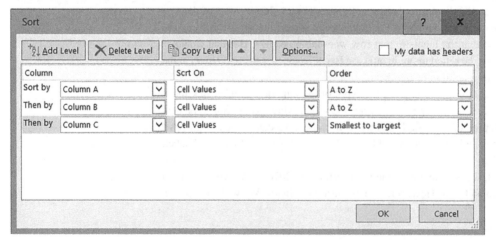

■ FIG. 6.6

(a) Copy the data A1:C101 from Sheet1 to Sheet3.
(b) Select any one of the cells in column A. Click on the *Filter* icon (it resembles a funnel) in *Data / Sort & Filter* group. This places drop-down arrows in the header row.
(c) Click on the arrow in cell A1 (Brand) to bring up the dialog shown in Fig. 6.7. Remove all the check marks except the one for Alpha, and click the OK button. Do the same in B1, selecting Brass. The result is as expected but note that the column headers now display a funnel and row headers are colored; this is to alert you to the fact that the data is filtered and that some rows are hidden.
(d) Clicking on the *Filter* icon in *Data / Sort & Filter* group again will unhide the data.
(e) Save the workbook.

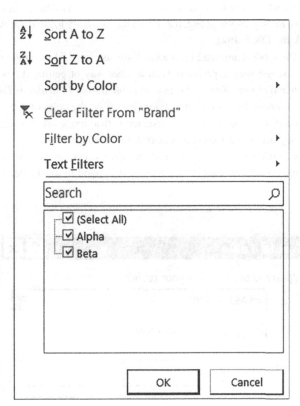

■ FIG. 6.7

EXERCISE 7: THE EXCEL TABLE

Any dataset consisting of two or more columns and several rows can be referred to as a table. However, within Excel, there is a feature to convert such data to an Excel Table which assists in managing and analyzing the data. To avoid any confusion between a simple dataset (aka "a table") and the Excel Table function, we will always refer to the latter with a capital T. Furthermore, Excel Tables should not be confused with the *data tables* that are part of a suite of what-if analysis commands; so be careful when using the Help facility.

We will be able to look at only a few features. The Help facility has some particularly useful information about Tables. One item is a link to an excellent Microsoft tutorial.

(a) Once again select and copy A1:C101 on Sheet1 and copy it to a new sheet—Sheet4.

(b) With the cursor anywhere in the range A1:C101 on Sheet4, use the command *Insert / Table* (*Insert / Tables / Table* on a Mac). This brings up the dialog shown in Fig. 6.8. Excel has correctly located our data, so click the OK button.

(c) Use the Undo command on the QAT (or use Ctrl+Z) to remove the Table so we may experiment with another way of getting it. Use the command *Home / Styles / Format as Table* (*Home / Format as Table* on a Mac) and select one of the styles (basically, the colors for the row banding). Once again, we are presented with a dialog box (Fig. 6.8) and we may accept its values by clicking the OK button.
As with filtering, the top row now has drop-down arrows. Also, some formatting (in this case, banded rows) is applied to the Table. To

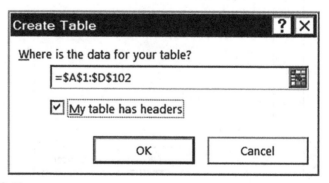

■ FIG. 6.8

manage the formatting click any cell in the Table and then open the contextual Ribbon tab *Table Tools*.

(d) Make C102 (the cell below the last number in the Hours column) the active cell. In *Home / Editing* (*Home* on a Mac) click the AutoSum tool (Σ). This inserts a formula showing the sum of Hours. You are not limited to just a Sum formula; there is a drop-down arrow that lets you select other functions such as Average and Max.

(e) Now we will add a "calculated column." Type a heading in D1 such as Days. In D2 type =ROUND(C2/24,1) and press [Enter↵]. Excel auto- matically fills the D column with the formula. If you make this formula using the pointing method and clicked on C2 then D2's formula would be =ROUND(Table1[[#This Row], [Hours]]/24, 1).

(f) Move to A101 and then tab across three times, moving to D101. Tab once more and you are taken to A102: Excel has automatically extended the Table.

This is a major advantage of Tables; they are dynamic. This means that data added to the bottom is automatically incorporated in the Table.

(g) Experiment with filtering. For example, have only the Brass items from Alpha displayed.

(h) Save the workbook.

You may convert the Table back to a simple range by selecting any cell in the Table and clicking the contextual *Table Tool* Table A new set of items appears on the ribbon. One of these is *Convert to Range*. However, the banded rows remain; to remove them select all the table with [Ctrl]+A then use the *Clear Format* command in the *Home / Editing* group. Experiment with other *Table Tools* commands.

IMPORTING DATA FROM THE WEB

Many web pages have tables of data. Frequently, you can copy and paste the tables directly from the web page. If this does not work, try selecting the table and a few lines past the table. Copy this region and paste the data into Word, as Word is better at recognizing web formatting. Once the table is in Word, you can copy the table from Word into Excel. If this does not work, you can save the webpage to your computer, and then open the webpage in Word. Word will recognize the web page table formatting, and you can copy the table from word into Excel.

For PDF files, if you try to copy a data table directly, it will paste as a single column in Excel. However, you can open the PDF file in Word, which will convert the PDF into a Word document. Once the file is in Word format, you can copy the table and paste it into Excel.

PROBLEMS

1. Your company specializes in copper, brass, and bronze plate in thicknesses of ¼″, ½″, ¾″, and 1″—so you have 12 products. Every day you receive a file with the orders from your five customers. The first few lines of the file may look like this:

Customer,Order Number,Metal,Thickness,Length,Width
Beta,BVL1000,Copper,0.75,4,6.5
Zeta,BVL1001,Brass,0.75,11,6.5
Kappa,BVL1002,Bronze,0.5,4,4.5

A file with this type of data is called a comma delimited file, and generally, it has the extension CSV (from Comma-Separated Variables).

Our aim is to generate a worksheet similar to that in Fig. 6.9.

(a) Download the file Chap6Problem1Data.csv from the companion website. Open it in Notepad to see its contents. Close Notepad and open the file in Excel. Note how Excel imports CSV files automatically putting records (rows) into cells. Rename the first worksheet as Sheet1 and save the file as an Excel file name Problem6.xlsx.

(b) Each product has a specific cost; the details of this are found in Chap6Problem1Pricing.xlsx on the companion website. Download the file and copy its data to Sheet2 of Problem6.xlsx.

(c) Using a VLOOKUP formula, find the cost for one square foot of the metal/thickness; thus for Bronze with thickness 0.5″, your formula

	A	B	C	D	E	F	G
1	Customer	Order Number	Metal	Thickness	Length	Width	Cost
2	Zeta	BVL1000	Bronze	0.5	12	3	$ 851.40
3	Zeta	BVL1001	Brass	0.5	11	4	$ 858.00
4	Beta	BVL1001	Bronze	0.25	10	6	$ 735.00
5	Delta	BVL1002	Copper	0.5	10	1	$ 184.50
6	Delta	BVL1002	Copper	0.25	9	5	$ 402.75
7	Zeta	BVL1003	Brass	0.5	9	0.5	$ 87.75
8	Delta	BVL1003	Bronze	0.5	14	6.5	$2,152.15
9	Delta	BVL1004	Brass	0.75	9	1	$ 256.05
10	Alpha	BVL1004	Bronze	0.25	4	3	$ 147.00
11	Kappa	BVL1005	Copper	0.5	6	4.5	$ 498.15
12	Kappa	BVL1005	Bronze	0.25	13	5	$ 796.25

■ FIG. 6.9

should return $23.65. Looking up the metal is easy: it is the normal VLOOKUP procedure. But what column to use? This requires you to locate the thickness value in row 4 of the pricing data. Do an internet search with Excel two-way VLOOKUP to get help. Now modify the formula for the cost of the order, taking length and width into account.

(d) Make the data A1:G101 into an Excel Table.

(e) Filter the data to show orders from Alpha. *Copy* this data and, on a new workbook, use *Paste Special* to get the data as values (i.e., not formulas). Now discover how to e-mail this to yourself from one of the tools on the File tab. Remove the filter from the data.

(f) Construct a pivot table as shown in Fig. 6.10. Use the fields Metal, Thickness for rows, Customer and Cost as the Value Field. Modify the Cost field to display average not sum.

(g) Download Chap6Generate.xlsx from the companion site to see how the hypothetical data was generated for this problem using the RANDBETWEEN and CHOOSE functions.

Before beginning this problem, open the Data tab and check if the *Analysis* group contains a command called *Data Analysis*. If not,

Row Labels ▼	Count of Customer	Average of Cost
⊟ **Brass**	**39**	**$ 944.67**
0.25	6	$ 214.40
0.5	11	$ 616.02
0.75	12	$ 1,098.88
1	10	$ 1,559.29
⊟ **Bronze**	**34**	**$ 1,077.81**
0.25	11	$ 497.24
0.5	8	$ 1,343.62
0.75	7	$ 1,526.50
1	8	$ 1,217.68
⊟ **Copper**	**27**	**$ 676.64**
0.25	8	$ 371.43
0.5	9	$ 523.78
0.75	6	$ 917.68
1	4	$ 1,269.45
Grand Total	**100**	**$ 917.57**

■ FIG. 6.10

use *File / Options / Add-ins* to load the command; if necessary do an internet search with *Excel 2013 load Analysis ToolPak*; the method is the same as in Excel 2010.

When completed, our worksheet will resemble that in Fig. 6.11. Because we are using a random function, each user will have different data in column A and hence different computed values in the rest of the worksheet.

(h) Enter the text shown in A1. Select A2:A501 (holding down ⌈⇧ Shift⌋ while tapping ⌈PgDn⌋ or ⌈↓⌋ will speed things up).

Enter =RANDBETWEEN(1100), and commit with ⌈Ctrl⌋+⌈Enter↵⌋ (no ⌈⇧ Shift⌋, this time). Select cell A1 and use ⌈Ctrl⌋+A to highlight A1:A501 and use the *Copy* command (or ⌈Ctrl⌋+C). With the data still selected, on a PC, right-click (on a Mac hold the ⌈Ctrl⌋ key and click) and use the *Paste Special Values* tool the data and use *Paste Special* to convert formulas to values; if you right-click the second icon in the *Paste Options* it has the label *123* indicating

	A	B	C	D	E	F	G	H	I
1	Mydata								Bin
2	50		*Mydata*			Bin	Frequency		5
3	79					5	19		10
4	55		Mean	51.944		10	20		15
5	58		Standard Error	1.2542		15	21		20
6	9		Median	52		20	24		25
7	91		Mode	63		25	27		30
8	17		Standard Deviation	28.044		30	23		35
9	70		Sample Variance	786.47		35	21		40
10	23		Kurtosis	-1.066		40	27		45
11	45		Skewness	0.0087		45	30		50
12	17		Range	99		50	30		55
13	77		Minimum	1		55	30		60
14	25		Maximum	100		60	27		65
15	45		Sum	25972		65	42		70
16	44		Count	500		70	25		75
17	98					75	11		80
18	46					80	22		85
19	36					85	15		90
20	50					90	22		95
21	97					95	31		100
22	23					100	33		
23	32					More	0		
24	65								

■ FIG. 6.11

that formulas are converted to values. (For a Mac, while hold down the Ctrl key and click …)

(i) Use *Data / Analysis / Data Analysis / Descriptive Statistics with Summary Statistics* in the dialog box to generate the results in columns C and D. Since the data was generated using random numbers your results will vary somewhat from those in the figure.

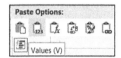

Paste Options:

Values (V)

(j) Enter the word Bin in I1. In I2 and I3 type 5 and 10, respectively. Select the two cells and drag the fill handle down to row 21 (you should see *100* displayed in a screen tip.)

(k) Use *Data / Analysis / Data Analysis / Histogram* (*Data / Data Analysis / Histogram* on a Mac) to generate the results in columns H and I. In the Histogram dialog leave the bottom three option boxes unchecked. You could make the worksheet look tidier by hiding column I since its numbers are reproduced in column H. All of the data generated in steps (b) and (d) could have been obtained with formulas. The Data Analysis tools provide a quicker way, especially for someone not totally familiar with the Excel functions. However, it must be noted that results from these tools are static: they will not be recalculated if the source data is edited. Unlike with pivot tables, there is no refresh command.

2. It is desired to select the tube for a 24 in. simply supported beam supporting a 300 lb load in the center. The resulting moment at the middle of the beam is 1800 in. lbs. It is decided to use 4130 Alloy Steel tubing to support the load. A table of available steel tubing options is available on the companion web page under *Steel Tubing Prices.txt.*[1] The design requires the stress to be less than 30,000 psi. The formula for stress is

$$\sigma = \frac{MR_o}{I}$$

where M is the moment, R_o is the outer radius of a tube, and I for a tube is

$$I = \frac{\pi}{64}\left(D_o^4 - D_i^4\right)$$

Calculate the I and the stress for each tube. Which tube should you use if you wanted to minimize (a) cost and (b) weight? Hint: put into a table and create

---

[1] https://www.aircraftspruce.com/catalog/mepages/4130tubing_un1.php.

separate columns which only have the price and weight if the stress is below the threshold.

APPENDIX: SUPPLEMENTARY MATERIAL

Supplementary material related to this chapter can be found on the accompanying CD or online at https://doi.org/10.1016/B978-0-12-818249-9.00006-6.

Chapter 7

Charts

CHAPTER OUTLINE

Scientists and engineers generally speak about making graphs. However, we shall use the term *charts* in this chapter since that is how Excel refers to these items. Perhaps the most important chart type for a technical person is the XY (scatter) chart where the data consists of ordered pairs (x, y) of numerical

Liengme's Guide to Excel 2016 for Scientists and Engineers. https://doi.org/10.1016/B978-0-12-818249-9.00007-8

data. We shall, therefore, concentrate on this type of chart (calling it an XY chart and avoiding the word *scatter* beloved of statisticians) and only briefly look at the other types.

Fig. 7.1 shows the major types of charts one can produce in Excel. Combination charts are also possible; for example, one can make a chart where one data series is displayed as a column chart while a second data series is displayed as a line chart. There are also so-called 3D versions of these charts, together with stock, surface, bubble, doughnut, and radar. Once you have mastered some basic concepts, you will be able to generate any of these with a little experimentation.

Line charts and XY charts: New Excel users often have trouble with the difference between Line and XY charts. The similarity between the two samples in Fig. 7.1 is misleading and is somewhat coincidental. To demonstrate this, look at the two charts in Fig. 7.2. They were made from the same data but are totally different. The Line chart treats its *x*-values (the values used to determine the horizontal position) as a *category* (textual label) with each category being equally spaced. The fact that these are numerical values is totally disregarded. This is true of all Line charts except when dates are used for *x*-values. In an XY chart, the numerical values of the *x*-data determine the horizontal positions of each data point.

> If the *x*-values are numbers, you most likely need an XY chart.

The use of the name *Line* is misleading. In Fig. 7.2 both the Line and the XY chart have their data displayed with lines and markers. A better name would be a *Category* chart.

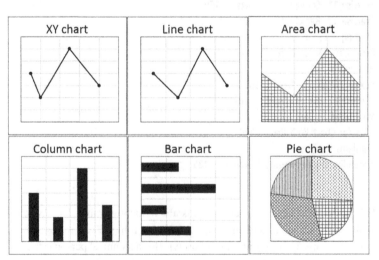

■ FIG. 7.1

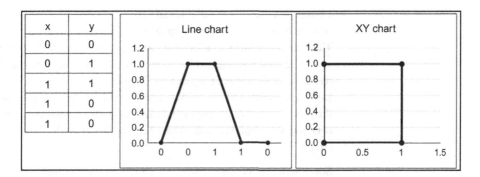

x	y
0	0
0	1
1	1
1	0
1	0

■ FIG. 7.2

Note: If you try to make the Line chart shown in Fig. 7.2, begin by removing the *x* label at the top of the first column; otherwise because both columns have numbers Excel will make a chart with two data series. The label will, of course, cause no problem when making an XY chart.

Guidelines when making a chart: One of the internationally acknowledged experts on charts is Edward R. Tufte.[1] His main points may be paraphrased by:

(i) Let the chart show the data clearly; do not add extraneous matter— what Tufte calls chart junk.

(ii) The chart should not distort the data.

(iii) The chart should show the data at both the broad and the detailed level. The general trend of the data and any fluctuations should be clear.

(iv) The viewer should be drawn to the chart's data, not the method used to construct it.

Tufte's advice may be summarized as "keep it simple." Of overriding importance is the avoidance of distorting the data. For this reason, he and others, deplore the use of the so-called 3D charts which are column charts with blocks in place of simple rectangles. Pie charts are similarly criticized for not faithfully depicting their data. So, we shall use simple charts in all the exercises.

Michael Alley recommends the use of assertion-based chart titling. Using the axis labels as the chart title (plot of *x* vs *y*) does not add value, as the reader can readily obtain this information. Rather, the title should be the point that you are trying to make with the chart.

---

[1]The Visual Display of Quantitative Information (Graphic Press, 1983).

Chart elements. Excel's dialogs and Help topics speak of the various parts of a chart as *elements*. The compulsory elements of a chart are as follows: The *Chart area* includes everything within the borders of the chart; the *Plot area* is generally delineated by the vertical and horizontal axes, and at least one *Data series* showing the data that is being charted. In an XY chart (and Line chart) a data series can consist of a *line* or *markers*, or both. In Fig. 7.3 the Plot Area has been given a light fill, but no border has been added. The data series is displayed with both a line and markers; the latter are hollow circles.

Some of the optional elements are the *Primary x-axis* which is generally the lower horizontal border, and the *Primary y-axis* forming the left vertical border. It is fairly common for a chart to have both a primary and a secondary y-axis—one on the left and the other on the right. Charts with secondary x-axes are less common. Each axis has a bounds range; in Fig. 7.3 the bounds for the x-axis is 0–6 while for the y-axis it is 5–11. Axes are also divided into major and minor units which specify where *tick marks* and *gridlines* may occur. In Fig. 7.3, the horizontal axis has both major and minor tick marks while the vertical has only major ones. *Labels* (which are numbers for XY charts) are optional but when used they align with the major units of the axis. In Fig. 7.3 only the major y-axis gridline has been added; these have been formatted to be dotted lines.

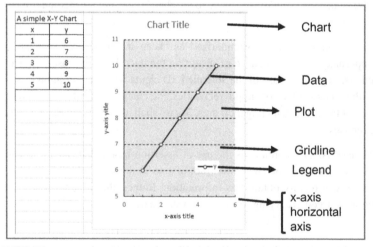

■ FIG. 7.3

A chart may have titles: *chart title*; *x-axis title*, and *y-axis title*. It may also have a *legend* box. The titles and legend may be dragged to anywhere on the chart, but axes titles should be close to their own axis line. The size of a title box is determined by its content; a legend box may be resized like a text box. A legend was added to Fig. 7.3 but since there is only one data series it is totally superfluous!

EXERCISE 1: AN XY CHART

In this exercise we will make a chart similar to that in Fig. 7.5. We will not have shading (this is called fill) in the plot area, nor will we have a legend that is entirely superfluous with only one data series. You may find this exercise long-winded, but the objective is to familiarize you with many chart tools.

(a) Open a new workbook. In A1 of Sheet1 enter a suitable title for the worksheet—something like *A Simple XY Chart*. Then starting in A3, enter the data (text and numbers) shown in Fig. 7.5.

(b) We could select all the data in preparation for making the chart, but since our data is surrounded by empty cells or the edge of the worksheet, we need only click on one cell (A6 for example) within the source data and let Excel find the data. Open the Insert tab on the ribbon. You will see various options in the Chart group. Select *Scatter* and from the drop-down menu (Fig. 7.4) select either of the examples with both lines and markers (the difference between them will be revealed later).

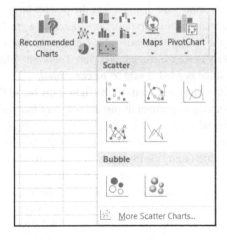

■ FIG. 7.4

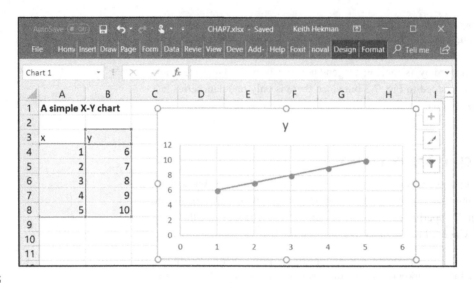

■ FIG. 7.5

We now have a fairly reasonable XY chart. We will proceed to make some formatting changes. But first, let us make some observations.

(c) When you first made the chart, it was selected; you could see this by the six fill handles around its border. If necessary, click on the chart to select it; see Fig. 7.5.

   (i) The data source is picked out by range finders—colored fill and borders. The cell B3 is included since it was used for the rather unimaginative chart title.

   (ii) Slowly move the mouse pointer around and observe how the element names are displayed as the pointer hovers above elements such as the chart title, the data series, a gridline, and so on.

   (iii) Note the two new Ribbon tabs: *Design* and *Format* which collectively make the contextual *Chart* Tools tab. A tab that appears only when needed is said to be *contextual*.

   (iv) Note also the three icons to the right (only on a PC). These give rapid access to commands that are also on either the Design or the Format tabs.

(d) With the chart selected, use the fill handles to alter its size. Click within the Chart Area avoiding the Plot area (so click somewhere near the chart title) and see how one can move the chart about the worksheet. Again click near the chart title and drag the mouse while holding down the Ctrl key; this is a quick way to make a copy of the chart.

(e) Now we will remove the vertical gridlines. Click on any of the vertical gridlines and press the Delete key. To experiment with another tool use

⌈Ctrl⌉+Z to undo the change. Let the mouse pointer hover over the top tool ⌊+⌋ to the right of the chart to see that its purpose is to add/remove chart elements. Click the tool, move down to Gridlines and click the right-pointing arrow. In the popup menu deselect Primary Major Vertical—see Fig. 7.6 (only works on PC). One can also use *Chart Tools / Design / Chart Layouts / Add Chart Element* (*Chart Design / Add Chart Element* on a Mac) to add/remove a chart element.

(f) Excel 2016 seems to favor light colors for chart elements, but this can make them less visible. We will make the gridlines darker (i.e., use black for its color) and change the style to a round dotted line. When you double-click on a chart element an appropriate formatting dialog opens to the right of the worksheet—see Fig. 7.7. The reader should experiment with this to discover how to perform our two tasks (darker, dotted line). Don't let the terminology Solid Line in the top part of the dialog confuse you; a dotted line is still solid! Since the chart is rather small you may need to use well-spaced dots or dashes. If you wish, you may close the formatting dialog by clicking the X in its top right corner.

(g) The labels likewise are somewhat faint. Double-click on the labels of the vertical axis. Ignore the formatting dialog this time but with the Home tab open use the last tool (icon showing the letter A) in the Font group to change the color of the font from a gray color to real black. And now for some magic! Click the horizontal axis labels and tap the ⌈F4⌉ key. This is the "repeat the last command" shortcut. It works for many Excel features but not for everything. With either of the axis labels selected, observe the Home tab; most of the commands are grayed out except those in the Font group and the orientation command in the Alignment group. These commands may be used to

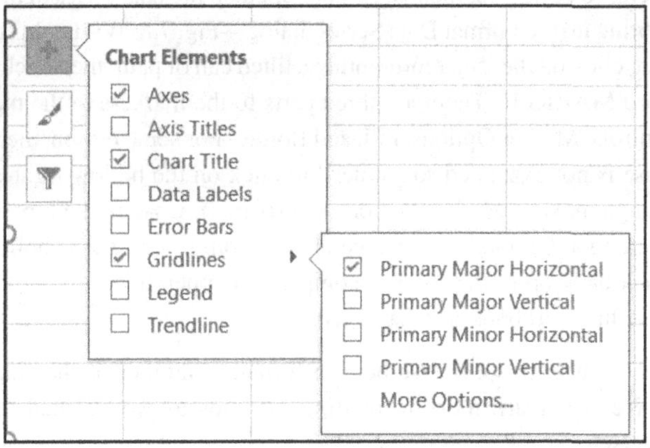

Chart Elements
- ☑ Axes
- ☐ Axis Titles
- ☑ Chart Title
- ☐ Data Labels
- ☐ Error Bars
- ☑ Gridlines ▸
  - ☑ Primary Major Horizontal
  - ☐ Primary Major Vertical
  - ☐ Primary Minor Horizontal
  - ☐ Primary Minor Vertical
  - More Options...
- ☐ Legend
- ☐ Trendline

■ FIG. 7.6

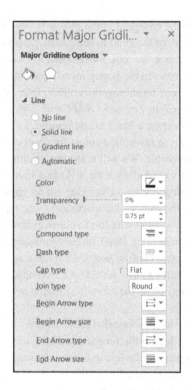

FIG. 7.7

format labels and titles in a chart. Unfortunately, the commands in the Number group are not available when a chart is selected.

(h) Next, we will format the markers changing them from the default filled circles to hollow diamonds. Double-click on the data series in the chart to bring up the Format Data Series dialog—Fig. 7.8. Within this dialog, first, click on the icon representing a tilted can of paint then click on the word MARKER. There are three parts to the markers to the marker controls: Marker Options, Fill, and Border. For some reason, the first of these is not expanded so we need to click on the hollow right-facing triangle next to Marker Options to expand it. Change the type from a circle to a diamond and change the fill from a solid fill to hollow fill. Click back on to the chart to complete the operation.

(i) Save the workbook as Chap7.xlsx.

We have seen in this exercise how to format a number of the chart elements. We will learn more formatting methods as we proceed through the exercise.

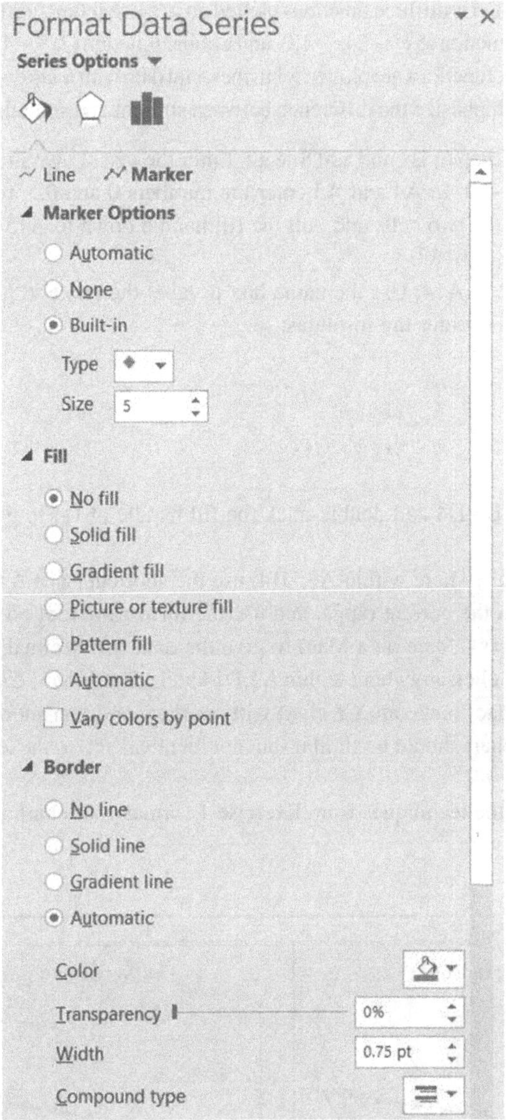

Format Data Series

Series Options ▼

~ Line   ∿ **Marker**

▲ **Marker Options**

○ A̲utomatic

○ N̲one

● Built-in

Type ◆ ▼

Size 5

▲ **Fill**

● N̲o fill

○ S̲olid fill

○ G̲radient fill

○ P̲icture or texture fill

○ Pa̲ttern fill

○ A̲utomatic

☐ V̲ary colors by point

▲ **Border**

○ N̲o line

○ S̲olid line

○ G̲radient line

● A̲utomatic

Color

Transparency 0%

Width 0.75 pt

Compound type

■ FIG. 7.8

EXERCISE 2: PLOTTING FUNCTIONS

In this exercise, we learn how to plot functions. We shall also see how to format an axis to change the way numbers are displayed and to adjust the range of an axis. We shall see when to use smooth lines in an XY chart.

We make a chart with three functions plotted on it: a linear function ($y = 2x + 4$), a quadratic function ($5x^2 - 24x + 12$), and a cubic function ($2x^3 - 13x^2 + 22x - 8$). In general, functions are plotted with lines, and data with markers, but we will use both to emphasize the difference between straight and smooth lines.

(a) Open Chap7.xlsx and add Sheet2. Enter the text shown in row 1 and 3 of Fig. 7.9. In A4 and A5 enter the numbers 0 and 0.5, respectively. Select the two cells and pull the fill handle down to A15 to generate the series 0–5.

(b) Select A4:A14. Use the name box to label the range x.

(c) In B4:D4 enter the formulas:

B4: `=2*x+4`
C4: `=5*x^2-24*x+12`
D4: `=2*x^3-13*x^2+22*x-8`

(d) Select B4:D4 and double-click the fill handle of D4 to generate all the data.

(e) Click anywhere within A3:D14, use the shortcut Ctrl+A to select all cells in the current range, and use the formatting tool on *Home / Numbers* (*Home* on a Mac) to give the data two decimal places.

(f) Again, click anywhere within A3:D14 and using *Insert / Charts* (*Insert* on a Mac) make an XY chart with markers and nonsmoothed lines. Your chart should be similar (but not identical yet) to the left hand one in Fig. 7.9.

(g) Using the techniques from Exercise 1 format your chart as you wish.

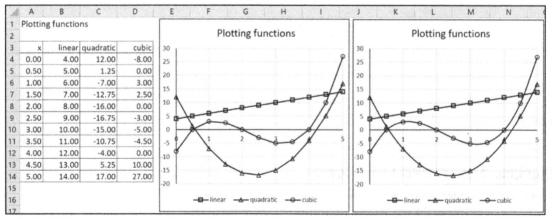

■ FIG. 7.9

(h) To link the chart title to cell A1: click on the chart title box, in the Formula bar type an equals sign ($=$) and then click on A1. In the Formula Bar, you should see $=$Sheet!\$A\$1. You should be aware that no formatting (e.g., italic, bold, superscript, etc.) in A1 will be carried over to the chart title.

(i) Your chart will have an *x*-axis range (or as Excel likes to call it *bounds*) 0–6. This makes the chart look a little asymmetrical, so we will change the range to 0–5. Furthermore, both axes have inherited to two decimal format from the source data and this looks strange.

(j) Since the *x*-axis is not at the bottom of the chart, selecting it by clicking can be problematical so we shall use another method.

(k) With the chart selected, open the *Chart Tools / Format* (*Format* on a Mac) tab, at the far left of the ribbon in the Current Selection group, tap the down arrow next to the words Chart Area (look to the left in Fig. 7.10) and select *Horizontal (Value) Axis*—see right side of Fig. 7.10. Then use the *Format Selection* command in the Current Selection group to open the dialog box shown in Fig. 7.11.

(l) If your dialog box differs from Fig. 7.11, click on the words AXIS OPTIONS and then click the fourth icon (picture with three columns). In the Maximum box enter the value 5.

(m) Observe the hollow arrow next to the word NUMBER in Fig. 7.11. Click this to open that part of the dialog, scrolling down if necessary, and adjust the settings to have zero decimal places—Fig. 7.12.

(n) If your chart is large enough, you will note that the markers are joined by line segments. This is most noticeable with the cubic

■ FIG. 7.10

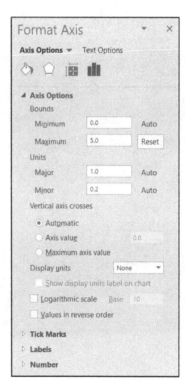

■ FIG. 7.11

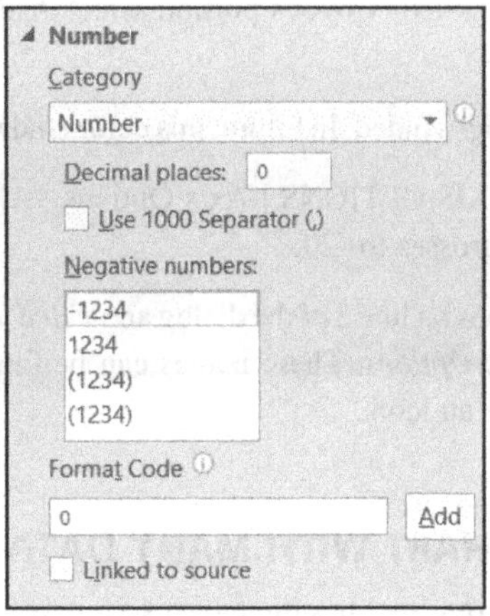

■ FIG. 7.12

function which changes more rapidly. This is not the appearance that we want. We will duplicate the chart and give the second one smooth lines.

(o) Using the trick of holding the [Ctrl] key while dragging the chart, or by the use of [Ctrl]+C followed by [Ctrl]+V make a duplicate of the chart. On a PC right-click (on a Mac hold down the [Ctrl] key and click) anywhere in the second chart and select *Change Chart Type*. From the gallery select the XY subtype with smooth lines.

(p) Save the workbook.

During this exercise, you may have asked yourself how many data points are needed to plot a function. There is no hard and fast rule. The author likes to have sufficient data points such that the chart with nonsmooth lines is almost indistinguishable from that with a smooth line. In the present context steps in the *x*-values of 0.2 rather than 0.5 would have achieved this.

CHANGING THE POSITION OF AXES CROSSING

In the charts of Exercise 2, the *x*-axis crosses the *y*-axis at the origin—at the Cartesian point (0, 0). There are occasions when we wish to change the crossing point. Let us say we want the x-axis to cross the y-axis at –20, the bottom of the chart; note that this requires a change to the *Y*-axis, not the *X*-axis. The steps to do this are as follows:

(i) Click on the *y*-axis to bring up the Format Axis dialog.

(ii) If necessary, click on AXIS OPTIONS on line 2 of the dialog.

(iii) Click on the fourth tool on line 3 of the dialog.

(iv) If necessary, expand the AXIS OPTIONS on line 4 of the dialog

(v) In the Horizontal axis Crosses portion select Axes Value and enter −20.

This is somewhat long-winded. In future, this command will be indicated by:

Format Axis | AXIS OPTIONS | Axis Options | AXIS OPTIONS | set Horizontal axis crosses to -20.

Note that the four icons on line 3 of the dialog are: *Fill & Line, Effects, Size & Properties*, and *Axis Options*. These names can be found by allowing the mouse to hover over an icon.

FILTERING A CHART WITH MANY DATA SERIES

The chart made in Exercise 2 has three data series. We may wish to have a copy of this chart which displays just one—the cubic function, for example.

The steps to do this are as follows:

(i) Click on the chart to bring up the three tools at the right.

(ii) Click the third tool (Filter) to bring up the dialog shown in Fig. 7.13.

(iii) Deselect the first two data series—the cubic data series is emphasized on the chart.

(iv) Click the *Apply* button. Now the chart has just one data series.

The process can be reversed to restore one or both of the other data series.

■ **FIG. 7.13**

FINDING ROOTS

In Chapters 11 and 12 we will show how to use Excel to find the roots of $f(x)=0$. Plotting can often be used to find approximate solutions, and these can serve as starting points to get a more accurate answer.

Looking at the right-hand chart in Fig. 7.9 we see that the cubic plot crosses the x-axis at points $(0, 0.5)$, $(0, 2)$, and $(0, 4)$. Clearly, the roots of $2x^3 - 13x^2 + 22x - 8 = 0$ are 0.5, 2, and 4. Actually, we could have seen this from the data table.

The linear and quadratic plots cross at approximately $(0.4, 5)$ and $(4.8, 14)$, so the roots of $5x^2 - 24x + 12 = 2x + 4$ are 0.4 and 4.8. Of course, the problem is the same as finding the roots of the quadratic $5x^2 - 22x + 8 = 0$.

EXERCISE 3: ADDING AND DELETING DATA SERIES

Let us imagine that we have made a worksheet similar to that in Fig. 7.9 but without column D (the cubic equation data), and that we have made an XY chart with two data series. Later we add column D and we wish to include it in the chart without starting all over again, as we may have made a lot of formatting changes!

(a) Make a copy of the right-hand chart on Sheet2 of Chap7.xlsx.

(b) Click on the cubic data series in the chart and use Delete to remove it. So, in our scenario, we have made a chart with columns A through C as the data.

(c) Imagine we now type in the data in column D. Select and copy D3:D14. Click on the chart to select it. On the *Home / Clipboard* group click the launch arrow under the Paste icon to bring up the dialog shown in Fig. 7.14. Note which selections have been made: we are adding a new series, this data is in a column, and the first cell in the selection has the name of the data series. We selected only one column, so the *Categories in First Column* option is not selected. Click the OK button and the third data series has been added to our chart.

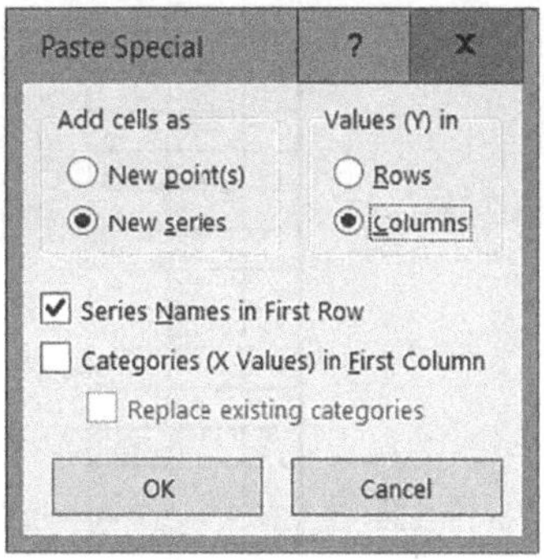

In the next exercise, we shall see an alternative way of adding or deleting a data series.

■ **FIG. 7.14**

EXERCISE 4: XY CHART WITH TWO *Y*-AXES

The scenario for this problem is as follows: A researcher has a recording thermometer and a recording light-meter, but they are not synchronized. Something interrupted a light-meter reading at time 8.5 h. The datasets are shown in Fig. 7.16. He requires a chart similar to that shown.

We are presented with three problems: (i) We have two sets of data with different *x*-values; (ii) the *y*-values of the two datasets have very different ranges and units. One dataset has an approximate range of 10–26, while the other's range is 0–110; (iii) one dataset has a missing value. We solve the first problem by making the chart with one dataset; then we use Copy followed by Paste Special to add the second dataset. Problem (ii) is solved by using a secondary *y*-axis so that the changes in both datasets are clear. The missing data point (iii) will generally by default result in the data series having a gap.

(a) Start by entering on Sheet 3 all the data shown in Fig. 7.15.
(b) Make an XY chart with a nonsmoothed line with the data in A3:B28. Smoothing would not be appropriate here since you are plotting sampled data. (Usually sampled data is plotted using markers only.)

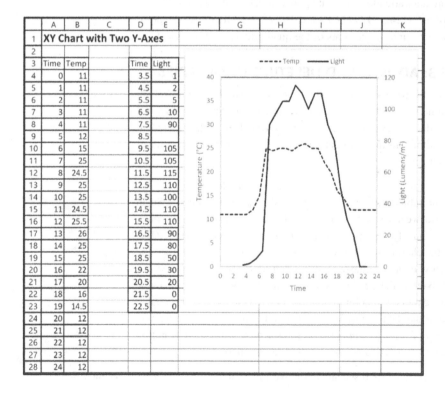

	A	B	C	D	E
1	XY Chart with Two Y-Axes				
2					
3	Time	Temp		Time	Light
4	0	11		3.5	1
5	1	11		4.5	2
6	2	11		5.5	5
7	3	11		6.5	10
8	4	11		7.5	90
9	5	12		8.5	
10	6	15		9.5	105
11	7	25		10.5	105
12	8	24.5		11.5	115
13	9	25		12.5	110
14	10	25		13.5	100
15	11	24.5		14.5	110
16	12	25.5		15.5	110
17	13	26		16.5	90
18	14	25		17.5	80
19	15	25		18.5	50
20	16	22		19.5	30
21	17	20		20.5	20
22	18	16		21.5	0
23	19	14.5		22.5	0
24	20	12			
25	21	12			
26	22	12			
27	23	12			
28	24	12			

■ FIG. 7.15

We now need to add the second data series. We could use the method demonstrated in the previous exercise: Select and copy D3:E23; with the chart selected use *Home / Clipboard / Paste / Paste Special* (*Home / Clipboard / Paste Special* on a Mac) to bring up a dialog similar to Fig. 7.14. The selected range has both an *x*-value and a *y*-value column, so the *Categories in First Column* option must be selected in the dialog. However, we will now try another way.

(c) On a PC right-click (on a Mac hold down the [Ctrl] key and click) the chart and click on *Select Data* in the popup menu to open the dialog shown in Fig. 7.16. (On a Mac, the dialog is shown in Fig. 7.17). Note the *Remove* command (– on a Mac) that would permit us to delete selected data series. Also, note the option box beside the Temp data series; this option box duplicates the function of the Filter tool—see the previous topic *Filtering a Chart with Many Data Series*.

(d) Click on the *Add* (+ on a Mac) button to bring up a dialog similar to that in Fig. 7.18 (on a Mac, the information in Fig. 7.19 is entered in the Select Data Source dialog). Fill this in as shown to specify the new data series name, *x*-values, and *y*-values.

(e) Format the new data series in the chart and use *Format Data Series / Series Options* (the icon with three columns) / *Plot Series On Secondary Axis*.

Our new data series has a gap corresponding to the empty cell E9.

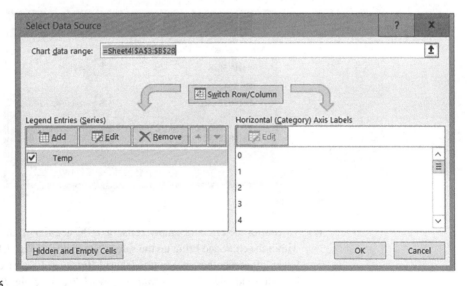

■ FIG. 7.16

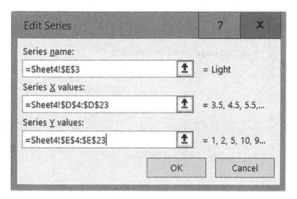

Select Data Source

Range Details

Chart data range: `=Sheet3!$A$3:$B$28`

Legend entries (Series): ⬆ ⬇

Temp		Name:	`=Sheet3!$B$3`
		X values:	`=Sheet3!$A$4:$A$28`
		Y values:	`=Sheet3!$B$4:$B$28`

`+` `−`   Switch Row/Column

Horizontal (Category) axis labels: [   ]

Hidden and Empty Cells

Show empty cells as: `Gaps`

☐ Show data in hidden rows and columns

Cancel   OK

■ FIG. 7.17

Edit Series   ?   X

Series name:
`=Sheet4!$E$3`   ⬆   = Light

Series X values:
`=Sheet4!$D$4:$D$23`   ⬆   = 3.5, 4.5, 5.5,...

Series Y values:
`=Sheet4!$E$4:$E$23`   ⬆   = 1, 2, 5, 10, 9...

OK   Cancel

■ FIG. 7.18

(f) To control the effect of empty cells, on a PC begin by right-clicking within the chart and bring up the select data dialog of Fig. 7.16. In the lower left corner there is a button labeled *Hidden & Empty Cells*; open its dialog and select *Connect data points with line*—see Fig. 7.20. (On a

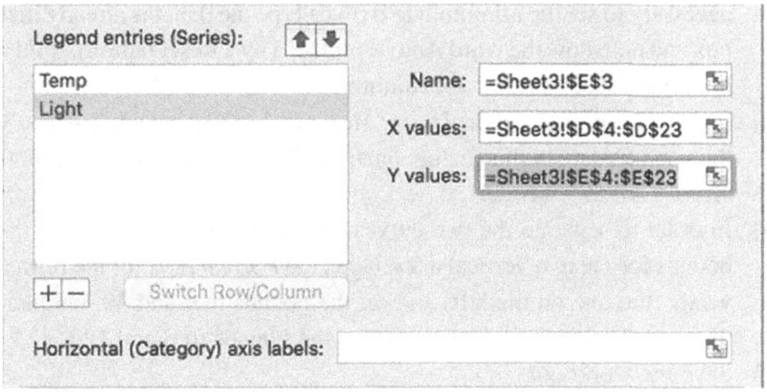

■ FIG. 7.19

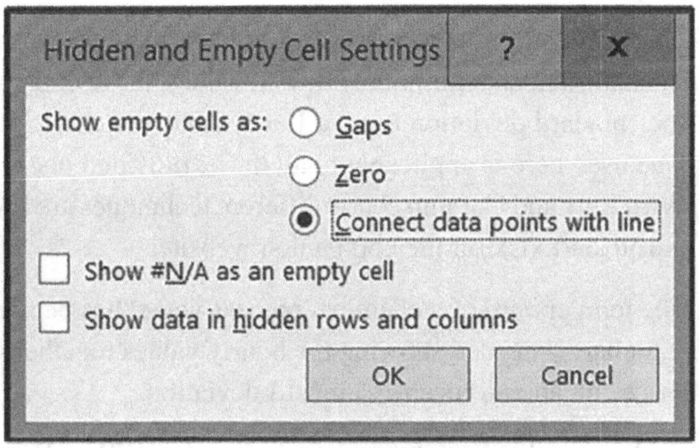

■ FIG. 7.20

Mac, this is part of the Select Data Source dialog—see Fig. 7.17.) This causes the charting engine to ignore the empty cell and the line joins the data points on either side of the missing one. This is sometimes referred to as interpolation.

(g) Use the command *Chart Elements* (the top icon at the right of a selected chart) (*Chart tools / Add Chart Element* on a Mac) to add a legend at the top of the chart, and to add axes titles. The degree symbol can be generated either from *Insert / Symbol* or by holding down Alt while typing 0176 on the numeric keypad (PC only). Similarly, for the superscript 2 use Alt + 0178 or by selecting the 2 and using the font dialog to make it superscript.

(h) The *x*-axis bounds are 0–30 but we want 0–24 to represent one day. Open the Format Axis dialog for the horizontal axis. Use *AXIS OPTIONS / Axis Options* and set the bounds maximum to 24. It is also necessary to set the minimum to 0 (over-type the 0 that is already in the box and note how the word Auto is replaced by a Reset button). If this is not done, Excel will set the minimum to −10.

(i) In the same dialog, set the *Units / Major* to 2 so that we get 2, 4, 6... on the x-axis. Also, in this dialog, open the *Tick Marks* section and set the *Major type* to "outside."

(j) In order to separate the two curves on the chart, we will change the bounds for the two vertical axes. Open the *Format Axis* for the primary y-axis (the one on the left) and set the bounds to 0 and 40. On like manner, for the secondary y-axis set the bounds to 0 and 120.

(k) Save the workbook.

EXERCISE 5: CHART WITH CONTROL LINES

A control chart generally includes one or more horizontal lines showing: a target value, maximum and minimum allowed values, the average value, the average ± the standard deviation (e.g., a Levey-Jennings chart), and so on. The technique used here is applicable to all these, provided one is using an XY chart. With Line and Column charts different techniques are needed; see the file ControlChart.xlsx on the companion website.

Scenario: The temperature of a chemical process vessel has been measured every hour. A chart is needed showing the hourly values together with control lines for the mean and mean ± standard deviation.

(a) In Sheet4, enter the data shown in columns A and B of Fig. 7.21. The degree symbol in °F is made with [Alt]+0176 on the numeric pad or using *Insert / Symbols / Symbol* (*Insert / Symbol* on a Mac). Make a nonsmoothed XY chart of this data.

(b) Enter the text in D3:G3. Enter these formulas:

```
D4:  =MIN(A4:A27)
D5:  =MAX(A4:A27)
E4:  =AVERAGE(B4:B27)
F4:  =E4+STDEV(B4:B27)
G4:  =E4-STDEV(B4:B27)
E5:  =E4 and drag this across to G5.
```

(c) Select D3:G5 and Copy. Activate the chart; use *Home / Paste / Paste Special* (*Home / Clipboard / Paste Special on a Mac*) as we did in the last exercise. Make sure to check the box Categories (*X* Values) in First Column before clicking the OK button.

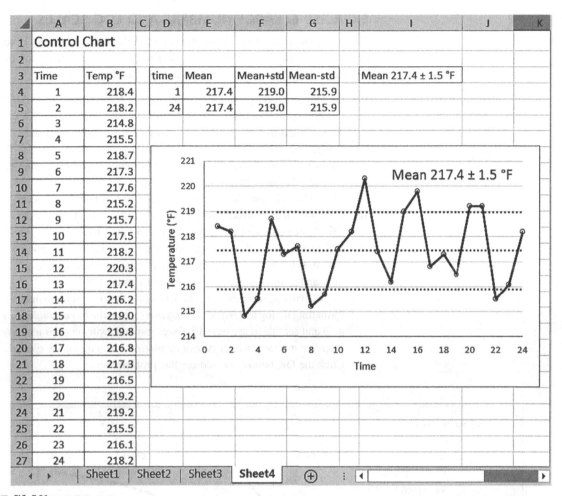

	A	B	C	D	E	F	G	H	I	J	K
1	Control Chart										
2											
3	Time	Temp °F		time	Mean	Mean+std	Mean-std		Mean 217.4 ± 1.5 °F		
4	1	218.4		1	217.4	219.0	215.9				
5	2	218.2		24	217.4	219.0	215.9				
6	3	214.8									
7	4	215.5									
8	5	218.7									
9	6	217.3									
10	7	217.6									
11	8	215.2									
12	9	215.7									
13	10	217.5									
14	11	218.2									
15	12	220.3									
16	13	217.4									
17	14	216.2									
18	15	219.0									
19	16	219.8									
20	17	216.8									
21	18	217.3									
22	19	216.5									
23	20	219.2									
24	21	219.2									
25	22	215.5									
26	23	216.1									
27	24	218.2									

Sheet1 | Sheet2 | Sheet3 | **Sheet4**

■ FIG. 7.21

(d) Format the chart to suit your requirements. For example, the *y*-axis needs no decimal in the labels, the control lines should be dotted without markers, and add the axes titles.

(e) In I3 enter `="Mean "& TEXT(E4,"0.0")& " ± " &TEXT(STDEV(B4:B27),"0.0 °F")`. You can insert the ± in a separate cell and then copy and paste it into the formula since Excel does not allow you to insert a symbol into a formula.

(f) With the chart selected use the top tool on the right of the chart to add a Chart Title. Link the title to cell I3 as you did in Exercise 2.
Move the chart title to a convenient position.

(g) Save the workbook.

EXERCISE 6: LARGE NUMBERS AND LOG SCALES

The y-axis labels can look cluttered when the values are very large. Excel provides a method of scaling these. When the arithmetic range of the y-values is large it can be advantageous to use a logarithmic scale.

(a) On Sheet5, enter the data shown in columns A and B (not column C at this point) as shown in Fig. 7.22. Recall that the series 0, 10...200 can be generated by entering the first two values, selecting both and dragging the fill handle. The formula in B4 is =A4^2*ROW(A1)/10. To see how this works, select B5 and use *Formulas Formula Auditing / Evaluate Formula* (not available on a Mac).

(b) Make Chart 1 from the data and format it to suit your preferences.

(c) Copy the chart to make Chart 2. You could use Ctrl+C followed by Ctrl+V or you might try dragging the chart area while holding down Ctrl. Double-click the y-axis to open the formatting dialog and within *Axis Options* locate the *Display Units* item and set this to thousands. Now the y-axis legends are in units of 1000.

(d) Copy Chart1 again to make Chart 3. Double-click the y-axis to open the formatting dialog and within *Axis Options* locate the *Logarithmic scale* item and set this to use base 10. Excel responds with an error message *Negative and zero values cannot be plotted correctly on log charts...* Click the OK button, we address this problem next.

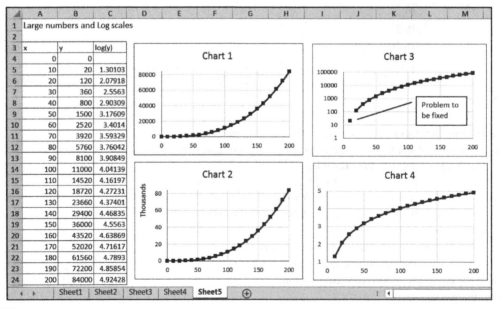

■ FIG. 7.22

The problem with Chart 3 is that the *y*-range includes cell B4 whose value is zero and the logarithm of 0 is mathematically undefined. This explains the there is one less marker in Chart 3, but it does not explain the missing line between the first two markers in the chart. This is an Excel bug; versions prior to Excel 2007 did not result in this. We will solve the problem by adjusting the range of values used for the chart so as to omit row 4. Once done, the first point will be linked to the others.

(e) There is a variety of ways of adjusting the ranges used by a chart data series. The reader may wish to experiment with one of these.

- **(i)** On a PC right-click (on a Mac hold down the Ctrl key and click) the chart and from the popup menu use *Select Data* to open a dialog box. Edit the ranges to exclude row 4—see Fig. 7.23.
- **(ii)** Click on the data series in Chart 3 and drag the range finder lines to exclude row 4—see Fig. 7.24.
- **(iii)** Click on the data series in Chart 3 and look in the Formula Bar which reads
  `=SERIES(Sheet5!$B$3,Sheet5!$A$4:$A$24,Sheet5!$B$4:$B$24,1).`
  Edit this to read `=SERIES(Sheet5!$B$3,Sheet5!$A$5:$A$24,Sheet5!$B$5:$B$24,1).`

(f) Enter the text shown in C3 of Fig. 7.22; in C5 enter `=LOG10(A5)` and fill this down the column by double-clicking the fill handle. Note that we have avoided the log-of-zero problem by starting in row 5. Make Chart 4—to specify the correct data begin by selecting A3:A24 then, while holding down Ctrl, use the mouse to select C3:C24. Note the similarity between Chart 3 and Chart 4; in the former, the *y*-axis has a log scale while in the later the *y*-values are logarithmic.

(g) Save the workbook.

■ FIG. 7.23

	A	B
1	Large numbers and L	
2		
3	x	y
4	0	0
5	10	20
6	20	120
7	30	360
8	40	800
9	50	1500
10	60	2520
11	70	3920
12	80	5760
13	90	8100

■ FIG. 7.24

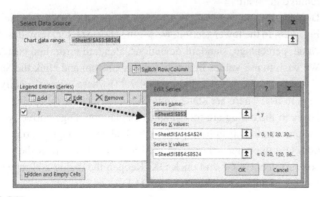

⯅	A	B	C	D	E	F	G	H	I	J	K	L	M
1	Error Bars												
2													
3	Voltage	Temp	Plus	Minus									
4	50	200	40	50									
5	60	250	60	50									
6	70	310	70	40									
7	80	375	60	90									
8	90	425	70	80									
9													
10													
11													
12													

■ FIG. 7.25

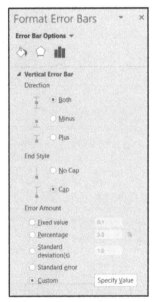

■ FIG. 7.26

■ FIG. 7.27

EXERCISE 7: ERROR BARS

Excel provides many options for adding error bars, but we have space to examine only one. Scenario: You have applied a voltage to a piece of equipment and measured the temperature ten times during an hour before increasing the voltage. Your data is as shown in Fig. 7.25. The Temp values are the hourly averages, each Plus value is the recorded maximum positive fluctuation from the average, and Minus is the negative fluctuation.

(a) On Sheet6, enter the data shown in A1:D8 in Fig. 7.25. Select A3:B8, make an XY chart (with title Error bars in Fig. 7.26) and format it as required.

(b) Activate the chart and using the Add Element tool displayed as a plus sign at the top right of the chart (see Fig. 7.6) add Error Bars. (On a Mac this can be done using *Chart Design / Add Chart Element / Standard Error*)

(c) Click on one of the horizontal error bars in the chart and use Delete to remove them all.

(d) Double-click on any vertical error bar to open the *Error Bar* dialog— Fig. 7.26. In the lower section, we see the choices Excel provides (fixed value, percentage, standard deviations, standard error, and custom). We wish to use values in cells so select Custom and click the *Specify Value* box to bring up a dialog (Fig. 7.27) where we can specify the cells to use. There are some minor bugs in Excel at this point: (i) the boxes in the dialog are too small to display the range correctly, and (ii) one cannot fill in both positive and negative ranges in one operation. Use the range selector (the red icon on the right) of the positive box to select C4:C8 and click OK. Reopen the dialog and repeat the process to add the negative range D4:D8.

(e) In the Format Error Bars dialog, the first tool (*Line & Fill*)—an icon resembling a tipped paint can—may be used to format the appearance of the error bars: line color and width for example.

(f) Make a copy of the chart; format the error bars such that only negative values are used and select Percentage set to 100% rather than Custom. This gives the second chart with 'drop lines' which are sometimes useful in a chart.

(g) Save the workbook.

EXERCISE 8: PLOTTING PARAMETRIC EQUATIONS

A point (x, y) on a circle of radius r will satisfy the equation $r^2 = x^2 + y^2$ so we might use $y = \sqrt{r^2 - x^2}$ to make an XY plot. However, it is simpler to realize that such a point will have polar coordinates r, θ and from simple trigonometry, we can convert to Cartesian coordinates with $x = r\cos(\theta)$ and $y = r\sin(\theta)$. In this pair of equations, we note that both x and y depend on the variable θ; such equations are called parametric equations.

In this exercise, we will plot a cardioid[2] given by $r = 1 + \sin(\theta)$. Clearly, in Cartesians coordinates, we have the parametric equations $x = (1 + \sin(\theta))$ $\cos(\theta)$ and $y = (1 + \sin(\theta))\sin(\theta)$. In Exercise 9 we see how an Excel radar chart may be used to plot simple polar curves.

(a) On Sheet7 of Chap7.xlsx begin by entering the text shown in rows 1 and 3 of Fig. 7.28.

(b) In A4 enter the value 0. In A5 enter =A4+22.5 and drag down to row 20.

(c) Select A3:B20 and use *Formulas / Defined Names / Create from Selection* (*Formulas / Create from Selection* on a Mac) and use the top row to label the columns.
Now that we have the polar coordinates we can generate the corresponding Cartesian coordinates.

(d) In B3 enter =1+SIN(RADIANS(theta)) and fill this down to B20 by double-clicking the fill handle.

(e) The formula in D4 is =r_*COS(RADIANS(theta)) and in E4 use =r_*SIN(RADIANS(theta)). Fill these down to row 20.

(f) Select any cell in D4:E20 and make an XY chart with a smooth line. Format the chart as required. Add a text box with the equation $r = 1 + \sin(\theta)$.

(g) Save the workbook.

---

[2]A cardioid is a plane curve traced by a point on the perimeter of a circle that is rolling around a fixed circle of the same radius.

	A	B	C	D	E
1	Cardioid				
2					
3	theta	r		x	y
4	0.0	1.000		1.000	0.000
5	22.5	1.383		1.277	0.529
6	45.0	1.707		1.207	1.207
7	67.5	1.924		0.736	1.777
8	90.0	2.000		0.000	2.000
9	112.5	1.924		-0.736	1.777
10	135.0	1.707		-1.207	1.207
11	157.5	1.383		-1.277	0.529
12	180.0	1.000		-1.000	0.000
13	202.5	0.617		-0.570	-0.236
14	225.0	0.293		-0.207	-0.207
15	247.5	0.076		-0.029	-0.070
16	270.0	0.000		0.000	0.000
17	292.5	0.076		0.029	-0.070
18	315.0	0.293		0.207	-0.207
19	337.5	0.617		0.570	-0.236
20	360.0	1.000		1.000	0.000

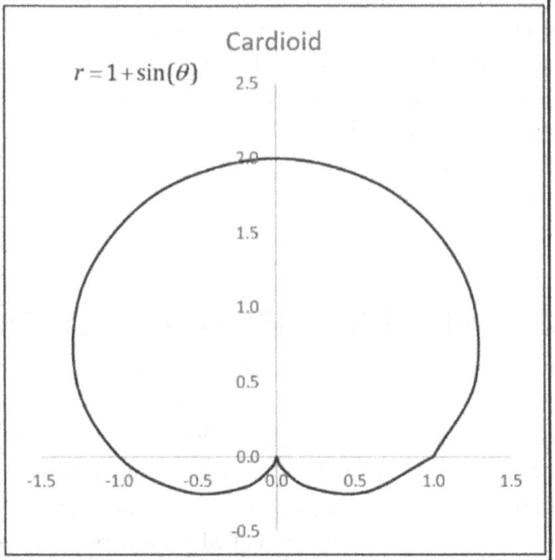

■ FIG. 7.28

OTHER CHART TYPES

We have concentrated on XY charts since these are the ones most frequently used by technical people. Most of the techniques we have covered are applicable to other Line, Column, and Bar charts. Furthermore, the reader now has enough knowledge to be able to work with other chart types (Radar, Area, etc.) with some experimentation. We conclude the chapter by looking at examples of some other types of charts.

EXERCISE 9: POLAR (RADAR) CHART

Excel does not provide a true polar chart option but in simple cases, one may use a radar chart to plot polar data.

> Excel's radar charts treat the first column as categories, not numeric values so it is essential to use a constant increment in that column when plotting a function

(a) On Sheet8 of Chap7.xlsx begin by entering the text shown in rows 1 and 3 of Fig. 7.29.

(b) Fill A4:A363 with values 0 through 359. You can enter 0 in A4 and with A4 selected use *Home / Editing / Fill / Series…* (*Home / Fill / Series…* On a Mac) with options shown in Fig. 7.30 to quickly do this.

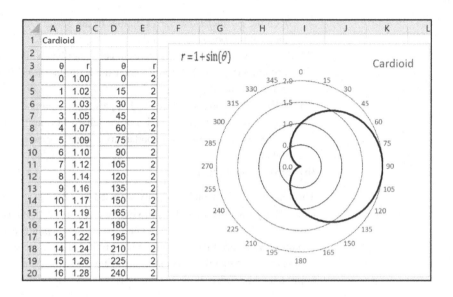

■ FIG. 7.29

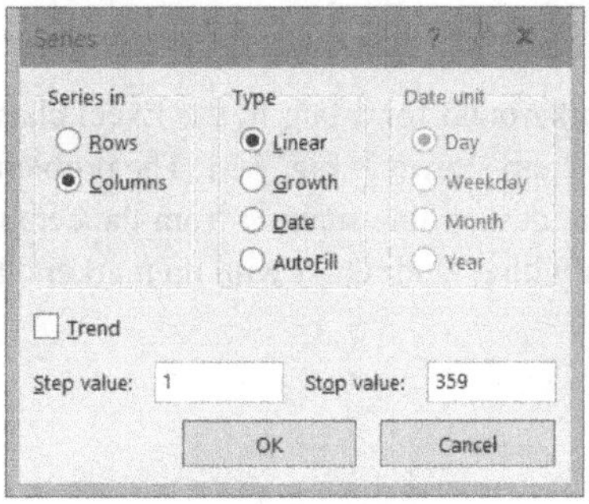

■ FIG. 7.30

(c) In B4 enter =1+SIN(RADIANS(A4)) and fill this down to B363 by double-clicking the fill handle.

(d) Select a cell such as A4 and create a radar chart. You may be surprised by the result. Because we have numeric values in the first column, Excel thinks we want two data series. On a PC right-click (on a Mac hold the Ctrl key and click) the chart and open the Select Data dialog. Two operations are needed: (i) remove the first series (Cardioid θ), and (ii) edit the Category Axis Labels specifying the range Sheet8!$A$4:$A$363.

(e) The large number (360) of labels around the chart's category axis makes them unreadable. Double-click the cardioid in the chart and open the *Format Data Series* dialog. In the *Series Option* section uncheck the box *Category labels*.

(f) Now we will add a dummy series to get neater labels. Enter the series 0 through 345 in D4:D27. In E3 enter the formula =MAX($B$3:$B$363) and drag this down to E27.

(g) Select D3:E7 and, using the techniques from Exercise 4, add the dummy data as a second series on the chart. Format this data series: (i) specify it is to use the secondary axis, and (ii) specify no line and no makers.

(h) Save the workbook.

The reader may be wondering why we used 360 data points for the radar chart. Return to Sheet7 and, following step (d) listed previously, make a radar chart of the range A3:B19. A chart with few data points is totally unacceptable. Be aware that the method shown in this exercise works only for simple curves; generally, it is more satisfying to use the parametric equation approach of Exercise 7.

RADAR CHARTS

This presents a workaround for a bug in the Excel chart engine. The user wishes to make the chart shown in Fig. 7.31. The problem is getting the category axis lines (the dashed line running from the center to the circumference). If you begin with a radar chart with no markers it seems impossible to do!

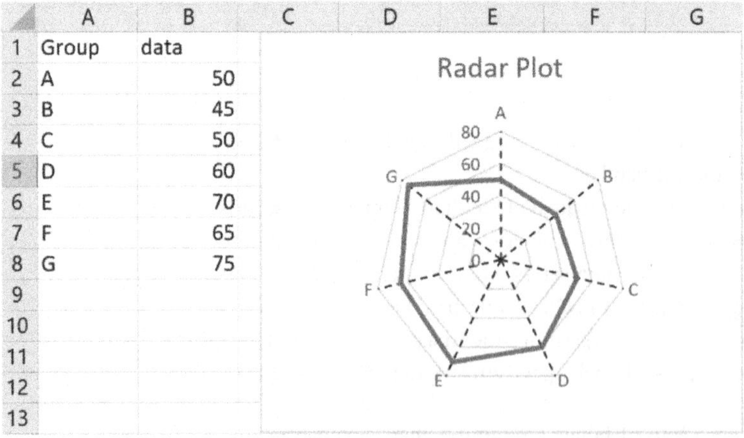

■ FIG. 7.31

The trick is to first make a chart with markers (marked with an arrow in Fig. 7.32) and then on a PC immediately right-click (on a Mac hold down the Ctrl key and click) on the chart, select Change Chart Type and select the no maker type (to the left of the one with the arrow in Fig. 7.32). Then you can click on the chart's value axis (the numbers 0–80), open the format dialog and specify a Line.

EXERCISE 10: SURFACE CHARTS

Excel can make surface plots, that is, a chart from a two-dimensional table. However, this has limitations in that the *x*- and *y*-axes are category axes, not value axes so it is essential to use equally spaced values.

Scenario: The table in Fig. 7.33 represents the result of an experiment in which a certain physical quantity was measured as parameters A and B were altered. We wish to show the data graphically.

(a) On Sheet9, enter the values shown in Fig. 7.33. The text Parameter B was typed into A6; the cells A6:A14 were merged and then formatted to have a 90° orientation. The gap in column B and row 4 helps define the chart area

(b) Select a cell in C5:H14 and use *Insert / Charts* (*Insert* on a Mac), open the waterfall chart drop-down and select the first Surface chart. Make a copy of the chart.

(c) On the second chart add Axis Titles for Primary Horizontal and Depth

(d) Format the Depth Axis Labels to have interval unit of 1 to get each value of parameter A.

(e) Use *Design / Chart Styles / Change Colors* (*Chart Design / Change Colors* On a Mac) drop-down and select the monochromatic grayscale option.

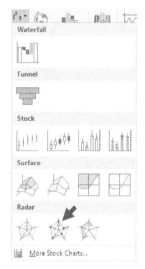

■ FIG. 7.32

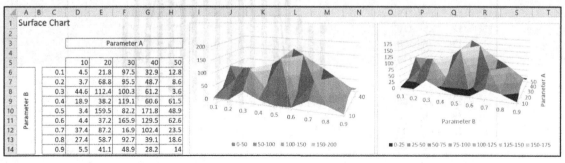

■ FIG. 7.33

(f) Compare the vertical axis with the legend of your first chart. The major units for the axis are set to 50 and the legend goes up in steps of 50. Double-click the vertical axis on the second chart and change the major unit to 25; your chart now resembles the right-hand chart in Fig. 7.33.

(g) On the second chart, first, click within the legend and then double-click on the legend item reading 0–25. This opens a dialog box with the title *Format Band*; this is where you can change the color of a single band.

(h) Save the workbook.

EXERCISE 11: COMBINATION CHARTS

It is possible to have a chart with one (or more) data series being displayed as a column chart while another is displayed as a Line or XY chart. In Exercise 3 we made an XY chart with two data series using two vertical axes. We will modify this to produce the chart shown in Fig. 7.34: in this, the temperature series has been changed to a column chart.

(a) While holding down the Ctrl key drag the tab of Sheet3 to the far right. Rename this new sheet as Sheet10.

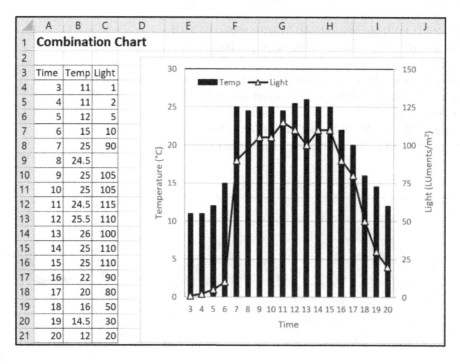

■ FIG. 7.34

(b) On a PC right-click (on a Mac hold down the Ctrl key and click) on the chart's plot area and from the popup menu select *Change Chart Type* to open the dialog shown in Fig. 7.35.

(c) Change the Temp data series to a clustered column chart. Click the OK button.

(d) If you look closely at the chart, you can notice that the light line starts at 2.5, rather than 3.5 as seen in Fig. 7.36. This is because the x-axis on a bar chart is categories, which Excel interprets as 1, 2, 3, … rather than their numerical values 0, 1, 2, …. In order to fix this, both should have the same categories. Go back to the *Change Chart Type* popup menu and make Light the type Lines with Markers.

(e) Now we want our data to match Fig. 7.34 so that both Temp and Light have the same categories. Delete columns C and D to make Light next to Temp. (Excel will warn you about changes in the chart data.) Move the light values down by cutting and pasting so that the first light reading matches hour 3. Now you can delete the rows with hours 0–2 and 19–24.

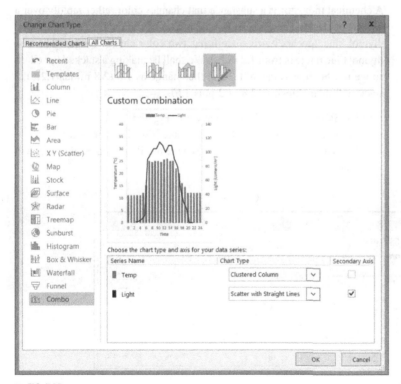

■ FIG. 7.35

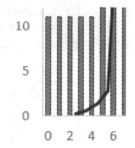

■ FIG. 7.36

(f) Format the Light data series to have markers with a white fill; this will make them visible against the columns. Note that if no fill is used the columns will show through the markers.

(g) Format the primary and secondary vertical axes to make the bounds 0–30 and 0–150, respectively. For the secondary axis, it is necessary also to change the major units to 25 (it needs to be a factor of 150–0 with the same spacing as the primary axis major units).

(h) Format the primary horizontal axis labels to have an interval unit of 1.

(i) Add the major horizontal gridlines. Note how the bounds we have chosen to make the gridline appropriate for either axis. Save the workbook.

EXERCISE 12: BAR CHARTS

We will make a stacked bar chart and hide one set of the bar segments to make the chart shown in Fig. 7.37. This technique can also be used to make a Gantt chart.

A chemical indicator is a substance that changes color rather rapidly over a narrow pH band. The chart shows the region of change for five indicators: thymol blue appears twice since it has two color change ranges. The first thymol blue range is from 1.2 to 2.8. We shall be making a stacked bar chart, so we use the low value as 1.2 and the change as 1.6 (2.8 minus 1.2); the stack will then extend to 2.8 (1.2 plus 1.6).

(a) On Sheet 11, enter the data shown in Fig. 7.37.

(b) Select any cell within A4:C9 and make a stacked bar chart.

(c) Format the Low pH data series to have no fill and no border; this will hide it.

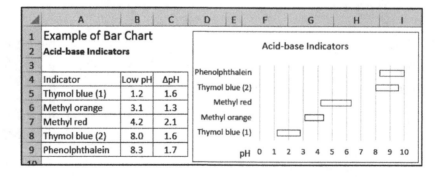

■ FIG. 7.37

(d) Format the Δ*pH* data series to have no fill but a black border.

(e) Add a horizontal axis title of *pH* and drag it to the left as shown in Fig. 7.37.

We have achieved our objective, but the chart tells us nothing about the actual colors. We could add ten text boxes, but it would be a chore dragging them into the correct positions to look tidy. In Fig. 7.38 we have another solution: make a chart with three series, add data labels to the first and last data series before hiding the bar segments. See Problem 8 for another approach.

(f) Insert a new sheet and call it Sheet11A. Enter the data shown in Fig. 7.38 (much of it can be copied from Sheet11).

(g) Make a stacked bar chart from A5:D10.

(h) On a PC right-click (on a Mac hold the Ctrl key and click) on the leftmost data series in the chart and select *Add Data Labels*. This will place the numbers in B6:B10 in each of the bar segments.

(i) On a PC right-click (on a Mac hold the Ctrl key and click) on one of the numbers in the chart and select *Format Data Labels*. In the Format Data Labels dialog: (i) Under Label Options clear all but *Value From Cells* in the Label Contains region, and using the mouse specify the range for the values as =Sheet11A!$F$6:$F$10, and (ii) in the Label Position area specify *Inside End* to cause the labels to appear to the right of the bar segment.

(j) Hide the bar segments for the first data series by formatting it to have no fill and no border.

(k) Repeat steps (i) and (k) for the third data series but this time the label position should be *Inside Base*. Save the workbook.

(l) If desired, experiment with a gradient fill for the center box to show the colors in the transition box as seen in Fig. 7.39.

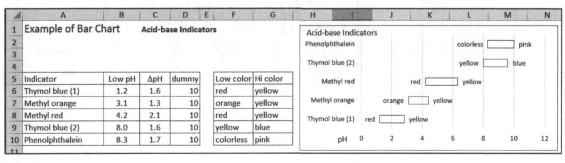

■ FIG. 7.38

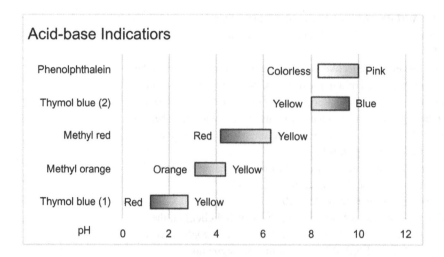

■ FIG. 7.39

PLOTTING SINE CURVES

Fig. 7.40 Shows two plots of $y = \sin(2\pi f t)$ with $f = 440\,\text{Hz}$. The solid curve has time increments of $1/12f$, giving 12 (actually 13) points for a full cycle while the dotted curve uses increments of $1/4f$. There are clearly not sufficient points in the second curve to accurately display the curve. We can show that the first curve is fine by adding a curve with time increments of $1/60f$ and show that this and the first curve exactly overlap. Moral: when plotting sine (and cosine) curve at least 12 points are needed for each full cycle.

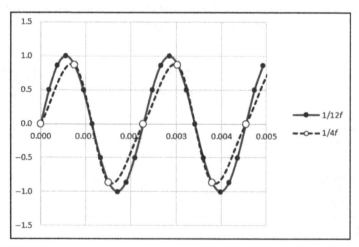

■ FIG. 7.40

EXCEL 2016: NEW CHARTS

Microsoft added a number of new chart types when Office 2016 was released. We shall look briefly at three of these: histogram, box-and-whisker, and waterfall. The quickest way to begin a histogram or box-and-whisker chart is it open the new Statistical Charts tool in the *Insert / Chart* group; see Fig. 7.41.

HISTOGRAM

[File Histogram.xlsx] Prior to Excel 2016 a user wishing to create a histogram began by setting up a range called the *bin*. Then one could either call upon the Histogram tool in Data Analysis[3] or use the FREQUENCY function[4] to count the number of occurrences for each bin value. The worksheet OldWay in Histogram.xlsx shows an example of this.

Generally, one uses bin values based on one's needs. But for those with statically sophistication may wish to use Scott's formula for the bin width:

$$bin\,width = 3.5 \times \sigma \sqrt[3]{\frac{1}{n}}$$

The first bin item will be the minimum of the dataset plus this bin value. Subsequent bin values will equal the previous bin value plus the bin width.

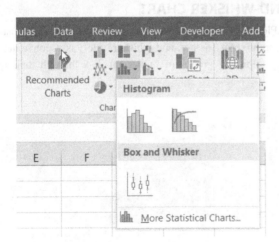

■ FIG. 7.41

---

[3]See, for example: http://www.excel-easy.com/examples/histogram.html.
[4]See, for example: http://www.eeo.ed.ac.uk/it/howto/Excel/g3histogram.html.

An example is shown on the OldWay.xlsx worksheet. Note that the column chart one makes (or the one made by the data analysis histogram tool) needs to be modified to close the gaps between the columns as traditionally histogram have no gaps.

The built-in histogram chart type in Excel 2016 makes life a little easier. It automatically works out an appropriate set of bin values without actually showing them on the worksheet. By default, it uses Scott's formula to compute the bin width. Using the same dataset as in *OldWay*, the worksheet *Numbers* shows the new histogram chart—see the left chart in Fig. 7.42. Note the values in category axis; these make the meaning of the histogram much clearer. To the right is the corresponding Pareto chart which is one of the choices when making a histogram in Excel 2016. By opening the Format Axis (for the category axis) the bandwidth or the number of bins may be altered to suit personal needs.

Fig. 7.43 shows a category histogram—see the worksheet *Category* in Histogram.xlsx. One begins by making an Excel 2016 histogram and then formatting the *x*-axis to Category. Unfortunately, the categories appear in the chart in the order of the first appearance in the data. To correct for this one would need to sort the data. The alternative is to make a column chart from a pivot table.

BOX-AND-WHISKER CHART

[File BoxPlot.xlsx] To quote from Wikipedia[5]: "In descriptive statistics, a **box plot** or **boxplot** [or box-and-whiskers plot] is a convenient way of graphically depicting groups of numerical data through their quartiles."

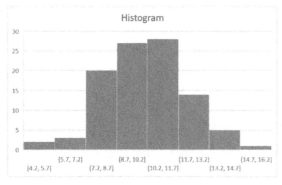
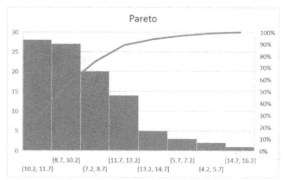

■ FIG. 7.42

---

[5]https://en.wikipedia.org/wiki/Box_plot.

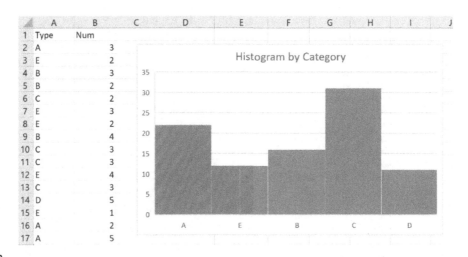

■ FIG. 7.43

Prior to Excel 2016, making such a plot in Excel was a rather lengthy business; see for example as shown in the clear webpage from Nathan Brixius.[6] Fig. 7.44 shows a box plot made in Excel 2016; this in on the worksheet *MM* in BoxPlot.xlsx. Just click within the data and use the *Insert / Charts* group. The only modifications made were as follows: (1) a legend was added, (2) the meaningless *x*-axis text was removed, (3) to make it easier to see the mean *X* and median line, the fill in some data items was changed to a lighter color, (4) after selecting on data point on the chart, the gap was adjusted to 10, and (5) the axis bounds were changed to 600–1100. Interestingly, there are things one cannot do with a box plot, including that the legend cannot be dragged around the chart area and the major and minor units on the *y*-axis cannot be adjusted.

Fig. 7.45 shows another boxplot—see worksheet *Ore* in BoxPlot.xlsx. In this one, the boxplot compiles data from two categories, Gold and Silver.

WATERFALL OR BRIDGE CHART

[File Waterfall.xlsx] On the AbleBits site,[7] Ekaterina Bespalaya shows how one made a waterfall chart in Excel 2013 and earlier versions. Using Ekaterina's data, a waterfall chart was made within a few minutes in Excel 2016 as shown in Fig. 7.46.

---

[6]https://nathanbrixius.wordpress.com/2014/03/10/beautiful-box-plots-in-excel-2013/.

[7]https://www.ablebits.com/office-addins-blog/2014/07/25/waterfall-chart-in-excel/#create-waterfall-chart.

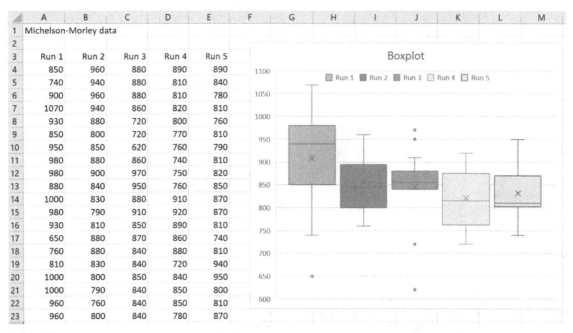

	A	B	C	D	E
1	Michelson-Morley data				
2					
3	Run 1	Run 2	Run 3	Run 4	Run 5
4	850	960	880	890	890
5	740	940	880	810	840
6	900	960	880	810	780
7	1070	940	860	820	810
8	930	880	720	800	760
9	850	800	720	770	810
10	950	850	620	760	790
11	980	880	860	740	810
12	980	900	970	750	820
13	880	840	950	760	850
14	1000	830	880	910	870
15	980	790	910	920	870
16	930	810	850	890	810
17	650	880	870	860	740
18	760	880	840	880	810
19	810	830	840	720	940
20	1000	800	850	840	950
21	1000	790	840	850	800
22	960	760	840	850	810
23	960	800	840	780	870

■ FIG. 7.44

	A	B	C
1	Box plot		
2			
3	Metal	Site A	Site B
4	Gold	0.5	0.6
5	Silver	0.4	0.6
6	Silver	1	1
7	Silver	0.6	0.2
8	Gold	0.7	0.1
9	Gold	0.5	0.8
10	Gold	0.35	0.5
11	Silver	1	0.8
12	Silver	1.1	0.3
13	Silver	0.5	0.1
14			
15			
16			
17			

■ FIG. 7.45

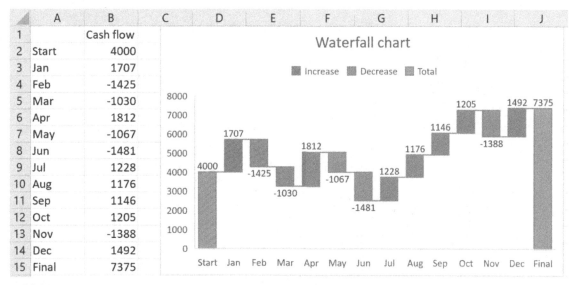

■ FIG. 7.46

Just click within the data, open Recommended Chart for the Insert | Charts group and select Waterfall. Some minor modifications can help: (1) click on the data series, open the Format dialog and check the Show Connectors box; (2) to make these a little more visible, click and delete the gridlines and open the Format Data Series dialog and increase the weight of the borders; (3) in turn click, on the first and last data point (Start and Final[8]) and in the Format Data Point dialog check the Set as Total box.

MAP CHART

If you have an Office 365 subscription and you have downloaded an updated version of Excel 2016, a map chart that can be used to compare values and show categories across geographic regions. To use a map chart you need categories of geographic regions, such as countries, states, counties, or postal codes. With the regions and values selected, the map chart is added using *Insert | Charts | Maps | Filled Map* (*Insert | Charts | Maps | Filled Map* on a Mac).

[File Minimum Wage by State.xlsx] Using data from the National Conference of State Legislatures[9] the minimum wage by state is plotted in Fig. 7.47. The federal minimum wage of $7.25 was used where the state minimum wage was

---

[8]All cells but B15 hold values; in B15 the formula =SUM(B2:B14) is used.

[9]http://www.ncsl.org/research/labor-and-employment/state-minimum-wage-chart.aspx.

Minimum wage in 2018

$11.50
$8.30
$7.25
$9.65
$10...
$10.75
$7.25
$7.25
$8.85
$7.25
$9.25
$10.40
$8.25
$7.25
$9.00
$7.25
$8.30
$7.25
$11.00
$7.25
$10.20
$8.25 $7.25
$...
$7.25
$7.25
$7.85
$7.25
$7.25
$7.25
$10.50
$7.50
$7.25
$8.50
$7.25
$7.25 $7.25 $7.25
$7.25
$9.84
$7.25
$8...

Powered by Bing
© GeoNames, MSFT, Navteq

MinimumWage
$7.25 $13.25

■ FIG. 7.47

lower than the federal minimum wage, or if the state did not have a minimum wage. Data labels were added to the chart to indicate the wage.

URLS FOR CHART WEBSITES

If you have a charting problem, one of the following sites will very likely have an answer for you.

Jon Peltier:	https://peltiertech.com/Excel/ChartsHowTo/
Andy Pope:	http://www.andypope.info/charts.htm
Tusha Mehta:	http://tushar-mehta.com/excel/charts/
Stephen Bullen:	http://oaltd.co.uk/Excel/Default.htm
Jan Karel	https://www.jkp-ads.com/articles/
Pieterse:	ChartAnEquation00.asp

PROBLEMS

1. *Using the information from Problem 5 in Chapter 4, make a chart similar to that shown in Fig. 7.48. To ensure that the shape with text is part of the chart, and will move with the chart, select the chart before using *Insert / Illustrations / Shapes*.
2. Make XY plots to (i) show that $\sin(x) - \cosh(x) + 1 = 0$ has roots at 0 and 1.3—see Fig. 7.49, and (b) to find the approximate roots of $\exp(-(x-2)^2 \cos(\pi x)) = 4\cos(x-2)$. In each case make a plot with two data series.

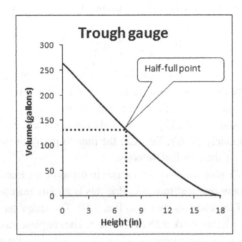

■ FIG. 7.48

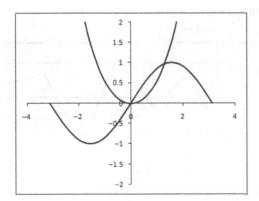

■ FIG. 7.49

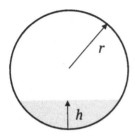

3. *The volume V of liquid in a hollow horizontal cylinder of radius r and length L is given by the following equation where h is the depth of the water. Using named cells and values $r = 6\,\text{m}$, $L = 5\,\text{m}$, make a plot of V against h.

$$V = \left[r^2 \cos^{-1}\left(\frac{r-h}{r}\right) - (r-h)\sqrt{2rh - h^2} \right] L$$

Can you find a way to avoid problems with the chart if you now give the radius a value of $3\,\text{m}$?

4. It can be shown that a body of mass m falling under the influence of gravity g and subject to a drag which is proportional to the square of the body's velocity (v), the velocity varies with time (t) according to the equation

$$v(t) = \sqrt{\frac{gm}{c_d}} \tanh\left(\sqrt{\frac{gc_d}{m}} t \right)$$

where c_d is the drag coefficient. Make a worksheet to enable you to chart the velocity from 0 to 25 s for a body starting from rest. The worksheet should allow the user to vary both m and c_d. Show that for $m = 75\,\text{kg}$ and $c_d = 0.25\,\text{kg/m}$, the body reaches a terminal velocity of approximately 54 m/s. Estimate the time to reach a velocity that is within 5% of the terminal velocity.

5. Make an XY plot with two data series to demonstrate that $x^3 - x + 2 = 0$ can have only one real root, and that this is approximately $x = -1.5$.

6. Make the Line chart as shown in Fig. 7.50. The technique used in Exercise 5 will not work with a Line chart. The simplest way is to make a column with the same number of points as in the main data series for each line to be added. Other methods include using secondary axes or using error bars. To explore these, do an internet search using `Peltier excel "line chart" horizontal line` to find Jon Peltier's advice on this topic.

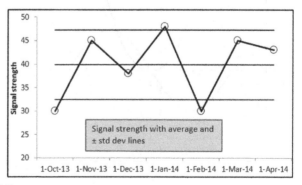

■ **FIG. 7.50**

7. Acme Tank Inc. will supply a customer with a storage tank in one of two ways: (i) the tank is constructed in their factory and transported to the site, or (ii) material is transported and the tank is constructed in situ. The formulas (albeit, rather complex formulas) for each method are as follows:

$$Cost1 = \text{Exp}\left(7.994 + 0.6637^* \ \ln V - 0.063088^*(\ \ln V)^2\right)$$
$$Cost2 = \text{Exp}\left(9.369 - 0.1045^* \ \ln V + 0.045355^*(\ \ln V)^2\right)$$
where V is the capacity in cubic meters.

Show that when $5 <= V <= 100$ (see Fig. 7.51) that the in situ construction cost is generally more than the factory-built cost, except for a very narrow band of V around $40\,\text{m}^3$ and even then the savings are minimal.

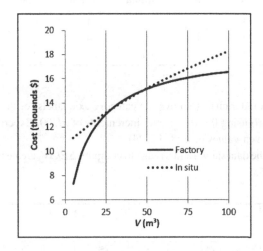

■ FIG. 7.51

8. Exercise 12 explored bar charts with hidden segments. Using the same techniques make the chart shown in Fig. 7.52. If your chart is made large enough, you can avoid the Phph abbreviation for phenolphthalein.
9. Following the directions in Exercise 8, make a parametric plot of the functions;

$$x = A\sin(at + \delta) \text{ and } y = B\sin(bt)$$

using t value 0–8 in increments (*inc*) of 0.01 to generate Lissajous curves—search the internet for more details. Try these values for the variables and experiment with others

Example of Bar Chart — Acid-base Indicators

Indicator	Colour-change	Low pH	ΔpH
Thymol blue (1)	red-yellow	1.2	1.6
Methyl orange	orange-yellow	3.1	1.3
Methyl red	red-yellow	4.2	2.1
Thymol blue (2)	yellow-blue	8.0	1.6
Phph	colorless-pink	8.3	1.7

■ FIG. 7.52

	A	B	Alpha	Beta	Delta	Inc
Simple	1	1	2	1	0	0.1
Pretty	1	1	−5	4	0	0.1
Knot	1	1	−2	4	5	0.6

10. Make a radar chart (remove the category axis labels) of the function $=\cos(n\theta)$ using $0 < \theta < \pi$ with increments of $\pi/200$. Experiment with large, even values of n (8, 12, 24).
11. Using the data shown and a range to compute $\Delta E/\Delta V$, create the chart in Fig. 7.53.

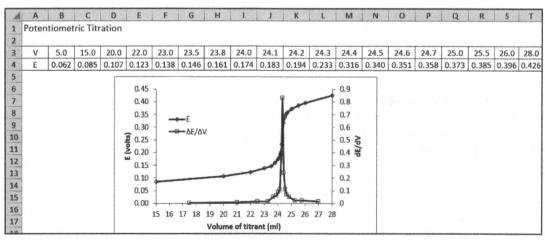

■ FIG. 7.53

▲	A	B	C	D	E	F	G	H	I	J	K
1	Gantt Chart										
2											
3											
4	Activity	Start	To complete								
5	Foundation	1-Jun-13	7								
6	Frame	8-Jun-13	21								
7	Plumb and Wire	29-Jun-13	6								
8	Wallboard	5-Jul-13	5								
9	Trim	10-Jul-13	4								
10	Roofing	29-Jun-13	3								
11	Windows	29-Jun-13	4								
12	Siding	3-Jul-13	5								
13											
14											

■ FIG. 7.54

12. Make the Gantt chart shown in Fig. 7.54.
13. A pictogram is a bar or column chart that uses pictures for the bars or columns as in the left-hand chart in Fig. 7.55. Make a regular bar chart of the data and open the format dialog. This contains an option *FILL / Picture fill* and a tool to search for, and insert, a picture either from the internet or from the user's hard drive. This method does not work with Line or XY charts. To use images for markers one must copy the image, select the chart data series and use Paste. The image must be correctly sized, so it is convenient to place it on the worksheet, resize it to be the same height as 1 or 2 rows, then copy and paste to the data series. The image may then be deleted from the worksheet.
14. This problem will demonstrate how beats are formed by the addition of sounds waves. The chart in Fig. 7.56 shows the addition of $A\sin(2\pi f_A t)$ and $B\sin(2\pi f_B t)$ with variables having the value shown in B3:C4,

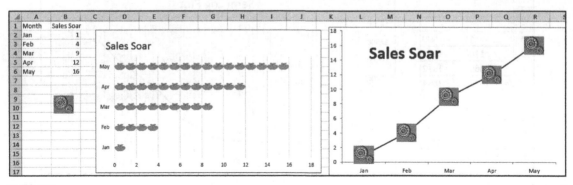

■ FIG. 7.55

◢	A	B	C	D	E	F	G	H	I
1	Adding Sine Waves								
2		Wave A	Wave B						
3	Frequency	440	500						
4	Amplitude	1	1.5						
5	delta	0.000189	0.000167						
6	Δt	0.000166667							
7									

■ FIG. 7.56

Cells B5:C5 compute $1/12f$ for the two sine waves while B6 identifies the smaller of these to use in the table of values to be plotted. Make a table with values of t, $A\sin(2\pi f_A t)$ and $B\sin(2\pi f_B t)$. Since there is no XY stacked option a final column is needed to add the two sine waves.

15. Make the semi-log plot shown in Fig. 7.57. The horizontal axis has major and minor units set to 100 and 10, respectively. If Excel uses a very light color for the minor grid lines, it may be necessary to locate them from the *Chart Tool / Format / Current Selection* menu. Somewhat confusingly, although these are vertical gridlines, Excel calls them the *Horizontal (value) Axis Minor Gridlines*.

16. The data shown in Fig. 7.58 comes from Exercise 3 in Chapter 6. Make the histogram using a column chart in which the gap is set to zero.

17. Construct a chart similar to that in Fig. 7.59 using the data in A1:E4.

18. In the summer of 2003, Paris suffered a terrible heat wave. From Fig. 7.60 one can see a correlation between temperatures and the death

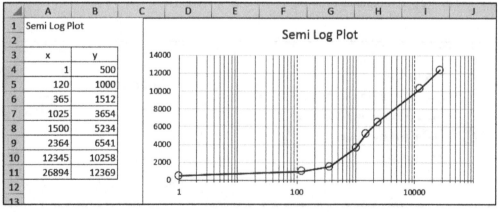

■ FIG. 7.57

◢	A	B	C	D	E	F	G	H	I
1	Frequency Histogram								
2									
3	Bin	Count							
4	300	1							
5	400	14							
6	500	16							
7	600	11							
8	700	18							
9	800	18							
10	900	10							
11	1000	12							
12									
13									

■ FIG. 7.58

◢	A	B	C	D	E	F	G	H	I	J
1		Block start	Block end	Low Δ	High Δ					
2	A	100	125	75	275					
3	B	150	50	25	325					
4	C	100	100	50	300					
5										
6										
7										
8										
9										
10										
11										

■ FIG. 7.59

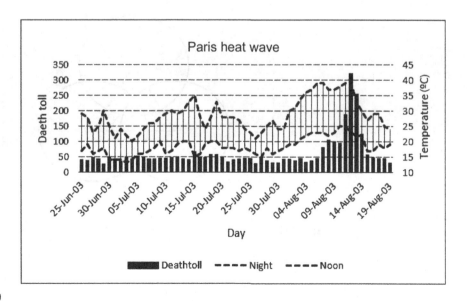

■ FIG. 7.60

rate. Download the file ParisHeatWave.xlsx from the companion website and construct a similar chart. [Data from Benedict Dousset, the University of Hawaii at Manoa.]

19. Fig. 7.61 shows a cylindrical vessel of mass M of 115 kg, height H of 0.8 m and radius r of 0.25 m. When empty the center of gravity of the vessel is 0.43 m from the base. Make a plot showing the position of the center of gravity as water is added.

20. Make a chart similar to that in Fig. 7.62. The three circles have the same radius and the triangle formed by joining their centers is equilateral.

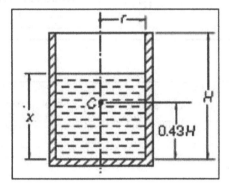

■ FIG. 7.61

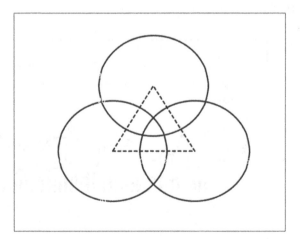

■ FIG. 7.62

21. A point on the rim of a rolling wheel follows a path known as the cycloid. Make a chart showing this curve. The coordinates are defined by

$$x = r(\theta - \sin\theta) \text{ and } y = r(1 - \cos\theta)$$

22. The temperature in Riverside, California was measured every 2 h as shown in Fig. 7.63.
    (a) Enter the data in columns A and B
    (b) Create a line graph based on columns A and B
    (c) Create an XY Scatter graph based on columns A and B. Change the axis limits to 0 and 1 (day), and the major unit to 0.166666 (4/24 of a day)

23. A bode plot is a plot of the gain (in decibels (dB)) and phase shift (in degrees) versus frequency (in log scale) for a circuit transfer function as seen in Fig. 7.64. The formulas are as follows

$$Gain = \frac{1}{\sqrt{1 + \left(\dfrac{\omega}{\omega_n}\right)^2}} \quad Gain(dB) = 20\log_{10}(Gain) \quad phase = \tan^{-1}\left(\frac{-\omega}{\omega_n}\right)$$

The scaling for the frequency is $10^{\,(1/(\text{points/decade}))}$. Each successive frequency is the previous frequency times the scaling. Create a user-friendly spreadsheet which plots the Bode plot as indicated using 2 y-axes and log scale on the x-axes. Note: The data continues beyond the end of the picture till $\omega = 1000$. Also, ATAN returns the angle in radians, so you need to convert it to degrees to match the chart.

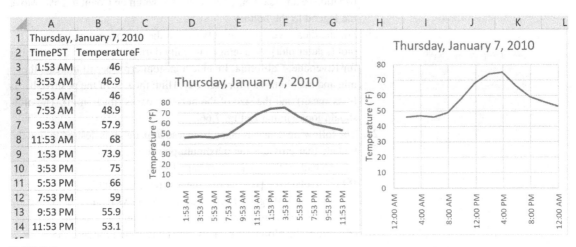

■ FIG. 7.63

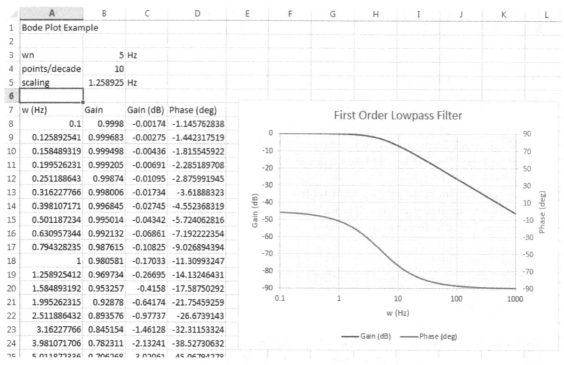

	A	B	C	D
1	Bode Plot Example			
2				
3	wn	5 Hz		
4	points/decade	10		
5	scaling	1.258925 Hz		
6				
7	w (Hz)	Gain	Gain (dB)	Phase (deg)
8	0.1	0.9998	-0.00174	-1.145762838
9	0.125892541	0.999683	-0.00275	-1.442317519
10	0.158489319	0.999498	-0.00436	-1.815545922
11	0.199526231	0.999205	-0.00691	-2.285189708
12	0.251188643	0.99874	-0.01095	-2.875991945
13	0.316227766	0.998006	-0.01734	-3.61888323
14	0.398107171	0.996845	-0.02745	-4.552368319
15	0.501187234	0.995014	-0.04342	-5.724062816
16	0.630957344	0.992132	-0.06861	-7.192222354
17	0.794328235	0.987615	-0.10825	-9.026894394
18	1	0.980581	-0.17033	-11.30993247
19	1.258925412	0.969734	-0.26695	-14.13246431
20	1.584893192	0.953257	-0.4158	-17.58750292
21	1.995262315	0.92878	-0.64174	-21.75459259
22	2.511886432	0.893576	-0.97737	-26.6739143
23	3.16227766	0.845154	-1.46128	-32.31153324
24	3.981071706	0.782311	-2.13241	-38.52730632
25	5.011872336	0.706268	3.02061	45.06794278

■ FIG. 7.64

24. Open normal.csv from the companion web page. Plot the Count and Normal distribution data as seen in Fig. 7.65. Change the normal distribution to an area plot. Change the gap width on Count to 25%. Move the legend to the bottom.

25. For the data given in Humidity.csv on the companion web page, plot (scatter plot) the average humidity during the month of December for Riverside, California. Include a custom error bar to indicate the min and max humidity for each day. Hint: the size of the positive bar is (max-mean) and the size of the negative bar is (mean-min). The plot should look similar to Fig. 7.66.

26. The following data was taken from a solar panel. It is desired to find the voltage that provides the maximum power.
    (a) Calculate the Current (Voltage/Resistance) in Amps (A) $= V/\Omega$.
    (b) Calculate the power (Voltage*Current) in Watts $= V*A$.
    (c) Plot the Voltage vs Resistance and Power vs Resistance. Use a log scale for the resistance. Plot the power on the 2nd axis.
    (d) From the plot determine the resistance that produces the maximum power.

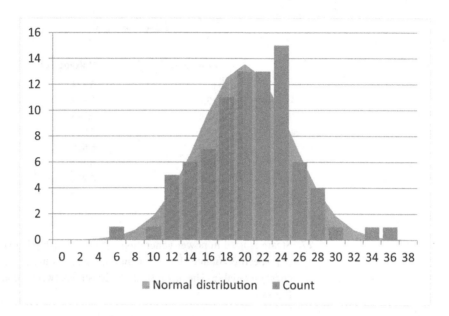

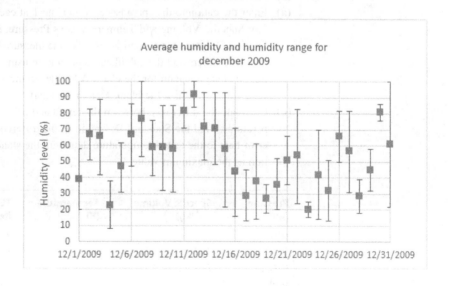

Collected Data	
Resistance (Ω)	**Voltage (V)**
1	0.124
15.9	1.872
33.4	3.745
49.3	4.934
71.4	5.566
165.2	6.113
314.7	6.294

27. Steam is used in power generation. The relationship between pressure
    (P), specific volume (V), and temperature (T) is used to calculate the
    energy available. The theoretical relationship between these values is as
    follows

$$T = \frac{M \cdot P \cdot V}{R}$$

where $M = 0.018\,\text{kg}\,\text{m}^3$, and $R = 8.312\,\text{mol}/(\text{kg}\,\text{K})$, P is in Pa, V is in m^3/
kg, and T is in K. In order to verify this relationship, the following
experimental data was recorded.

(a) Enter the data into the spreadsheet. Calculate T at each pressure.
    Plot Specific Volume and Temperatures vs Pressure. Put the
    Specific Volume on a second y-axis. Plot the measured data using
    distinct markers and the calculated temperature using a smooth
    line. Include a legend for the chart. Add titles to the x and y-axes
    with units. Minimize the blank area in the chart

(b) Calculate the difference between the tested and calculated tem-
    peratures. Find the Standard Deviation of the differences. Add
    error bars to the tested temperature data with the standard devia-
    tion as the maximum value.

Pressure (MPa)	Specific Volume (m³/kg)	Temperature (K)	TEMP (tested) (K)
0.01	14.8		317.2
0.04	4.01		348.9
0.08	2.11		366.5
0.14	1.25		382.3
0.22	0.81		396.3

28. It is desired when bungee jumping off a bridge to just touch the water which is 60 m below. The rope has an elasticity such that the spring constant is as follows

$$k = \frac{C_r}{l_0}$$

where $C_r = 3000\,\text{N}$ and l_0 is the unstretched length of the chord in meters. Using conservation of energy, the peak stretch of the chord (x) can be calculated using

$$mg(x + l_0) = \frac{kx^2}{2}$$

where $g = 9.81\,\text{m/s}$, and m is the mass of the person in kg. This can be solved for your peak displacement x to get

$$x = \frac{mg + \sqrt{(mg)^2 + 2kmgl_0}}{k}$$

The total distance fallen h can be calculated using

$$h = x + l_0$$

Create a spreadsheet to plot the total distance fell (h) vs. the initial rope length (l_0) for $0\,\text{m} < l_0 < 40\,\text{m}$ with a step size of 5 m. Have the mass of the person $(m = 70\,\text{kg})$ and the elasticity of the rope (C_r) be easily changed for different ropes and people.

Add a marker indicating the solution to the problem. It can be made by plotting another series with a single point with an x value of what you observe from your plot and a y value of 60. Also, add an x data label to your point to show the required chord length. Add error bars in 1 direction down and to the right that are 60 units in length to get the extension lines. On a 2nd y-axis, add the chord stiffness to the plot.

A plot for different values of m and C_r can be seen as follows to give you an idea of what the plot should look like. You need to add a descriptive title with the proper rope length for the person.

Hint: create a separate column for k and x and h calculations.

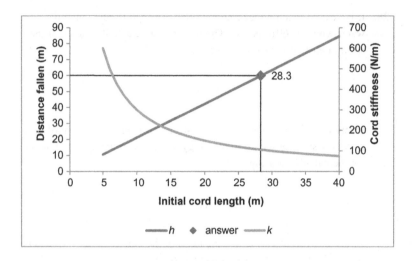

29. In vibrations, the position of a vibrating mass follows the exponential decaying curve

t_m (s)	x_m (m)
0	1.91
0.2	0.39
0.4	−1.11
0.6	−0.50
0.8	0.43
1	0.43
1.2	−0.064
1.4	−0.33
1.6	0.027
1.8	0.14
2	−0.045

$$x(t) = X_0 e^{-\zeta \omega_n t} \cos(\omega_d t)$$

The measured data shown on the right corresponds to $X_0 = 2\,\text{m}$, $\zeta = 0.2$, $\omega_n = 7\,\text{rad/s}$, and

$$\omega_d = \omega_n \sqrt{1 - \zeta^2}$$

(a) Create a user-friendly spreadsheet to calculate $x(t)$ for $0s \leq t \leq 2s$ with a step size of 0.1. Allow X_0, ζ and ω_n to be easily changed.

(b) Calculate the bounding curves

$$x_b(t) = \pm X_0 e^{-\zeta \omega_n t}$$

(c) Plot the measured data and the calculated values as follows.

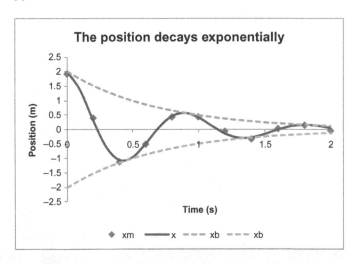

APPENDIX: SUPPLEMENTARY MATERIAL

Supplementary material related to this chapter can be found on the accompanying CD or online at https://doi.org/10.1016/B978-0-12-818249-9.00007-8.

Chapter **8**

Regression Analysis

CHAPTER OUTLINE

In this chapter, we seek answers to the question: *What equation fits my experimental data?* The general terminology for this type of activity is *regression analysis*. The reader may wish to Google or Bing to find how this term came to be used.

We begin by looking at simple linear functions—functions that can be recast as $y = mx + c$ where m is the slope of the line and b the intercept. Consider an experiment in which some *independent* variable such as temperature is changed while another, *dependent* variable (maybe the electrical resistance of an object) is measured. The independent variable is generally represented by x and is sometimes called the *predictor* or *explanatory* variable; the dependent variable is generally represented by y and may be referred to as the *response* or *predicted* variable. It is common in regression analysis to assume that the dependent variable, being under the control of the experimenter, is error free but that the dependent variable will often have experimental errors associated with it. The task of regression analysis is to find the

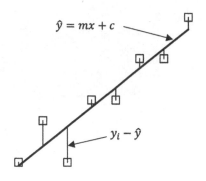

Liengme's Guide to Excel 2016 for Scientists and Engineers. https://doi.org/10.1016/B978-0-12-818249-9.00008-X
© 2020 Elsevier Inc. All rights reserved.

values of the regression *parameters* or *coefficients* (in $y = mx + c$, these are m and c) such that the equation correctly describes the data.

Most historians[1] credit Gauss with the development of the fundamentals of least-squares analysis in 1795 at the age of 18. One speaks about *the line of best fit*. In this instance, we will restrict ourselves to linear fits. Let the experimental data consist of pairs of x- and y-values. We write the equation of the line of best fit as $\hat{y} = mx + c$ ($\hat{y}$ is read as "y hat"). This is the predicted value for a given x. The vertical displacement between the actual y-value and the predicted for a given x (see diagram) is called the *residual*. The least-squares criterion requires that we adjust the constants m and b such that the sum of the squares of the residuals $\sum(y_i - \hat{y})^2$ is as small as possible. There are formulas for finding these parameters, but we shall let Excel do the work.

We explore the use of chart *trendlines* and the Excel functions SLOPE, INTERCEPT, TREND, FORECAST, and LINEST. Then we explore some nonlinear functions, again using trendlines, and the Excel function LINEST and LOGEST. We conclude by showing the use of Excel's Data Analysis tools. In Chapter 12 we will see how Solver can be used for curve-fitting problems, especially for problems where trendlines and Excel functions cannot be used.

In this chapter, we use the equation $y = mx + b$. You should be aware that there are other conventions. In the United Kingdom it is $y = mx + c$ and m is called the *gradient*. Statisticians, and the authors of some entries in Excel's Help, like to use $y = a + bx$. Hence b may not be the b you are thinking of when you flip through a textbook or glance at Help. Always check what convention is being used.

EXERCISE 1: TRENDLINE, SLOPE, AND INTERCEPT

Scenario: A physics student is tasked with finding the thermal coefficient of resistance of a sample. Her experimental results[2] are shown in Fig. 8.1. The textbook told her to work with Eq. (8.1), where R_0 is the resistance at $0°C$, R_t is the resistance at temperature $t °C$, and α is the required coefficient.

$$R_t = R_0(1 + \alpha t) \tag{8.1}$$

Of course, this can also be written as follows:

$$R_t = \alpha R_0 t + R_0 \tag{8.2}$$

---

[1]There are others who give the credit to Legendre. However, it was Gauss who expanded the basic method to encompass probability and normal distribution of errors.

[2]The data is shown in column form but we could just as well put the x-values in one row and the y-values in another row. See Exercise 6.

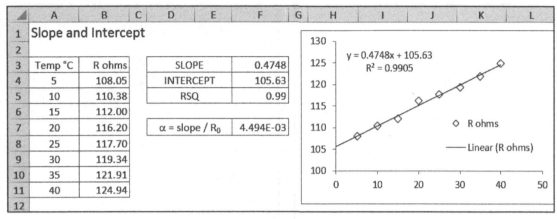

■ FIG. 8.1

This has the form of the well-known equation of a straight line $y = mx + b$. The slope will be αR_0 and the intercept R_0; hence α can be found from slope divided by intercept.

(a) Open a new workbook and on Sheet1 enter the text and data shown in columns A and B of Fig. 8.1.

(b) Construct an XY chart using the first examples shown in the gallery—markers only.
Now we are ready to add the trendline. We could select the chart (by clicking on it) and use *Chart Tools / Design / Chart Layouts / Add Chart Element / Trendline / Linear Trendline* (*Chart Design / Add Chart Element / Trendline / Linear* on a Mac) to quickly add a trendline. However, this just adds the trendline; we want more. The same steps, but ending with *More Trendline Options*, will open the required dialog but we shall use the shortcut menu.

(c) On a PC right-click (on a Mac hold down the Ctrl key and click) a marker on the chart and select Insert Trendline from the shortcut menu to open the Trendline dialog (see Fig. 8.2).
Clearly, we want a linear trendline, so make that selection. For this demonstration also check the boxes to give us the equation of the best fit and the *R*-squared value. Our data starts at 5°C but it will be interesting to have the trendline start at 0°C (then it will hit the *y*-axis), so in the Backwards box of the Forecast group, enter a value of 5. To format the appearance (line color and thickness, etc.) of the trendline, click on the icon representing a tipped paint can. The trendline can be formatted at any time by clicking it to open its formatting dialog.

(d) Note that the Trendline Label box can be dragged around the chart. As we see in Fig. 8.3, we can change numbers format—the number of

Format Trendline
Trendline Options
Trendline Options
Exponential
Linear
Logarithmic
Polynomial
Order
2
Power
Moving Average
Period
2
Trendline Name
Automatic
Linear (R ohms)
Custom
Forecast
Forward
0.0
period
Backward
0.0
period
Set Intercept
0.0
Display Equation on chart
Display R-squared value on chart

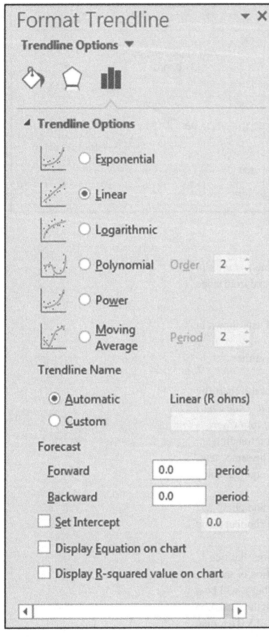

■ FIG. 8.2

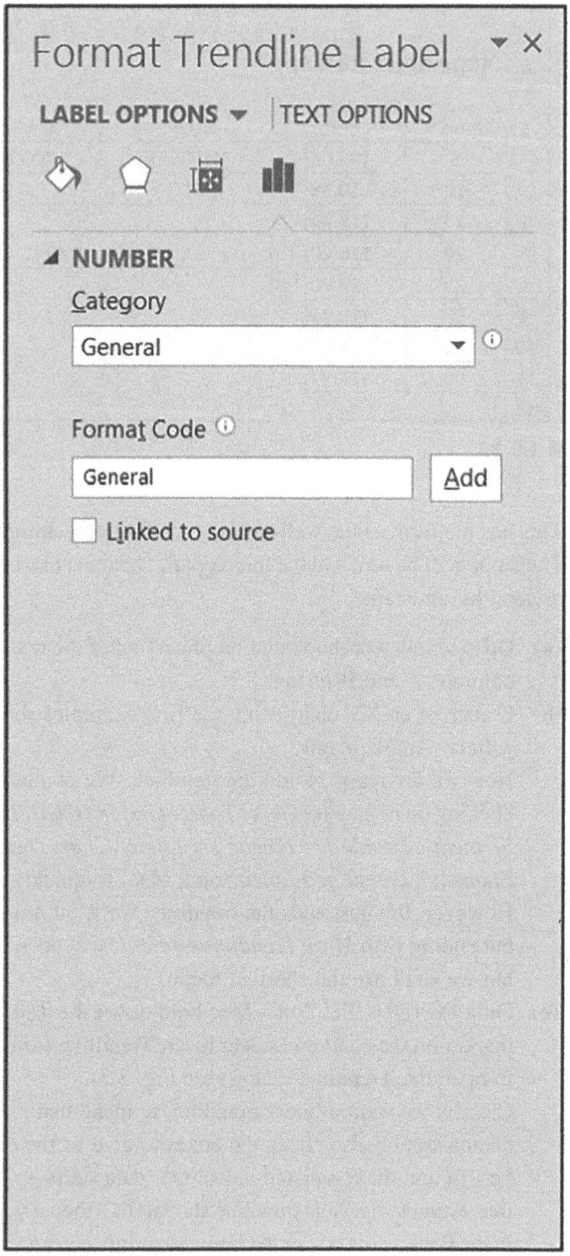

Format Trendline Label
LABEL OPTIONS
TEXT OPTIONS
NUMBER
Category
General
Format Code
General
Add
Linked to source

■ FIG. 8.3

decimals, scientific notation, and so on. Again the paint can icon opens a dialog for changing the font and fill colors, and so on.

Remember that we want both slope and intercept in order to compute the coefficient α. There are three good reasons not to just copy the trendline values into cells on the worksheet: (i) We may not use enough significant figures (the Trendline equation can be formatted to show more or fewer digits); (ii) should we make a correction to the data used to make the plot, we may forget to recopy the Trendline coefficients; and (iii) it is an error-prone operation. Rather, we shall get the coefficients of the line of best fit using functions.

R^2 (the coefficient of determination) gives a measure of the goodness of the fit. In a sense, it is a measure of how much of the variability in the *y*-values can be accounted for by changes in the *x*-values. Here it is 99%; the rest may be attributed to experimental errors.

(e) Enter the text in columns D and E.

(f) The formulas we need in column F are as follows:

```
T3:   =SLOPE(B4:B11, A4:A11)
F4:   =INTERCEPT(B4:B11, A4:A11)
F5:   =RSQ(B4:B11, A4:A11)
F6:   =F3/F4 (this computes α)
```

Format the cells to display the number of decimals shown in Fig. 8.1. Note the agreement between the trendline label and the functions. The syntax for the three regression formulas is FUNCTION(*y*-value, *x*-values). Engineers and scientists generally use *x* before *y* in this context, so be careful.

(g) Save the workbook as Chap8.xlsx.

EXERCISE 2: INTERPOLATION AND FORECAST

In the previous exercise, we fitted data for temperatures in the range 5–40°C in 5-degree intervals. How would we compute the expected resistance of the sample at (i) 22°C, (ii) 0°C, (iii) 30°C, and (iv) 100°C? The first three tasks are called interpolation (we want a value within the known range); the last one is extrapolation (we want a value outside the known range). It is generally safe to interpolate. Extrapolation is risky especially when the value lies far from the known range. Many physical systems appear to behave in a linear fashion over a short range but actually obey more complex laws. A gas obeys the Ideal Gas Law at low pressure and high temperatures, but not under other conditions.

◢	N	O	P
1	Interpolation, Extrapolation and FORECAST		
2		R ohms	
3	Temp °C	y=mx+b	Forecast
4	22	116.08	116.08
5	0	105.63	105.63
6	30	119.88	119.88
7	100	153.11	153.11
8			

■ FIG. 8.4

(a) Open Chap8.xlsx on Sheet1 and enter the text and data shown in rows 1 through 3 of Fig. 8.4. Note the use of Merge and Center with O2:P2.

(b) In O4 we compute the expected R value at 22°C using the formula =$F$3*N4+$F$4. Format it to show 3 decimal places since that is the precision of the experimental data. We use absolute cell references for the slope and intercept so that we can readily copy the formula down to O7.

(c) In some cases, we might wish to perform these types of calculation without the bother of finding the slope and intercept. That is where the FORECAST functions come in. Its syntax is =FORECAST(x-value, Known_y's, Known_x's) so in P4 we enter =FORECAST(N4, $B$4:$B$11, $A$4:$A$11). Copy this formula down to P7.

(d) Save the workbook.

Behind the scenes, FORECAST does compute both slope and intercept so it is of no surprise that the two methods agree. Of course, we already knew the answer for 0°C—it is the value of the intercept. Doing such a calculation is not necessarily a waste of time as it helps us ensure we are doing things correctly. Why does the result for 30°C in O6 (and P6) not exactly agree with the value in B9? In O6 we have computed the expected value (the one that sits on the trendline) while B9's value is experimental. Since our experimental data ended at 40°C we cannot be too sure of the result in O7 unless we are very sure that the resistance of the material does vary linearly with temperature. In general, this is not the case over a wide temperature range so we should treat it as an approximation.

EXERCISE 3: THE LINEST FUNCTION

In this exercise we use LINEST rather than SLOPE, INTERCEPT, and RSQ to get the parameters for a linear fit. LINEST is more flexible and can give more data, as we shall see in this and subsequent exercises.

In Fig. 8.5 we have the results of a chemistry experiment to measure the enthalpy of solution (ΔH) of ℓ-ascorbic acid at various mole fractions (χ). We will see how well this data can be fitted to a linear equation.

(a) Temporarily ignore E5:F9 and enter all text and values as shown in Fig. 8.5 on Sheet2 of Chap8.xlsx. The *l* is *l* in the Gigi font.
(b) Construct the chart. Add the trendline. Format the chart to suit your taste.

	A	B	C	D	E	F	G
1	Enthalpy of Solution of ℓ-ascorbic acid						
2	Dallos et al., *J. Chem. Therm*. 263, **30** (1998)						
3							
4	X	ΔH		LINEST			
5	0.00102	25.44		slope	-12.8034	25.4563	intercept
6	0.00526	25.33		std error of slope	0.692287	0.027008	std error of intercept
7	0.00510	25.32		R^2	0.957988	0.058887	std err for y
8	0.01034	25.33		F	342.0406	15	df
9	0.01465	25.30		SSreg	1.186091	0.052015	SSresid
10	0.02127	25.16					
11	0.02406	25.25					
12	0.02891	25.16					
13	0.03428	25.06					
14	0.03910	25.05					
15	0.04199	24.86					
16	0.04654	24.79					
17	0.05119	24.78					
18	0.05377	24.75					
19	0.05779	24.66					
20	0.06309	24.70					
21	0.06455	24.61					

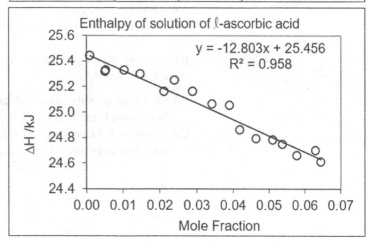

■ FIG. 8.5

(c) LINEST is an array function in that it returns more than one value so it needs to be committed correctly. Select the range E5:F9, type the formula =LINEST(B5:B21,A5:A21,TRUE,TRUE), and use [Ctrl] + [⇧ Shift] + [Enter ↵] to commit it. Note the braces around the formula when viewed in the formula bar.

(d) Save the workbook.

When used with four arguments, LINEST returns the slope, intercept, and R^2 value as well as a number of other statistics that we address in a later chapter. You will see that arguments one and two are the y- and x-values as used with SLOPE. When the third argument is TRUE or omitted, LINEST computes the intercept; otherwise, it sets the intercept to zero. The statistics in rows below the fit coefficients are not returned if the fourth argument is either set to FALSE or is omitted. A two-argument formula such as =LINEST (B5:B21, A5:A21) would just give the slope and intercept.

EXERCISE 4: FIXED INTERCEPT

Occasionally, one may wish to get a fit with a specified (fixed) intercept. You may, for example, want an intercept of zero or of some other value. If you look at Fig. 8.2, there is a setting *Set Intercept* where you can specify the required intercept value.

We will somewhat arbitrarily decide to fit some data with an intercept value of 0 and some other data with an intercept value of 5. The trendlines on the chart present no problem—you just enter the required values in the dialog. Getting a zero intercept value with LINEST is simple; you just enter FALSE for argument three. Specifying a value such as 5 needs a "workaround." Lines with equations $f(x) = 1.5x + 5$ and $g(x) = 1.5x$ are parallel. For a given x, say x_i, $f(x_i) = g(x_i) + 5$. So if we subtract 5 from each $f(x)$ value, we get the $g(x)$ line and its intercept is 0. Let's see how to implement that in Excel.

(a) On Sheet3, enter the values and text in columns A, B, and C, and the text in columns E and F as shown in Fig. 8.6.

(b) Create an XY chart (markers only) from A3:C11. Add a trendline to each data series specifying intercept 0 for the y data series and 5 for the z data series. Make sure to require that the trendline equations are displayed.

(c) Next, we will see how to use LINEST to get the slopes. In E5:F5 the LINEST formula for the y-line is =LINEST(B4:B11,A4:A11,FALSE); remember to select both cells, type the formula and use

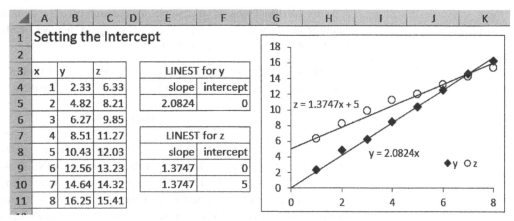

■ FIG. 8.6

[Ctrl] + [⇧ Shift] + [Enter↵] to commit it. The resulting slope agrees with the trendline for that data series.

(d) In E9:F9 the formula for the z-line is `=LINEST(C4:C11-5,A4:A11,FALSE)`. Note how this gives the correct slope but reports the intercept as 0.

(e) If we enter `=LINEST(C4:C11-5,A4:A11,FALSE)` in E10 we will compute the slope. Because we are returning only one value we may commit it with a simple enter. This gives the slope value since it is the first value in the returned LINEST array. In F10 type the value 5, since that is the intercept we demanded!

(f) Save the workbook.

EXERCISE 5: MULTILINEAR REGRESSION

In the term *linear regression,* the word *linear* is not used to denote a straight line. It means that the function to be fitted is in the form $F(x) = a_1 * f_1(x) + a_2 * f_2(x) \dots$ where $f_i(x)$ is a simple or complex function of x. Other authors[3] state that F is linear when the partial derivatives of each coefficient are not functions of other coefficients.

So far we have worked with the straight line equation $F(x) = mx + c$ where there is just one predictor variable function. When there is more than one $f(x)$, we may speak of *multilinear* (or *multiple*) regression. The Excel Help

---

[3]For example, E. Joseph Billo, *Excel for Chemists*, Wiley, New York, 2001.

topic for LINEST[4] uses, as an example, the assessed value (y) of an office building as predicted by four predictors: floor space, number of offices, number of entrances, and age of the building.

We shall look at the use of multilinear regression to fit data for the vapor pressure (P) of a liquid at various temperatures (T) to the equation:

$$\log_{10}(P) = b_1 + b_2 \bullet \frac{1}{T} + b_3 \bullet \log_{10}(T) + b_4 \bullet T^2$$

Our predictors are all related to T but not in a linear or a simple logarithmic way. However, using a simple workaround we can have LINEST do the regression. As can be been in Fig. 8.7, the workaround is to use formulas to generate a column for the Log(P) values and separate columns for the four predictor values. Make a worksheet similar to Fig. 8.7 on Sheet4, following the notes given as follows.

(a) Enter the heading and values in A3:B13.
(b) Select column A and label it T using the name box. Similarly, label column B as P. Labeling the whole column allows the label to be used in both the curve fitting and calculation.
(c) The formulas in this worksheet are as follows:

    D4:   =LOG10(P)
    E4:   =1/T
    F4   =LOG10(T)
    G4:   =T^2

These formulas in row 4 are filled down to row 13.

(d) In D17:G17 a LINEST formula is used: =LINEST(D4:D13,E4:G13). In the interests of simplicity, we have generated just the first row of LINEST results and have omitted the statistics. Do not forget to use [Ctrl] + [⇧ Shift] + [Enter⏎] to commit the cells, as it is a matrix formula.
(e) The cells in D7:G7 are named by the text in row 16; but we must take note that, since b4, b3, b2, and b1 are all valid cell references, the actual names are b4_, b3_, b2_, and b1_.
(f) The range A20:B29 merely repeats the values in A4:B13.
(g) In C20 the formula =10^(b1_+b2_*(1/T)+b3_*LOG10(T)+b4_*T^2) is used to generate the predicted P values.
(h) The chart shows that the P values in B20:B29 fit very nicely on the prediction curve P(cal) made from C20:C29. P is plotted on a log

---

[4]See http://office.microsoft.com/en-ca/excel-help/linest-function-HP010069838.aspx.

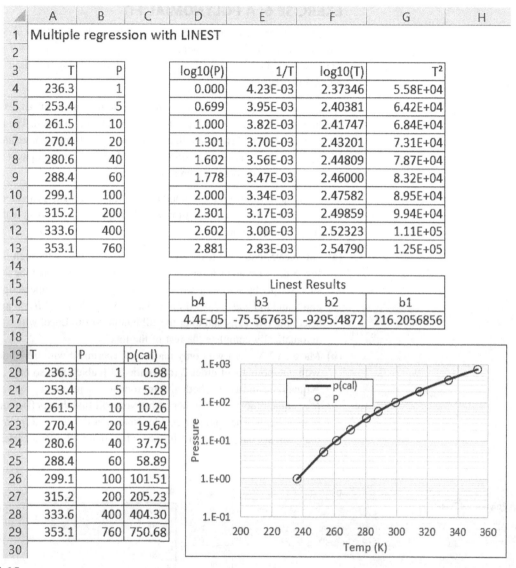

The spreadsheet shown contains the following:

	A	B		C	D	E	F	G	H
1	Multiple regression with LINEST								
2									
3	T	P			log10(P)	1/T	log10(T)	T²	
4	236.3	1			0.000	4.23E-03	2.37346	5.58E+04	
5	253.4	5			0.699	3.95E-03	2.40381	6.42E+04	
6	261.5	10			1.000	3.82E-03	2.41747	6.84E+04	
7	270.4	20			1.301	3.70E-03	2.43201	7.31E+04	
8	280.6	40			1.602	3.56E-03	2.44809	7.87E+04	
9	288.4	60			1.778	3.47E-03	2.46000	8.32E+04	
10	299.1	100			2.000	3.34E-03	2.47582	8.95E+04	
11	315.2	200			2.301	3.17E-03	2.49859	9.94E+04	
12	333.6	400			2.602	3.00E-03	2.52323	1.11E+05	
13	353.1	760			2.881	2.83E-03	2.54790	1.25E+05	
14									
15					Linest Results				
16					b4	b3	b2	b1	
17					4.4E-05	-75.567635	-9295.4872	216.2056856	
18									
19	T	P		p(cal)					
20	236.3	1		0.98					
21	253.4	5		5.28					
22	261.5	10		10.26					
23	270.4	20		19.64					
24	280.6	40		37.75					
25	288.4	60		58.89					
26	299.1	100		101.51					
27	315.2	200		205.23					
28	333.6	400		404.30					
29	353.1	760		750.68					
30									

■ FIG. 8.7

scale to show that the fit matches across the entire range with the primary minor horizontal gridlines visible to emphasize that it is in log scale.

Another way to fit this data is by using matrix algebra. This is shown in the workbook RegregionWithMatrix.xlsx on the companion website. We shall see yet another method in Chapter 12 when we experiment with Solver.

EXERCISE 6: A POLYNOMIAL FIT

If we compare the multilinear regression formula $F(x) = a_1 \times f_1(x) + a_2 \times f_2(x)$ to that of a polynomial such as $y = ax^3 + bx^2 + cx + d$ clearly $f_1(x)$ is x^3, $f_2(x)$ is x^2, and so on. So LINEST may be used to fit polynomial data.

Scenario: An engineer has measured the temperature of an extruder machine's die at various settings of the screw revolution speed. The results are shown in Fig. 8.8. He would like a cubic equation to summarize the data. We will find the regression coefficients with LINEST and compare them to those given by a third-order trendline.

We might have followed the method used in Exercise 5 and generated rows for x, x^2, and x^3. However, this time we will explore a rather neat feature of Excel: the use of an array.

(a) On Sheet5, enter the text and values shown in rows 1 through 6 of Fig. 8.8. In B6 and C6 type x3 and x2, respectively; make the 3 and 2 superscripts by selecting the number and locating the superscript command in *Home / Font* drop-down menu (*Format / Cells* on a Mac); select the two cells and drag the fill handle to E6. Excel will automatically complete the rest of the text.

(b) Make an XY chart with only markers. Experiment with trendlines with polynomials of order 2 (quadratic), 3 (cubic), and so on. A quadratic and a cubic both give reasonable fits judging by the value of R^2. A quadratic gives a higher R^2 but you will find the coefficient of x^4 is very small. The engineer decides to stop at the cubic. Remember that with six data points, a fifth-order function will give a perfect fit.

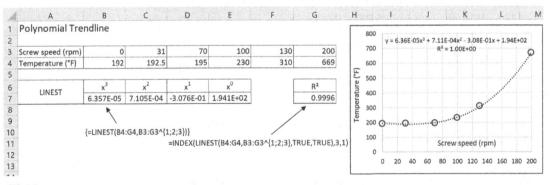

■ FIG. 8.8

There is, however, no justification in using this, as the leading coefficients are very small. In the chart, format the curve fit parameters to be in scientific notation.

(c) Select B7:E7, enter the formula =LINEST(B4:G4,B3:G3^{1;2;3}) and use [Ctrl] + [⇧ Shift] + [Enter↵] to commit it. Here we have used the array {1; 2; 3} to raise the *x*-values to powers, thus making it unnecessary to generate these values on the worksheet.

(d) Save the workbook.

> NOTE: In the formula used in step (c) we have an array of constants: {1;2;3}, the elements of which are separated by semicolons. This is because our *x*-values are in columns. Had we made a vertical table, the array would have used commas to separate elements: {1,2,3}.

EXERCISE 7: A LOGARITHMIC FIT (LOGEST)

A simple model for the growth of bacteria predicts that if the initial population is N_0, the population N_t at time t will be given by the following equation, in which B is the reproduction rate.

$$N_t = N_0 \exp(Bt) \tag{8.3}$$

We can linearize (which means to give it the form $y = mx + b$) by taking natural logs on both sides:

$$\ln(N_t) = Bt + \ln(N_0) \tag{8.4}$$

Before computers, the normal practice was to convert equations to linear form since fitting to a straight line is relatively simple.

For this exercise, we will find the fitting parameters of some exponential data both with and without linearization. Fig. 8.9 rows 3 and 4 show the experimental data for the population size (N_i) of a bacteria colony at

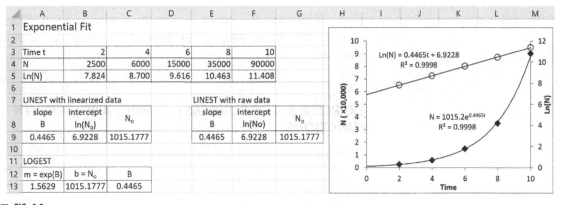

■ FIG. 8.9

various times (t_i). In row 5 we have the values for $\ln(N_i)$, so B4 has the formula =LN(B4). Referring to the previous equations, we wish to estimate N_0 and B.

For a change of pace, the reader is asked to develop this worksheet on Sheet6 without detailed instructions. Make a chart with two vertical axes, a linear trendline for the N data and an exponential one for the $\ln(N)$ data. Note that "display units" is set to 10,000 for the N values, but "Show display units label on chart" is not checked. Extend the trendlines back to the origin.

Comparing Eq. (8.4) with the equation for a straight line ($y = mx + b$) we see that B will be the slope and $\ln(N_0)$ the intercept, meaning that N_0 equals EXP(intercept). We may compute N_o and B with:

A9:B9   =LINEST(B5:F5,B3:F3)
C9:   =EXP(B9)

We could avoid having to make a row of Ln(N) data by taking the logarithms within the LINEST formula. We get the same results as before with:

E9:F9   =LINEST(LN(B4:F4),B3:F3)
G9:   =EXP(F9)

An alternative method is to use the LOGEST function. It fits x and y values to the equation $y = bm^x$ or in our case $N = bm^t$. Compare the logarithm of this with Eq. (8.4) to show that $B = \ln(m)$ and $N_0 = b$. The LOGEST function returns m in the first cell and b in the second. Use the formulas:

A13:B13   =LOGEST(B4:F4,B3:F3)
C13:   =LN(A13)

As expected, the results of the three methods are in agreement with each other and with the trendline values. Remember to save the workbook when done.

EXERCISE 8: THE FORECAST, TREND, AND GROWTH FUNCTIONS

The functions FORECAST and TREND are used to predict values from data with a linear fit. The FORECAST(x, known_y's, known_x's) function returns the predicted value of the dependent variable (represented in the data by known_y's) for the specific value, x, of the independent variable (represented in the data by known_x's) by using a best fit (least squares) linear regression to predict y values from x values. We saw an example of its use in Exercise 2. While FORECAST is generally used for a single predictor, TREND(known_y's, known_new_x's, x, const) is more suited to data points in a series such as a time series and is capable of computing multiple y values for multiple new x values; it is then used as an array function. In practice, either function can be used interchangeably for a first-order curve fit. Since TREND is an array function, it can also be used to calculate values from higher order curve fits as well. The optional *const* argument in TREND is used to return additional statistics (similar to the last argument in LINEST) but we will not explore its use. The GROWTH(known_y's, known_x's, x's, constant) function is used to perform a regression analysis where an exponential curve is fitted. So LINEST and TREND are complementary, as are LOGEST and GROWTH.

Fig. 8.10 is to be made on Sheet7 of Chap8.xlsx. The data in rows 4 and 5 was copied from Sheet4 while rows 11 and 12 were copied from Sheet5. The formulas to be used are shown in the figure. Remember to select ranges and use [Ctrl]+[⇧ Shift]+[Enter ↵] for the array formulas in rows 7, 8, and 13. Why did we need absolute references in B6 but not in B7 or B13?

In each case, we have used the same x-value for the predictors as we have for the fitting process. So we have computed the values that would be displayed in trendlines. The reader should experiment to show that both TREND and GROWTH may be used with a single predictor: predict the temperature when the screw speed is 150 rpm (interpolation), and the value of N at $t = 12$ (extrapolation.). Remember to save the workbook.

	A	B	C	D	E	F	G	H	I	J
1	FORECAST, TREND and GROWTH Functions									
2										
3	FORECAST and TREND									
4	Screw speed (rpm)	0	31	70	100	130	200			
5	Temperature (°F)	192	192.5	195	230	310	669			
6	FORECAST	97.38	167.69	256.13	324.16	392.20	550.94		B6	=FORECAST(B4,$B$5:$G$5,$B$4:$G$4)
7	TREND (1st order)	97.38	167.69	256.13	324.16	392.20	550.94		B7:G7	{=TREND(B5:G5,B4:G4)}
8	TREND (3rd order)	194.12	187.16	197.87	234.02	305.79	669.55		B8:G8	{=TREND(B5:G5,B4:G4^{1;2;3})}
9										
10	GROWTH									
11	Time t	2	4	6	8	10				
12	N	2500	6000	15000	35000	90000				
13	GROWTH	2479.67	6056.84	14794.44	36136.89	88267.94			B13:G13	{=GROWTH(B12:F12,B11:F11)}

■ FIG. 8.10

RESIDUALS

Recall that we have defined *residuals* as the difference between the actual and the predicted values in a curve-fitting problem. If the residuals are the result of normal experimental errors, we would expect them to be distributed randomly above and below the $x = 0$ line. If, on the other hand, the residuals display an observable trend, then one should question the fit.

In the example shown in Fig. 8.11 (albeit a very contrived set of data), a linear fit seems very appropriate, but the residuals appear to follow a parabolic rule. This is a good indication that we should look further at the fitting function. The reader is encouraged to construct a similar worksheet on Sheet9.

Two trendlines were added to the original chart; both give R^2 as 1. In B14 we have =RSQ(x,y)—the ranges A3:A12 and B3:B12 have been named x, and y, respectively; in B15 we have =INDEX(LINEST(y,x^{1,2},TRUE, TRUE),3,1). However, but the results from REQ (linear fit) and LINEST (for a quadratic fit) show very slightly different values. This indicates that the quadratic is a better fit. It is a little disturbing that the linear trendline's R^2 value does not exactly match the REQ result. We could have used =INDEX (LINEST(y,x,TRUE,TRUE),3,1) in places of RSQ; show that it gives the same result as RSQ.

The meaningfulness of the small x^2 coefficient (often called by others a lurking variable) will depend very much on the circumstances of the experiment.

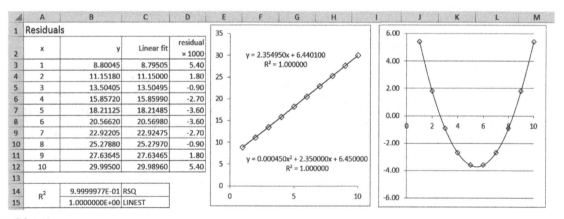

■ FIG. 8.11

EXERCISE 9: SLOPE AND TANGENT

In this exercise, we see how to compute the slope of a polynomial for a specified x-value and how to display a tangent line on a chart. Suppose we find the slope m at a point x_0, y_0, then the tangent is the line that obeys $y_0 = mx_0 + b$. Hence, $b = y_0 - mx_0$ and we can find a second point (x, y) on the tangent using $y = m(x) + b$.

The next table gives the formulas needed to compute approximations to the first and second derivatives of tabulated data. The central formula is generally more accurate but requires that we have a point each side of the point of interest.

Order	Forward	Backward	Central
First	$\frac{dy}{dx} = \frac{y_1 - y_0}{h}$	$\frac{dy}{dx} = \frac{y_0 - y_{-1}}{h}$	$\frac{dy}{dx} = \frac{y_1 - y_{-1}}{2h}$
Second	$\frac{d^2y}{dx^2} = \frac{y_2 - 2y_1 + y_0}{h^2}$	$\frac{d^2y}{dx^2} = \frac{y_0 - 2y_{-1} + y_{-2}}{h^2}$	$\frac{d^2y}{dx^2} = \frac{y_1 - 2y_0 + y_{-1}}{h^2}$

(a) On Sheet9 of Chap8.xlsx, enter all the text shown in Fig. 8.12. Enter the values shown in A3:B13.

(b) The formula to compute the slope using the Central Difference method in C5 is =(B6-B4)/(2*($A$5)), and this must be copied down to row 12.

(c) Enter a number between 2 and 9 inclusive in G3. We will use this as an index to the x-values. With the central difference method we cannot use

> Data Validation is the best way to prevent users from entering inappropriate data in cells. We have barely touched on all it can do. The reader is encouraged to experiment.

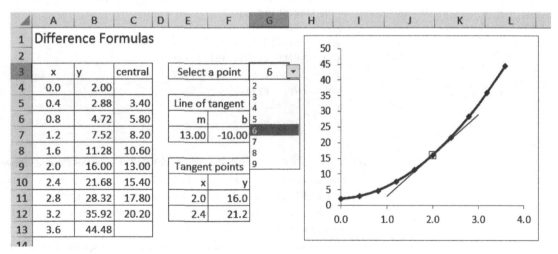

■ FIG. 8.12

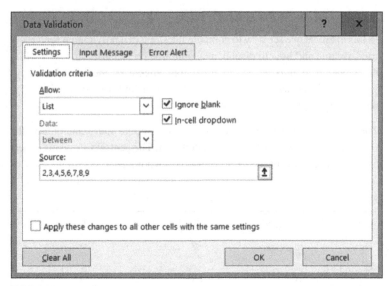

■ **FIG. 8.13**

points 1 or 10, so we need to prevent the user from entering invalid data here. Select G3 and use *Data / Data Tools / Data Validation (Data / Data Validation* on a Mac) to open and complete the dialog box shown in Fig. 8.13.

(d) Cells E11 and F11 hold our x_0, y_0 data pair; this is the point on the curve where we want the tangent. Cells E12 and F12 hold the second point on the tangent. In E7:F7 we compute the slope and intercept values of the tangent line.

```
E11:    =INDEX($A$4:$A$13,$G$3)
F11:    =INDEX($B$4:$B$13,$G$3)
E7:     =INDEX($C$4:$C$13,$G$3)
F7:     =F11-E11*E7
E12:    =E11+0.4
F12:    =F11-E11*E7
```

(e) Make an XY chart with the data in A3:B13. Using the *Copy-&-Paste Special* method from Exercise 11 in Chapter 7, add the points defined by E10:F11 as a second series. We want only the first point to be visible, so click on the second point twice and format it to have no marker line or fill.

(f) Add a linear trendline to the new series with appropriate forward and backward projections to make the tangent line visible.

(g) Save the workbook.

EXERCISE 10: THE ANALYSIS TOOLPAK

Excel has an add-in feature called the Analysis ToolPak,[5] which has a variety of tools that enable the user to generate results without using formulas and formatting. In this exercise, we will see the use of the Regression Tool by repeating the problem set out in Exercise 3 for comparison purposes.

(a) Copy A1:B20 from Sheet3 to A1 in Sheet10.
(b) Use the command *Data / Analysis / Data Analysis (Data / Data Analysis* on a Mac) and from the resulting dialog select Regression, which opens the dialog shown in Fig. 8.14.
(c) The *x* range is B3:B20, and the *y* range is A3:A20. Ensure you have checked the Labels box. A suitable output range for our purposes is E5, but you will note that you could output to a new worksheet or workbook. Check the box *Line Fit Plots* to generate a chart. Click the OK button.

If you compare the results in F21 and F22, shown in Fig. 8.15, you will see that the slope and intercept are the same as were generated with LINEST in Exercise 3. You will also see that the statistics are in agreement. None of this is surprising as the Tool uses the LINEST function.

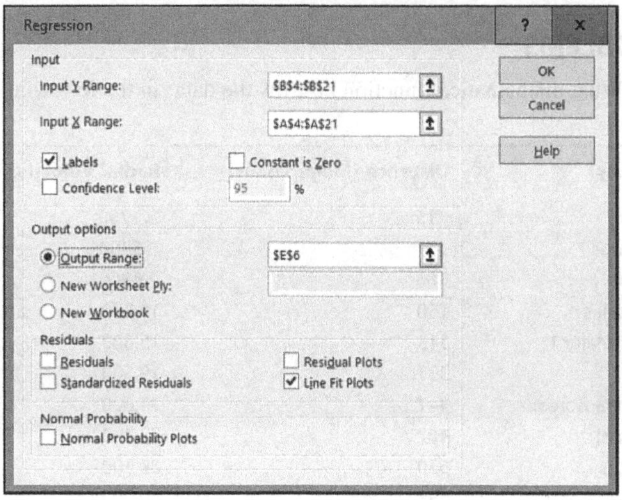

■ FIG. 8.14

---

[5]If not visible, it can be added using *File | Options | Add-ins* and then clicking the *Go...* button next to *Manage Excel Add-ins* option (*Tools | Excel Add-ins...* on a Mac).

	E	F	G	H	I	J	K	L	M
5	SUMMARY OUTPUT								
6									
7	*Regression Statistics*								
8	Multiple R	0.978768601							
9	R Square	0.957987974							
10	Adjusted R Square	0.955187173							
11	Standard Error	0.058887087							
12	Observations	17							
13									
14	ANOVA								
15		*df*	*SS*	*MS*	*F*	*Significance F*			
16	Regression	1	1.186090546	1.186091	342.0406	9.77619E-12			
17	Residual	15	0.052015336	0.003468					
18	Total	16	1.238105882						
19									
20		*Coefficients*	*Standard Error*	*t Stat*	*P-value*	*Lower 95%*	*Upper 95%*	*Lower 95.0%*	*Upper 95.0%*
21	Intercept	25.45630329	0.027008461	942.5307	3.26E-37	25.39873611	25.51387046	25.39873611	25.51387046
22	X	-12.80338931	0.692286884	-18.4943	9.78E-12	-14.27896387	-11.32781475	-14.27896387	-11.32781475

■ FIG. 8.15

There are two major drawbacks to using this Tool. The user has no control over the positioning of the various resulting values and, like all Data Analysis Tools, the results are static. This means that if you make a change in the input data you must remember to rerun the Tool.

PROBLEMS

1. *What mathematical function best fits the data[6] in the following table?

Galaxy	Distance (Megaparsec)	Radial Velocity (km/s)
Virgo	15	1200
Perseus	71	5400
Coma	83	6600
Hercules	150	10,500
Ursa Major I	313	15,600
Leo	337	19,500
Corona Borealis	347	21,600
Gemini	402	23,400
Bootes	650	39,300
Ursa Major II	653	40,200
Hydra	831	60,600

[6]www.astro.indiana.edu/catyp/activities/hubble.doc.

2. *A chemical engineer is studying the rate at which compound X reacts under certain conditions. The following table gives the percentages of X remaining after measured times. Fit the data to $(1 - X) = \exp(-kt)$ to determine k using (a) a graphical method, and (b) a single cell with a LINEST or LOGEST formula.

T	200	400	600	800
X	18%	29%	42%	51%

3. *In Problem 3 of Chapter 4, we used numerical differentiation formulas to find di/dt for same tabulated data. Another approach is to use LINEST to get the polynomial coefficients; then from $f(x)$, we can find the coefficients of $f'(x)$. Compare the results from each method.

4. The solution to Problem 9 of Chapter 2 consisted of a table giving the amount of solute m_0 remaining in the water after extraction n.

   (i)   Plot this data and add an exponential trendline in the form
         $m_0 = 5\exp(-An)$.

   (ii) Fit the data using the LOGEST function to get parameter B and 5.

   (iii) Clearly, the 5 results from the fact we started with 5g. How are A and B related to each other and to the data in the experiment?

   (iv) Do a mathematical analysis of the experiment to explain the exponential fit.

5. Fit the following data[7] to the equation $N = aPk$ by (i) making a plot and adding a power trendline; (ii) plotting $\ln(N)$ against $\ln(P)$ and adding a linear trendline, and using LINEST. Ensure you understand the

P	N	P	N	P	N
0.46	24.80	10.00	84.50	55.00	195.00
0.53	26.50	17.70	115.00	58.50	193.00
0.63	28.50	18.60	115.00	70.30	189.00
0.74	30.00	25.30	150.00	93.00	245.00
3.00	58.40	31.60	127.00	95.00	245.00
4.20	60.30	32.00	140.00	185.00	315.00
5.00	70.70	37.00	165.00	340.00	380.00
5.60	69.00	41.00	170.00	590.00	480.00

---

[7]W. L. Friend and A. B. Metzner, American Institute of Chemical Engineering Journal 4, 393, 1958.

relationship between the various fitting parameters. Note that you can plot P vs N and give both axes a logarithmic scale to get a straight line, but this does not help with regression analysis.

6. A chemist measured the partial pressure of a decomposing gas at various times; see the following table. Make an appropriate chart to show that this data follows the equation $\ln(p_0/p) = kt$ where p_0 is the pressure at time $t = 0$. What value of k is reported by the trendline? Can you get the same result with a LINEST formula

t	0	600	1200	1800	2400	3000	3600
P	350	247	185	140	105	78	58

7. * The heat of vaporization of a liquid (ΔH_v) may be found by measuring the liquid's vapor pressure at various temperatures and applying the Clausius-Clapeyron equation, which chemists generally write as follows:

$$\ln\left(\frac{P_1}{P_2}\right) = \frac{\Delta H_v}{R}\left(\frac{1}{T_2} - \frac{1}{T_1}\right) \quad \text{or} \quad P = A\exp\left(\frac{\Delta H_v}{RT}\right)$$

A plot of $1/T$ against $\ln(P)$ where T is measured in Kelvin and P in torr will give a straight line with a slope $-\Delta H_v/R$ where R has the value $8.3145\,\mathrm{J\,mol^{-1}\,K^{-1}}$. From the following data[8] find ΔH_v for water.

T (K)	313	323	333	343	353
P (torr)	55.364	92.592	149.51	233.847	355.343

8. Make a plot of the following data and add two trendlines, one quadratic and the other cubic. Format the cubic trendline as a dotted line. Hint: Remember the *Selection* group in *Chart Tools / Format* (*Chart Design / Add Chart Element* on a Mac).

x	573	534	495	451	395	337	253
y	1000	800	600	450	300	200	100

---

[8]H. F. Stimson, Journal of Research of the National Bureau of Standards, 73A, 493, 1969.

9. *A sociological study in 1976 tested the hypothesis that the larger the city the more rushed were the inhabitants. Google with `Pace of Life` for more details. The table that follows lists some results. Which model best fits the data: (i) a power model $V = kP^a$ or (ii) a logarithmic model $V = m\ln(P) + c$?

Location	Population	V (ft/s)
Brno, Czechoslovakia	341,948	4.81
Prague, Czechoslovakia	1,092,759	5.88
Corte, Corsica	5,491	3.31
Bastia, France	49,375	4.90
Munich, Germany	1,340,000	5.62
Psychro, Crete	365	2.76
Itea, Greece	2,500	2.27
Iraklion, Greece	78,200	3.85
Athens, Greece	867,023	5.21
Safed, Israel	14,000	3.70
Dimona, Israel	23,700	3.27
Netanya, Israel	70,700	4.31
Jerusalem, Israel	304,500	4.42
New Haven, United States	138,000	4.39
Brooklyn, United States	2,602,000	5.05

10. *The following data[9] records the observations of the number of cricket chirps per 20 s, as a function of temperature. What relationship do you find? A search of the Internet found two comments that if you count the chirps in 15 s and add 40 (one said 37) you get a good estimate of the temperature in °F. Does this data agree with these comments?

T (°F)	46	49	51	52	54	56	57	58	59	60
chirps	40	50	55	63	72	70	77	73	90	93
T (°F)	61	62	63	64	66	67	68	71	71	72
chirps	96	88	99	110	113	120	127	137	137	132

11. The table that follows shows the results of an enzyme kinetics experiment. The quantity V is the velocity of the reaction, while $[S]$ is the concentration of the substrate S. Ideally, this data should be fitted to the Michaelis-Menten equation to find K. Traditionally, biochemists

---

[9] F. E. Croxton et al., *Applied General Statistics*, Prentice-Hall, Englewood Cliffs, N J, 1967; page 390.

linearize the M-M equation to give the Lineweaver-Burke equation and then plot $1/V$ against $1/[S]$. What value of K is obtained using a trendline and using LINEST? We revisit this problem in Chapter 12 and use Solver to make a direct fit.

$$\text{Michaelis} - \text{Menten Eqn}: \quad V = \frac{V_{max}[S]}{[S] + K}$$

$$\text{Lineweaver} - \text{Burke Eqn}: \quad \frac{1}{V} = \frac{K}{V_{max}}\frac{1}{[S]} + \frac{1}{V_{max}}$$

[S] (mM)	8.33	5.55	2.77	1.38	0.83
V (mM/s)	3.62E – 06	3.39E – 06	2.75E – 06	1.99E – 06	1.49E – 06

12. Use LINEST to fit the following data to the van Deemter equation[10] $y = Ax + B/x + C$. Attempt this using Exercise 5 as a guide and then repeat using Exercise 6 as a guide. The data can be entered row-wise or column-wise. Make a plot showing the experimental data and the line of best fit. Attempt to make the latter using the TREND function.

x	3.4	7.1	16.1	20.0	23.1	34.4	40.0	44.7	65.9	78.9	96.8	115.4	120.0
y	9.59	5.29	3.63	3.42	3.46	3.06	3.25	3.31	3.50	3.86	4.24	4.62	4.67

t (min)	T(t) (°F)
0	40.0
0.5	38.3
1	37.0
1.5	36.0
2	35.3
2.5	34.8
3	34.3
3.5	34.0
4	33.8
4.5	33.6
5	33.4

13. The temperature in a refrigerator can be modeled as a first-order system as a function of time t. The basic time response is as follows

$$T(t) = T_{ss} + (T_0 - T_{ss})*\exp(-t/\tau) = T_{ss} + (T_0 - T_{ss})e^{\frac{-t}{\tau}}$$

where T_0 is the initial value, T_{ss} is the steady-state value, and τ is the time constant of the system.

The data to the right was measured experimentally from the system:

(a) Enter the data into excel. Plot the measured $T(t)$ vs t. Looking at the plot, determine the steady-state value, T_{ss}, which is the value of temperature if you kept on measuring forever. (HINT: it is an integer). For each value of T, plot calculate $|T - T_{ss}|$.

(b) On a separate plot, plot $|T - T_{ss}|$ vs t. This is just for curve fitting, so you do not need to make it look nice. Add an exponential curve fit to the curve. Show the equation on the chart. From the exponential curve fit parameter, determine the time constant τ of the system.

---

[10]The van Deemter equation is used in chromatography. The data in the table comes from H. W. Moody, *Journal of Chemical Education*, 1982, 59, page 290.

(c) From the times in the data, calculate $T(t)$ from the equation. Add the calculated $T(t)$ to the plot of the measured $T(t)$ you made in part b. Adjust the axis limits and legend position to clearly show the data. Title your axes.

14. Diodes express a voltage-current relationship of

$$I = I_0(\exp(\alpha V) - 1) = I_0(e^{\alpha V} - 1) \qquad (8.5)$$

For an experiment, the voltages and currents in the table were measured.
(a) Enter this data into Excel. Plot Current vs Voltage for the last four data points. Have the y-axis in log scale. You do not need to label your axes. Add an exponential curve fit to the data. Display the equation and R^2 on the chart. Use scientific notation in the equation. The α value in Eq. (8.5) can be determined from the curve fit. For the last four data points calculate

$$I_{0_{est}} = \frac{I}{e^{\alpha V} - 1} = I/(\exp(\alpha V) - 1)$$

where α is what was calculated from the curve fit. Use I_0 as the average of the calculated numbers.
(b) Calculate the theoretical value for I using Eq. (8.5) based on the I_0 and α you determined. The load current can be calculated using

$$I_{load} = \frac{V_0 - V}{R}$$

where $V_0 = 1$ V and $R = 200$ Ω. Calculate it for each voltage. Plot measured current, theoretical current, and load current vs voltage. Your plot should look similar to Fig. 8.16 but have axis numbers and labels. Also, the intersection point could be at a different value, depending on V_0 and R. Have the title of the plot explaining the operating current flowing through the diode.

15. Using the PASCO "picket fence," the acceleration of gravity can be calculated. The time that each dark line breaks the photogate is recorded, as well as the approximate velocity. (The time of the velocity is halfway between each position measurement.) The data is shown to the right.
(a) Plot position vs time using markers. Do a second-order curve fit for the position curve. Display the equation and R^2 value on the chart. It should be in the form of

$$y = At^2 + Bt + C$$

Voltage(V)	Current (ma)
0.00	0
0.26	0.001
0.40	0.01
0.47	0.05
0.63	1
0.72	5
0.76	14

Time (s)	Position (m)
0	0
0.0414	0.05
0.0734	0.1
0.1005	0.15
0.1244	0.2
0.146	0.25
0.1659	0.3

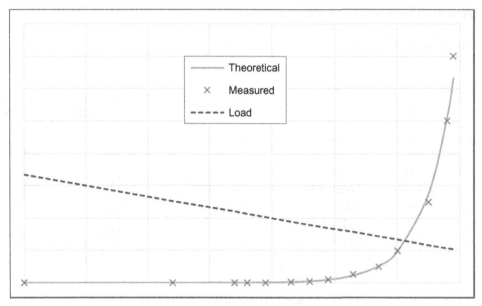

■ FIG. 8.16

(b) From the curve fit values, calculate the velocity at each of the times for your position measurements. The velocity can be calculated using

$$V = 2At + B$$

Have the values of A and B in separate cells to make them easy to change.

(c) Add both your calculated velocity from (b) and the approximate velocity from the data file to your chart as shown on the plot. Plot both of these on a 2nd y-axis. Add appropriate axis labels and title to your chart.

16. In the radial direction in a cylinder, that is, cylindrical coordinates, the temperature (T) is a function of the logarithm of the radius of the cylinder.

(a) Enter this data into an Excel sheet. Plot the temperature as the y-axis and the radius as the x-axis, using a scatter plot do not connect the symbols representing the data points. Title the y-axis **Temperature** and the x-axis **Radius**, and include units. Adjust the crossing of the axis to minimize the areas that do not include data

(b) Add a logarithmic trend line to the data. Show the equation and R^2 value on the chart. Based on the logarithmic equation, using formulas determine the temperature at a radius of 1 mm. Include that calculated point in your plot using a separate marker.

17. Steam is used in power generation. The relationship between pressure (P), specific volume (V), and temperature (T) is used to calculate the energy available. The theoretical relationship between these values is as follows

$$T = \frac{M \bullet P \bullet V}{R}$$

where $M = 0.018\,\text{kg m}^3$, and $R = 8.312\,\text{mol/(kg K)}$, P is in Pa, V is in m^3/ kg, and T is in K. In order to verify this relationship, the following experimental data was recorded.

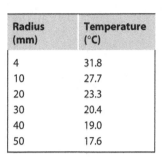

Radius (mm)	Temperature (°C)
4	31.8
10	27.7
20	23.3
30	20.4
40	19.0
50	17.6

Pressure (MPa)	Specific Volume (m³/kg)	Temperature (K)	TEMP (tested) (K)
0.01	14.8		317.2
0.04	4.01		348.9
0.08	2.11		366.5
0.14	1.25		382.3
0.22	0.81		396.3

(a) Enter the data into the spreadsheet. Calculate T at each pressure. Plot Specific Volume and Temperatures vs Pressure. Put the Specific Volume on a second y-axis. Plot the measured data using distinct markers and the calculated temperature using a smooth line. Include a legend for the chart. Add titles to the x- and y-axes with units. Minimize the blank area in the chart.

(b) Calculate the difference between the tested and calculated temperatures. Find the Standard Deviation of the differences. Add error bars to the tested temperature data with the standard deviation as the maximum value.

18. Shear stress across a member can vary by height. It is desired to compare the theoretical shear stress with the shear stress calculated by a finite element (FEA) program. The theoretical shear stress (τ_{xy}) can be calculated using

$$\tau_{xy} = \frac{VQ}{wI} = \frac{Vy(y-h)}{2I}$$

where V is the vertical force, and Q is calculated by

$$Q = wy\frac{(y-h)}{2}$$

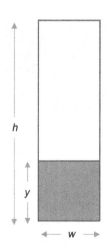

and I is the second area moment of inertia:

$$I = \frac{wh^3}{12}$$

The finite element analysis was done with a part ($h = 1''$, $w = 0.2''$) with $V = 1000$ lb. The result (FEA Tau) from the program is given as follows

Y (in.)	FEA Tau (psi)	τ_{xy} (psi)
1.0	−6.4E+01	
0.9	−2.8E+03	
0.8	−4.9E+03	
0.7	−6.4E+03	
0.6	−7.3E+03	
0.5	−7.6E+03	
0.4	−7.3E+03	
0.3	−6.4E+03	
0.2	−4.9E+03	
0.1	−2.8E+03	
0.0	−6.0E+01	

(a) Enter the data into excel. Calculate τ_{xy} for each of the values of y. Have h, w, and V be cells to make them easy to change. You can calculate a separate column with Q and use the 1st formula or just use the 2nd formula for τ. Plot both the FEA and the theoretical values of τ_{xy} vs y. Use markers for the FEA values and a smooth line for the theoretical. Label your axis including units. Choose axis limits and legend location to limit blank areas on the chart.

(b) Add a polynomial trendline to the data with a zero intercept. Choose the order which reasonably fits the data. Include the equation on the graph as well as the R^2 value.

19. Plot x vs $\sqrt{x}$ for $x = 1, 2, ..., 20$. Choose axes (log or linear) to have the plot look like a line and choose the appropriate curve fitting function. Show the equation on the chart and that it is a perfect fit.

20. For a tensile test, the load is recorded verses stretched length as shown in the table to the right. The initial length is 2 in. and the diameter is 0.5 in. Create a user-friendly spreadsheet calculating stress and strain,

allowing the user to input initial length and diameter. The formulas are as follows:

$$Strain = \frac{change\,in\,length}{inital\,length}$$

$$Stress = \frac{load}{area}$$

$$Area = \frac{\pi(diameter)^2}{4}$$

Load (lb)	Change in Length (in)
0	0
2200	0.0008
4300	0.0016
6400	0.0024
8200	0.0032
8800	0.0048
9600	0.0096
10,600	0.02
12,600	0.06
13,900	0.12
14,500	0.2
14,500	0.28
14,300	0.36
13,000	0.44

(a) Create a plot with the x-axis for the plot is strain, and the y-axis for the plot is stress as shown as follows.

(b) On a second x- and y-axis on the plot in part (b), plot another copy of the stress vs strain, and also plot only the first 5 data points. Have the scale of both y-axes be the same. Adjust the scale of the 2nd x-axis to match the figure. And move the crossing to the bottom. Now you can hide the secondary y-axis. Add the appropriate axis labels.

(c) Add a first-order curve fit to the first 5 points with an intercept at 0. Extend the curve fit line 0.001 periods. Display the equation in scientific notation and the R^2 value on the chart. The slope of this line is the Modulus of Elasticity of the material.

(d) Calculate the slope using LINEST as well.

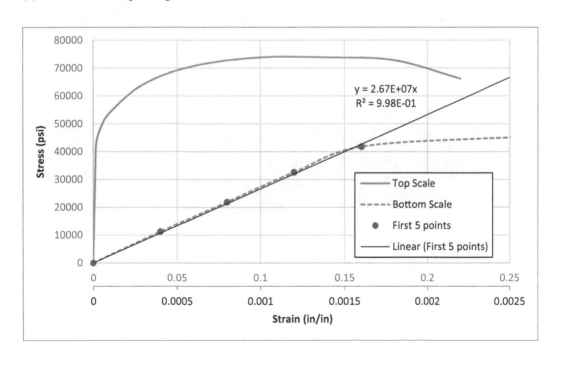

APPENDIX: SUPPLEMENTARY MATERIAL

Supplementary material related to this chapter can be found on the accompanying CD or online at https://doi.org/10.1016/B978-0-12-818249-9.00008-X.

VBA User Defined Functions

CHAPTER OUTLINE

In this chapter and Chapter 10 we will briefly look at Visual Basic for Applications (VBA) which is an important part of Microsoft Office. When used within Excel it enables us to write functions and subroutines. A function returns a value to a cell (or a range) in the same way as a worksheet function. A subroutine performs a process; we look at these in the next chapter. Collectively, subroutines and functions are called *modules* or *macros*. To add some confusion, the word *module* is also used for the place where one or more macros are coded.

Liengme's Guide to Excel 2016 for Scientists and Engineers. https://doi.org/10.1016/B978-0-12-818249-9.00009-1

If you have experience with any programming language, you will be familiar with many of the topics covered in this chapter. If you are not yet a programmer, VBA is a great way to begin. The emphasis in this chapter is on
coding, so we will use simple examples. Later chapters make use of this skill
to code more useful functions.

Why and when do we use user-defined functions (UDF)? Just as it is more
convenient to use =SUM(A1:A20) rather than =A1+ A2+...+A20, a user-
defined function may be more convenient when we repeatedly need to perform a certain type of calculation for which Microsoft Excel has no built-in
function. Once a user-defined function has been correctly coded, it may be
used in the same way as a built-in worksheet function.

Before you write a user-defined (or custom) function, make sure that it is not
already provided by Excel. The built-in functions are more efficient than
user-defined functions. After you have written a function, you must test it
thoroughly with a wide range of input values.

The Developer Tab: In order to work with the VBA editor, we shall
need to have the Developer tab on the ribbon. If it is not present on
PC right-click (on a Mac hold the [Ctrl] key and click) any of the tabs
on the ribbon and select *Customize Ribbon* and in the right-hand panel
enter a check mark in the *Developer* box. (On a Mac this is added using
Excel / Preferences / Ribbon and Toolbar) Click the OK button to close
the dialog.

Security Note: While macros are indispensable, they are also a source of
danger. A macro (primarily subroutines) may contain malicious code.
Office 2016 incorporates various security features, but it is the user's
responsibility to protect his work. If you unexpectedly receive a file, do
not open it even if it appears to come from a friend or colleague. Check with
the sender first.

One of the new security features results in Excel files containing macros
being given a different extension. When the newly created file is saved,
we need to specify that it is a *macro-enabled* file. The file is then saved with
the extension .XLSM (note the M).

You may need to adjust the security setting before you can work with modules. Open the *Developer tab* and in the *Code* group click the *Macro Security* tool to open the *Trust Center* dialog shown in Fig. 9.1. In the *Macro
Settings* tab, if you choose *Disable all Macros with Notification*, you will
be presented with a security question each time you open a file containing
a macro.

Trust Center

Trusted Publishers	**Macro Settings**
Trusted Locations	
Trusted Documents	○ Disable all macros without notification
Trusted App Catalogs	◉ Disable all macros with notification
Add-ins	○ Disable all macros except digitally signed macros
ActiveX Settings	○ Enable all macros (not recommended; potentially dangerous code can run)
Macro Settings	**Developer Macro Settings**
Protected View	☐ Trust access to the VBA project object model
Message Bar	
External Content	
File Block Settings	
Privacy Options	

■ FIG. 9.1

EXERCISE 1: THE VISUAL BASIC EDITOR

The editor (VBE) is where we do the coding of subroutines and user-defined functions. We are just scratching the surface of VBA, so we do not have time to explore the VBE window in depth. In this exercise, we begin to develop a function to compute the area of a triangle given the lengths of two sides (a and b) and the included angle (θ). The formula is $Area = \frac{1}{2}ab\mathrm{Sin}(\theta)$

(a) Open Excel and use the command *Developer / Coding / Visual Basic* (or the shortcut [Alt]+[F11]) (*Developer / Visual Basic* on the Mac) to open the VBE—see Fig. 9.2. The top part of the VBE window displays a menu and a toolbar. To the left, we have the *Project* window. The top part of the right-hand side is the *Module* window (yours will be empty at this point), and the lower part is the *Immediate* window. If the Immediate window is not visible, use the menu command *View / Immediate Window*.

(b) Move the mouse pointer into the Immediate window and click, or use the shortcut [Ctrl]+G. Type ?3*4 and tap [Enter↵]. The result 12 is displayed. Hence the name *Immediate*; we use this area mainly for testing short pieces of code or issuing brief VBA commands.

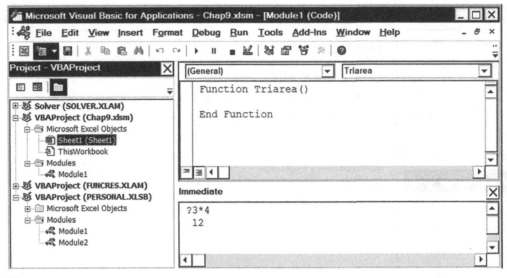

■ FIG. 9.2

(c) Your module window is most likely empty, so we will open a new module. Ensure that one of the items in the Chap9.xlsm project is selected. Use the menu command *Insert / Module*. Note how Module 1 is added to the project tree.

(d) In preparation for the next exercise, in the module window type Function TriArea() and hit [Enter←]. Note how VBE helpfully adds End Function. VBE gives reserved words (like *Function* and *End Function*) one font color and user words (like the name *TriArea* we elected to use for the UDF) another color.

(e) Return to Excel (you can click the appropriate item in the Windows taskbar or the Excel icon on the VBE toolbar) and save the workbook. Since this workbook contains a macro, in the Save dialog you must open the *Save as type* box and select *Excel Macro-Enabled Workbook (*.xlsm)*. Once the file is given the *xlsm* extension future saves will not require any additional action.

SYNTAX OF A FUNCTION

To successfully code a function, you need two skills. The first is the ability to compose, in English and mathematical symbols, the set of rules that will yield the desired result. This is called the *algorithm*. The second is the ability to translate the algorithm into the Visual Basic language. Like all languages, both natural and computer, Visual Basic has a set of rules known as the language syntax. Following we have an outline of the syntax for a user-defined function; for simplicity, optional items (Private / Public) have been omitted from before the word *Function*.

```
Function name [(arglist)][As type]
     [statements]
     [name = expression]
     [Exit Function]
     [statements]
     [name = expression]
End Function

name                        The name you wish to give to the function.
arglist                     List of arguments passed to the function.
                            Arguments are separated from each other by commas,
type                        The data type of the value returned by the function.
statement                   A valid Visual Basic statement.
expression                  An expression to set the value to be returned by the function.

Items shown within square brackets [...] are optional.
Words in bold must be typed as shown (reserved words).
```

The name used for a function must not be a valid cell reference such as AB2, nor may it be the same as the name assigned to a cell or a region. If you make this mistake, the cell that calls your misnamed function will display #REF!

Each statement must begin on a new line. If a statement is too long for one line, type a space followed by an underscore character and complete the statement on the next line. Do not split a word using this method.

EXERCISE 2: A SIMPLE FUNCTION

In this exercise, we write a user-defined function to calculate the area of a triangle given the length of two sides and the included angle: $Area = \frac{1}{2}ab\mathrm{Sin}(\theta)$. A worksheet formula is also used to confirm the VBA result.

(a) Open Chap9.xlsm and on Sheet1 type the entries shown in A1:E3 and A4:C6 of Fig. 9.3. The formula in D4 is =0.5 * A4 * B4 * SIN(RADIANS (C4)) and computes the area so that we may test our function. Copy this down to row 6. Leave E4:E6 empty for now.

(b) Use `Alt`+`F11` to open the VBE window. Click on Module1 in the Project window. The window title should read Chap9.xlsm—[Module1 (Code)]. One of the most common errors for VBA beginners is entering

If you edit a UDF that is already used in a worksheet, then the worksheet must be recalculated (using `F9`) to have the function report its new value.

◢	A	B	C	D	E
1	Test function to compute area of triangle				
2					
3	SideA	SideB	Angle	Formula	Function
4	1	2	90	1	1
5	2	2	45	1.414214	1.414214
6	2	2	60	1.732051	1.732051

■ FIG. 9.3

the code in the wrong place. For our purposes, the only correct place is on a general module, not a worksheet or workbook module.

(c) Enter this code exactly as shown using Enter← at line ends and Tab⇆ to indent:

```
'Computes the area of a triangle given
'top sides and included angle in degrees
Function TriArea(SideA, SideB, Theta)
    Alpha = WorksheetFunction.Radians(Theta)
    TriArea = 0.5 * SideA * SideB * Sin(Alpha)
End Function
```

The indentations in the code are used only for readability; they are not required by syntax. A UDF returns a value to the cell(s) containing the formula that calls it. A UDF cannot change the values in other cells. So the cells referenced by *a*, *b* and *c* do not change in the worksheet. A UDF cannot format a cell.

(d) Return to the worksheet and in E4 enter the formula =TriArea(A4, B4, C4). Note that as you type =Tri, your function appears in the popup window in the same way that worksheet functions do. Copy the formula down to row 6. The values in the D and E columns should agree. If they differ, return to the module sheet and correct the function. Remember to press F9 to recalculate the worksheet after editing a function.

(e) Save the Chap9.xlsm file.

Although this is a simple function, it demonstrates some important Visual Basic features. We now examine each line of the TRIAREA function. The numbers refer to the function line numbers.

(1) This is a comment; the initial single quote (apostrophe) ensures this. A statement may end with a comment: for example,

$$x = \text{srt(b)}' \text{ find the square root of x.}$$

(2) Another comment.

(3) The *Function* is displayed in blue in the VBE to indicate a key or reserved word. We chose the name TriArea; a function name can be anything but a keyword. Our function has three arguments. Arguments are passed from the formula in the worksheet to the function heading by their position, not by their names.

(4) This is an *assignment statement*: we give a value to the variable *Alpha*. We do so using an Excel function, so we need to use the Worksheet-Function before Radians. The complete syntax for referencing a worksheet function is *Application.WorksheetFunction.FunctionName*, but the first or second word may be omitted; we have omitted the first as it is in our function.

Did you notice that when you had typed the period after WorksheetFunction, VBE offered a choice of functions by a process that Microsoft calls *IntelliSense*? Did you also notice that you could have typed a lowercase *f*? VBE would fix that when you pressed [Enter↵].

(5) This is another assignment statement. There must be at least one statement that assigns a value to the function. In this statement, we use the VBA sine function. We might have typed `VBA.Sin(Alpha)`, but this would be redundant.

(6) The *End Function* statement is required as the final line.

NAMING FUNCTIONS AND VARIABLES

Try to use short but meaningful names for variables, functions, and arguments. These three simple rules must be followed:

(1) The first character must be a letter. Visual Basic ignores uppercase and lowercase. If you use the name *term* in one place and *Term* elsewhere, Visual Basic will change the name to match the last used form.

(2) A name may not contain a space, a period (.), exclamation point (!), or these symbols: @, $, or #.

(3) A name may not be a VBA restricted keyword. A full list of reserved keywords is hard to find. However, it is not necessary to know them because, if you try to use one, VBA highlights the word and displays an error message. Generally, this will read "Identifier expected," but certain keywords generate other messages. Note that, generally, VBA displays keywords in blue.

Some cautionary notes on naming variables, functions, and modules are in order. If you avoid dictionary words like *Range,* you are less likely to run into conflicts with keywords. Variable names such as *myRange* are safe. If you name a function *Extract*, you get no warning until you run it and then the message is terse: That function is invalid. Using *View / Properties Window,* you can name a module something other than *Module3*. Use "odd" names.

The author once named a module *Pi*, which caused every UDF in the workbook that used that name for a variable to report an error!

WORKSHEET AND VBA FUNCTIONS

The mathematical functions available within VBA are shown in the following table. Unfortunately, at this time, the Help feature for VBA in Office 2016 (unlike that in earlier versions) is unsatisfactory. The website *http:// msdn.microsoft.com/en-us/library/office/jj692811.aspx* may be consulted for details on VBA functions. Alternatively, the reader could perform an internet search.

You cannot use a worksheet function when VBA provides the equivalent function even when the name is not the same. So, none of the worksheet trigonometric functions SIN, COS, or TAN may be used, but ASIN and ACOS are permitted. The worksheet function SQRT cannot be used since VBA includes the equivalent SQR function. You may, however, use the worksheet function MOD because Mod in VBA is an operator, not a function.

Abs(x)	The absolute value of *x*
Atn(x)	Inverse tangent of *x*. Other inverse functions may be computed using trigonometric identities such as: `Arcsin (X) = Atn(X / Sqr(-X * X + 1))`. For more information search Visual Basic Help for Derived math functions
Cos(x)	The cosine of *x*, where *x* is expressed in radians
Exp(x)	The value e^x
Fix(x)	Returns the integer portion of *x*. If *x* is negative, Fix returns the first negative integer greater than or equal to *x*; for example, `Fix(-7.3)` returns −7. See also Int
Int(x)	Returns the integer portion of *x*. If *x* is negative, Int returns the first negative integer less than or equal to *x*; for example, `Int(-7.3)` returns −8. See also Fix
Log(x)	The value of the natural logarithm of *x*. Note how this differs from the worksheet function with the same name which, without a second argument, returns the logarithm to base 10 In VBA, the logarithm of *x* to base *n* may be found using the statement y = `Log(x)/Log(n)`
Mod	In Visual Basic this is an operator, not a function, but it is similar to the worksheet MOD function. It is used in the form `number Mod divisor` and returns the remainder of a number divided by the divisor after rounding floating-point values to integers. The worksheet function and the VBA operator return different values when the number and divisor have opposite signs; see Help for details
Rnd(x)	Returns a random number between 0 and 1
Sgn(x)	Returns −1, 0, or 1 depending on whether *x* has a negative, zero, or positive value
Sin(x)	The sine of *x*, where *x* is expressed in radians
Sqr(x)	Square root of *x*
Tan(x)	The tangent of *x*

If you try to code a user-defined function that references an unavailable function (e.g., `WorksheetFunction.Sin(Alpha)`), the worksheet cell in which your user-defined function is called will display #VALUE!

The VBA Round function differs from the Excel ROUND: When the last digit to be rounded is 5, an even number is always produced. Round (4.5,1) gives 4, while Round(5.5,1) gives 6. The method VBA uses is sometimes called Banker's Rounding, but there is no evidence of bankers using this method!

Should one use, for example, `Application.Radians(theta)` or `WorksheetFunction.Radians(theta)`? The two forms are equivalent as long as the worksheet function does not return an error value. It is easier to trap errors when you use the first form, but this deprives you of the convenience of IntelliSense—see diagram to the side. The companion website file WorksheetFunction.xlsm gives some details on this topic. Note that older books/websites will use the rather long-winded `Application.WorksheetFunction.<whatever>`; this works exactly the same as the simpler `WorksheetFunction.<whatever>`.

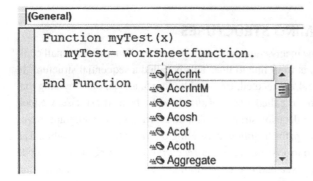

EXERCISE 3: WHEN THINGS GO WRONG

Recalling the adage "To err is human, but to really mess up you need a computer," we will make an error in a module and lock our worksheet. Do not worry, we can fix the problem. It is very likely that you will accidentally make such an error, so it is good to know what is needed to correct it.

(a) Open the VBE and change line 5 of TriArea by replacing the equal sign by a minus sign. Do not press `Enter↵` and do not use *Debug / Compile*. Just return to the worksheet.

(b) Press `F9` to recalculate the worksheet. Excel returns you to the VBA Editor window and displays an error dialog box. Click the OK button. Note that the function header is highlighted in yellow. Correct the error by replacing the minus sign by an equal sign. The yellow highlighting does not disappear.

(c) Return to the worksheet. Try to do something like changing the active cell. Nothing works; the worksheet (indeed the whole workbook) is locked.

(d) Return to the VBE and use the command *Run / Reset* to remove the highlighting. Alternatively, we could use the reset tool; this is the 12th item on the toolbar and is displayed as a solid square. Now when you go to the worksheet all is well again.

In short, an error in a module can cause a worksheet to lock. The remedy is to use the command *Run / Reset* to reset the module.

This exercise should help you with syntax errors. Another type of problem is the logical error. This occurs when you have coded your function correctly as far as Visual Basic is concerned (the syntax is correct), but the wrong answer results from an error in the algorithm. Some techniques for solving this type of problem are explored in a future exercise.

PROGRAMMING STRUCTURES

The normal flow in any computer program (and our function is a small computer program) is from line to line. This is called a sequential structure. In Exercise 2, line 4 is executed, followed by line 5, and so on. Anything that changes this flow is called a *control structure*. In the next exercise, we look at a branching or decision structure. This structure gives the program two or more alternative paths to follow depending on the value of a variable. The other type of control structure is the *repetition* or *looping* structure; code within a loop is executed one or more times. We will explore this in later exercises.

EXERCISE 4: THE IF STRUCTURE

In Chapter 5 we looked at conditional formulas using the IF function to make a selection based on a test. We do the same in a macro using the IF...ELSE structure whose syntax is shown as follows. The items enclosed in square brackets are optional. There is a slightly simpler syntax for a one-line IF statement in which statements are separated by colons. For details on logical expressions (aka *conditions*) refer back to Chapter 5.

<table>
<tr><td colspan="3">Syntax for IF…ELSE</td></tr>
<tr><td>If condition Then</td><td>condition</td><td>Expression that is True or False</td></tr>
<tr><td>[statements]</td><td>Statements</td><td>One or more statements</td></tr>
<tr><td>[ElseIf condition-n Then</td><td>condition-n</td><td>Expression that is True or False.</td></tr>
<tr><td>[elseifstatements]] …</td><td>elseifstatements</td><td>One or more statements executed associated</td></tr>
<tr><td></td><td></td><td>Statements executed if no previous condition- n expression are True</td></tr>
<tr><td>[Else</td><td>elsestatements</td><td></td></tr>
<tr><td>[elsestatements]]</td><td></td><td></td></tr>
<tr><td>End If</td><td></td><td></td></tr>
</table>

Note: A condition is the same as what we called a logical expression in Chapter 5.

EXERCISE 5: BOOLEAN OPERATORS

In this exercise, we will use the short and the long form of the IF statement and show the use of the AND and OR Boolean operators.

We will construct a function that reports what type of triangle is formed when given the length of the three sides. Before coding this, we need to think more about the algorithm. We might make a list of the things we know about triangles and their sides:

 (i) One side is always shorter than the sum of the other two.
 (ii) In an equilateral triangle, all the sides are equal.
(iii) In an isosceles triangle, two sides are equal.
(iv) Pythagoras' theorem is true with a right-angle triangle.

How do we know which is the hypotenuse if we are given just three values? How can we see if two, and only two sides are equal? These questions are readily answered if the values for the sides are in descending order. The reader may wish to complete the algorithm before proceeding.

(a) On Sheet2 of Chap9.xlsm construct the worksheet shown in Fig. 9.4. The cell D4 contains the formula =Tritype(A4, B4, C4). It will return the error value #NAME? until we have coded the function.

(b) Open the VBE, add a second module to the Chap9.xlsm project using *Insert / Module*, and code the following function.

```
Function Tritype(a, b, c)
'Sort the three sides in ascending order
 If b > a Then holder = a: a = b: b = holder
 If c > a Then holder = a: a = c: c = holder
 If c > b Then holder = b: b = c: c = holder
'Determine triangle type
```

◢	A	B	C	D
1	Triangle tester			
2				
3	Side A	Side B	Side C	Type
4	1	2	3	Scalene

■ FIG. 9.4

```
If a > b + c Then
              Tritype = "None"
      ElseIf a * a = b * b + c * c Then
              Tritype = "Right"
      ElseIf (a = b) And (b = c) Then
              Tritype = "Equilateral"
      ElseIf (a = b) Or (b = c) Then
              Tritype = "Isosceles"
      Else
              Tritype = "Scalene"
    End If
  End Function
```

(c) Return to the worksheet and press [F9]. Experiment with other values for the three sides to test the function.

Make sure you follow the logic in the sorting section. We use a variable called *holder* to temporarily store a value when we need to exchange the values of two other variables.

We made a second module for the user-defined function of this exercise. This was not essential; we could have added it to Module1. However, there is a problem with having more than one function on a single module. If any one of the functions contains an error, then all functions on that module return error values on the worksheet. This can be confusing, especially for beginners. There are other considerations that help you decide whether to put more than one function on a single module, but these relate to the use of keywords *Public* or *Private* in the Function header—a topic we will not be exploring.

THE SELECT...CASE STRUCTURE

Visual Basic for Applications provides another branching structure called the SELECT CASE structure. Its syntax is shown as follows.

Syntax for SELECT... CASE		
Select Case testexpression	testexpression	Any numeric or string expression
[**Case** expressionlist-n	expressionlist-n	A list of one or more of expression types separated by commas.
[statements-n]] . . .		Valid expression types are: expression, expression To expression,
[Case Else		Is comparisonoperator expression
[elsestatements]]		One or more statements executed if testexpression matches any
End Select	statements-n	part of expressionlist-n
		One or more statements executed if testexpression does not
	elsestatements	match any of the Case clauses

Examples of an expression include:

(i) A simple value such as 100;

(ii) Smaller-*value To larger-value*; as in 0 To 20.

(iii) *Is > some-value*; as in Is > 20. Clearly all comparison operators ($>$, $<$, $>=$, $<=$) are permitted.

When *testexpression* matches one of the expressions, the statements following that Case clause are executed up to the next Case (or End Select) clause. Control then passes to the statement following End Select. When *testexpression* matches more than one expression, only the statements for the first match are executed. The statements following Case Else are executed if no match is found in any of the other Case selections. It is advisable always to use a Case Else statement to handle unexpected *testexpression* values.

For comparison, here are two functions give the same results:

```
Function Test1(a, b)              Function Test2(a, b)
  If a > b Then                     Select Case (b - a)
    Test1 = "A is larger"           Case Is > 0
  ElseIf b > a Then                   Test2 = "B is larger"
    Test1 = "B is larger"           Case Is < 0
  Else                                Test2 = "A is larger"
    Test1 = "A & B are equal"       Case Else
  End If                              Test2 = "A & B are equal"
End Function                        End Select
                                  End Function
```

EXERCISE 6: SELECT EXAMPLE

The number of real roots of the quadratic $ax^2 + bx + c = 0$ is determined by the value of the discriminant $d = b^2 - 4ac$. In this exercise, we write a function to return a value indicating the number of real roots for a quadratic equation.

◢	A	B	C	D
1	Number of real root roots of a quadratic			
2				
3	a	b	c	Roots
4	1	3	-15	2
5				

■ FIG. 9.5

(a) Using the file Chap9.xlsm, open the VB Editor, insert another module, and enter the function shown here.

```
Function RootCount(a, b, c)
    d = (b * b) - (4 * a * c)
    Select Case d
        Case 0:             RootCount = 1
        Case Is > 0:        RootCount = 2
        Case Else:          RootCount = 0
    End Select
End Function
```

Note: The expression $d = (b * b) - (4 * a * c)$ could have been coded without the parentheses. But they do make it easier to read the expression because of the way VBE spreads out arithmetic expressions.

(b) Set up Sheet3 as in Fig. 9.5 to test the function. The function is called in D4 with =RootCount(A4, B4, C4).

(c) Save the workbook.

THE FOR...NEXT STRUCTURE

In a looping structure, a block of statements is executed repeatedly. When the repetition is to occur a specified number of times, the FOR...NEXT structure is used. The syntax for this structure is given as follows.

The reader is strongly advised never to alter the value of the counter within the For...Next loop, as the results are unpredictable.

For our example, we will write a function to find the sum of the squares of the first n integers. While we are learning VBA, it is good to use an example where the answer is known. In this case, the analytical solution to the

	Syntax for FOR...NEXT	
For counter = first **To** last [**Step** step]	counter	A numeric variable used as a loop counter.
[statements]	first	The initial value of counter.
[Exit For]	last	The final value of counter.
[statements]	step	The amount by which counter is changed each time through the loop.
Next [counter]	statements	One or more statements that are executed the specified number of times.

The *step* argument can be either positive or negative. If *step* is not specified, it defaults to 1.
After each execution of the statements in the loop, *step* is added to *counter*. Then it is compared to *last*. When *step* is positive, the loop continues while *counter* <= *last*. When *step* is negative, looping continues while *counter* >= *last*.
The optional Exit For statement, which is generally part of an IF statement, provides an alternate exit from the loop

problem is given by $(n/6)(n+1)(2n+1)$. Open the VBE, insert another module, and enter the following function.

```
Function SumOfSquares(n)
  SumOfSquares = 0
  For j = 1 To n
    SumOfSquares = SumOfSquares + j ^ 2
  Next j
End Function
```

(a) Set up Sheet4 as in Fig. 9.6 to test the function. Name cell A4 as *N*. In B4 and C4 enter the functions as shown in the figure. As expected, the two results are in agreement.

(b) Save the workbook.

	A	B	C
1	Sum of the Squares of first N integers		
2			
3	N	Function	Formula
4	12	650	650
5			
6	Function call	=SumOfSquares(N)	
7	Formula	=(N/6)*(N+1)*(2*N+1)	

■ FIG. 9.6

THE EXCEL OBJECT MODEL: AN INTRODUCTION

From a nontechnical point of view, the Excel Object Model is a detailed blueprint of how Excel operates behind the scenes. The model consists of *objects*. Examples of objects are a workbook, a worksheet, a range. A group of objects of the same type is called a *collection*. The *Workbooks* collection contains all the open workbook objects. The term *Workbooks(1)* refers to the first open workbook, while *Workbooks("Data.xlsx")* refers to a workbook by name.

Objects have *methods* and *properties*. Thus we might see subroutine statements with terms such as *Workbook(1).Close* and *Activesheet.Delete* where Close and Delete are methods. An expression such as *Worksheet(1).Name* is a reference to a property of Worksheet(1).

For this chapter, our main interest is in ranges. A range may be a cell, a row, a column, a selection of cells containing one or more contiguous blocks of cells, or a 3D range.

A range is a strange object. Consider the range A1:C10. It contains other ranges such as A2:B4 and C9, but each of these is itself a range. So, the range object is also a range collection; this is the only collection in Excel that does not use a plural name. Surprisingly, there is no *Cell* object. You may see code using a term such as *Cells(1,1)*, but this is just another way of referencing Range ("A1").

When working with Functions, we need to know about two range properties: *Count* and *Value*. Very often the term *Value* is omitted since this is the default property.

EXERCISE 7: FOR EACH ... NEXT—RESISTORS REVISITED

The *For Each* structure may be used with any collection. This structure references each member of the collection in turn with code such as: *For Each MyCell in MyRange...Next*.

In Exercise 5 of Chapter 2, and again in Exercise 3 of Chapter 5, we found the equivalent resistance of a number of resistors in parallel. For comparison, we will use both *For Next* and *For Each* to find the equivalent (or effective) *R* value.

(a) With file Chap9.xlsm, open the VBE and enter the two functions shown as follows one under the other on a new module.

```
Function EquivR(myRange)
  recipR = 0
  For j = 1 To myRange.Count
    If myRange(j).Value > 0 Then
        recipR=recipR+(1 / myRange(j).Value)
    End If
  Next j
  If recipR > 0 Then
    EquivR = 1 / recipR
  Else
    EquivR = "Error"
  End if
End Function
```

```
Function EffectR(myRange)
  recipR = 0
  For Each myCell In myRange
    If myCell.Value > 0 Then
        recipR = recipR + (1 / myCell.Value)
    End If
  Next
  If recipR > 0 Then
    EffectR = 1 / recipR
  Else
    EffectR = "Error"
  End If
End Function
```

(b) Set up Sheet5 as in Fig. 9.7 to test the functions. The cells B5:C5 have been merged with the tool in the *Home* / Alignment group; as have B6:C6. The formulas used are as follows:

D5: =ROUND(equivR(B3:E3),-1)
D6: =ROUND(effectR(B3:E3),-1)

◢	A	B	C	D	E
1	**Resistors in Parallel**				
2					
3	R	1240	1800	2000	0
4					
5	R_e	EquivR		540 ohms	
6	R_e	EffectR		540 ohms	

■ FIG. 9.7

Both use the Custom Format of *0 "ohms."*

The rounding could have been done within the VBA code. For example:

```
EffectR = WorksheetFunction.Round(EffectR, -1)
```

It is left as an exercise for the reader to experiment with the correct place to enter this code.

(c) Save the workbook.

EXERCISE 8: THE DO…LOOP STRUCTURE

Whereas the FOR…NEXT structure is used for a specific number of iterations through the loop, a DO…LOOP structure is used when the number of iterations is not initially known but depends on one or more variables whose values are changed by the iterations. There are two syntaxes for this structure: the Pretest and the Posttest forms. These are shown as follows.

Syntax 1 for the DO statements	Syntax 2 for the DO statements
Do {While | Until} condition	**Do**
[statements]	[statements]
[Exit Do]	[Exit Do]
[statements]	[statements]
Loop	Loop {While | Until} condition

We have the option of looping until a condition *becomes* true or *while* a condition is true. In the Pretest form (Syntax 1), the condition is tested *before the first statement* within the loop is executed while in the Posttest (Syntax 2) the test occurs *after the last statement* has been executed. This means that with syntax 2, the statements within the loop are executed at least once regardless of the value of the condition at the start of the structure.

If you inadvertently end up with an infinite loop, your worksheet will "hang."

Use [Esc] or [Ctrl]+[Break] to terminate the function.

Typically, the condition in the UNTIL or WHILE phrase refers to one or more variables. The programmer is responsible for ensuring that the variables eventually have values such that the terminating condition is satisfied. The only exception is when a conditional EXIT statement is used to terminate the loop. Should the terminating condition never be reached, you have an *infinite loop*.

In Chapter 10 we will use the Do Loop structure to find the root of an equation using an iterative method. But for the example here, we show its use to compute Exp(*x*) using the Maclaurin series.

$$\exp(x) = \sum_{k=0}^{\infty} \frac{x^k}{k!} = 1 + x + \frac{x^2}{2!} + \frac{x^3}{3!} + \cdots$$

This is known to be a convergent series. Also, we observe that:

$$term_k = \frac{x^k}{k!} \ and \ term_{k+!} = \frac{x^{k+1}}{(k+1)!}$$

$$\therefore term_{k+1} = term_k \times \frac{x}{k}$$

We will make use of this recursive relationship. But how many terms shall we use? Since nothing in Excel can be more precise than $1E-15$, we will keep looping until term k is within this precision of term $k-1$.

(a) Open a new VBE module and enter this function:

```
Function MacExp(x)
  Const Precision = 0.000000000000001
  MacExp = 0
  Term = 1
  k = 1
  Do While Term > Precision
  MacExp = MacExp + Term
  Debug.Print k; Term; MacExp
  Term = Term * x / k
  k = k + 1
  If k > 100 Then
     MsgBox "Loop aborted, k > 100"
     Exit Do
   End If
   Loop
  End Function
```

(b) The Debug.Print statement is to check the operation of the function; it can be deleted or "commented out" when not needed. It prints results in the Immediate Window. If not visible, the Immediate Window can be opened using *View / Immediate Window* or Ctrl + G. Note that we have an "escape hatch"; if k gets above 100, we jump out of the loop after displaying a message. It is a good idea to use this technicality to avoid unending loops. The 100 limit will be too small for values of x much over 25.

(c) Set up Sheet6 as in Fig. 9.8 to test the function.

(d) Save the workbook.

Did you find any problems? How about $x=-4$? Our function gives an answer of 1 for any negative x value. Look at the loop condition. What will be the sign of *Term* after the first iteration? We can solve this by using `Do While Abs(Term) > Precision`. Conclusion: always give macros a very extensive testing.

◢	A	B	C
1	Maclaurin Series for Exp(x)		
2			
3	x	=MacExp(x)	=EXP(x)
4	2	7.389056098930650	7.389056098930650
5	Difference	0.000000000000000	

■ FIG. 9.8

VARIABLES AND DATA TYPES

Unlike most computer languages, all dialects of BASIC allow the programmer to use variables without first declaring them. While this feature slightly speeds up the coding process, it has a major disadvantage that typo errors can go undetected. What would happen in the last exercise if you had mistakenly typed *MacExp=MacExp+Team* where the variable *Term* is misspelled as *Team*. Since *Team* is not mentioned elsewhere, its value is zero and the final result would have been zero. It is easy to make silly typos. The problem is avoided by making variable declarations mandatory. We can do this with the use of the *Option Explicit* statement at the start of the module, or by opening the VBE *Tools / Options (Excel / Preferences...* on a Mac) dialog and on the *Editor* tab checking the *Require Variable Declaration* box. You are encouraged to use the latter. With this in place, all variables must be declared before being used. We use the DIM statement for this purpose. The code for the last exercise would then begin with:

```
Function MacExp(x)
    Dim Term, k
    Const Precision = 0.000000000000001
```

MacExp is already defined by the function header, and Precision gets defined by the *Const* statement. Now if you use *Team* when *Term* was needed, VBA will issue a warning.

There is yet another difference from other programming languages. In languages such as FORTRAN and C, it is not sufficient merely to name the variable; the programmer must also state its data type. In this example, we would need to define k as an integer variable and term as a floating-point variable. We could do this by coding `Dim Term As Double, k As Integer`. We would use *Double* rather than *Single* to get the maximum precision. When the data type of variables is not declared, VBA uses a special data type called the variant. This is acceptable for the simple functions shown in these examples, but in general, one should declare data types. It should be noted that variant data types are memory hogs. You may wish to use Help to find out more about this topic, especially the permitted range of values for Integer, Short, Long, Single and Double.

INPUT-OUTPUT OF ARRAYS

In the function EquivR in Exercise 7, we saw how an input array may be processed by a VBA function. That was a one-dimensional array. Fig. 9.9 shows the use of a function that processes a two-dimensional array. The objective is to find $\sum a_i/b_i$ (the sum of quotients a/b for each row in the table). So, we have $16/8 + 25/5...$ giving 32. The function is called with =SUMQUOT(A4:B10). Note that it will work with any number of rows.

Note the use of two index values; with the use =SUMQUOT(A4:B10) we have defined *myrange* as A4:B10. When *j* has the value 1, *myrange(j,1)* references the first row and the first column of the range, i.e., A4, and *myrange(j,2)* references B4. The reader is encouraged to experiment with this function on Sheet7.

EXERCISE 9: AN ARRAY FUNCTION

In this exercise we look at a function that outputs an array: it is used to put results in a number of cells so it is an array function. Within the function we have a 3-by-1 array called *temp* into which data is placed, and then we pass that array to the function-name in the final statement. For our example, we will write a function to find the real roots of a quadratic equation $ax^2 + bx + c = 0$.

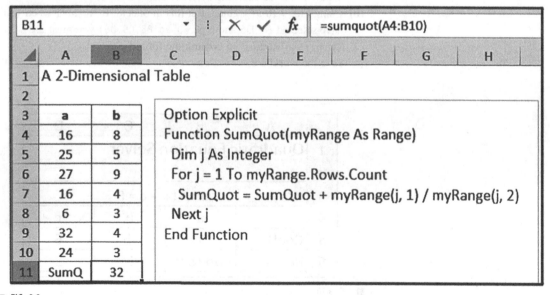

■ FIG. 9.9

(a) In Chap9.xlsm, insert Module 7 and enter this code:

```
Function Quad(a, b, c)
  Dim Temp(3)
  d = (b * b) - (4 * a * c)
  Select Case d
    Case Is < 0
      Temp(0) = "No real"
      Temp(1) = "roots"
      Temp(2) = ""
    Case 0
      Temp(0) = "One root"
      Temp(1) = -b / (2 * a)
      Temp(2) = ""
    Case Else
      Temp(0) = "Two roots"
      Temp(1) = (-b + Sqr(d)) / (2 * a)
      Temp(2) = (-b - Sqr(d)) / (2 * a)
  End Select
  Quad = Temp
End Function
```

The statement *Dim Temp(3)* establishes a 3-by-1 array. Note that VBA arrays use an indexing system that begins with zero. If this is not acceptable, you can add the code Option Base 1 before the function header to use an indexing system starting from 1.

(b) Using Fig. 9.10 as a guide, test this in Sheet8. Remember when you have selected B6:D6 and type =Quad(B5,C5,D5) you must use
Ctrl + ⇧ Shift + Enter↵ to commit the formula.

(c) Save the workbook.

	A	B	C	D
1	Quadratic Equation Solver			
2				
3		$ax^2 + bx + c = 0$		
4		a	b	c
5	Coeff	1	-5	6
6	Roots	Two roots	3	2

■ FIG. 9.10

USING FUNCTIONS FROM OTHER WORKBOOKS

The user-defined functions we have created have been used in the workbook in which they were coded. Every function in a workbook is available from any sheet in it. A number of procedures permit us to use a function from another workbook.

(a) The least efficient method is to copy the function to the new workbook using either *Copy&Paste* or, within the VBE File menu, *Export,* and *Import.*

(b) The user-defined functions of an open workbook are available in other workbooks. Thus if Chap9.xlsm is open, then in a second workbook we may code, for example, =Chap9.xlsm!Quad(...). We could either type this formula or use the Insert Function tool to locate the function in the User-defined category. Frequently used macros may be conveniently kept in a file called Personal.xlsb stored in the Xlstart folder.[1] Normally, one hides this workbook in Excel (with View / *Window / Hide*) before it is saved. Files in Xlstart automatically open when Excel starts.

(c) The most versatile way is to convert a macro-enabled workbook (extension xlsm) into an add-in workbook with extension xlsa. Excel can be configured to load an add-in every time the program is started thereby making all the macros in the add-in accessible from any file the user opens or creates.

We will quickly look at the steps needed to make an add-in from Chap9.xlsm once all the functions have been coded and thoroughly tested. So that Chap9.xlsm is preserved for future reference, save it as Chap9Add.xlsm for this exercise.

i. Rename and lock the project: Open the project properties dialog either from the shortcut menu found when you right-click a member of Chap9.xlsm in the Project Window or by using the Tools menu command. By default, all projects are named VBAProject; it is better to give it a unique name such as *Chap9*. If you are distributing the add-in, you may wish to lock the code and add a password, so others cannot see or alter it.

ii. Add file properties: While the project properties dialog is open, in the Project Description, enter descriptive information about your add-in.

iii. Save in Add-in format: Go back to the Excel window. Use *File / Save As* to open the Save As dialog. In the Type box, near the end of the list, select Excel Add-in (.xlam). You could also save an Excel 97–2003 Add-in.

---

[1]Generally this folder will be stored as C:\Users\[user_name]\AppData\Roaming\Excel \Xstart.

The file will be saved with the extension xla. Note that the directory changes to the \AppData\Roaming\Microsoft\AddIns directory when you change the file type. This is the default directory for Add-ins. You can change this to make the file easier for you to share.

iv. Install the Add-in: You, and others to whom you give the file, can install the add-in using the Add-in Manager found in *File / Excel Options / Add-ins*. The Manager box should read *Excel Add-Ins*; click the Go button (*Tools / Excel Add-ins* on a Mac). Use the Browse command to locate and install the add-in. The add-in will display the descriptive information you added in step (ii). Now you will be able to use all the functions in the add-in as if they were in the current file.

A number of companies market Microsoft Excel add-ins. For example, one add-in contains functions for performing mass-mole chemistry calculations. You may find some shareware add-ins by searching the internet.

PROBLEMS

1. Alter the Quad function in Exercise 9 in such a way that it may be called with =Quad(A1:C1).

2. *The pressure drop in a pipe of length L ft and diameter D ft with an average water velocity of V ft/s is given by the Darcy-Weisbach equation

$$\Delta P = f \frac{L}{D} \frac{V^2}{2g}$$

where f is the friction factor (generally taken as 0.02) and g is the acceleration due to gravity (32.2 ft/s^2). Write a UDF with the header Function DW(Length, Diameter, Flow, Friction) where *Flow* is the average volume flow in ft^3/s. The velocity V is related to the flow rate Q by $V = Q/(\pi R^2)$ or $V = 4Q/(\pi D^2)$. Test it against results obtained with worksheet formulas and/or an Internet site.

3. *Write a UDF to calculate the resultant of two vectors; see Fig. 9.11. Cells B6:C6 call the function with =ForceVector(B4:C4,B5:C5) as an array function. The formulas to solve this are as shown in the figure.

4. *Write an array function with the header Function SciNum(Number) that accepts a number and returns the significand and exponent in two separate cells as shown in Fig. 9.12.

5. Refer to Problem 5 in Chapter 4. Write a UDF with the header Function Trough(length, radius, height) that returns the volume of water in the trough. The arguments *length* and *radius* are in feet, while *height* is in inches. The result should be rounded to the nearest gallon. Write another UDF where *height* refers to the water depth—the length of the wet part of the dip stick rather than the dry part.

▲	A	B	C	D	E	F	G
1		Resultant Force Vector					
2							
3		Force (lb)	Angle(deg)				
4	V_1	120	25				
5	V_2	200	42				
6	V_r	316.71	35.64				
7							

$$F_x = F_1 \cos(\theta_1) + F_2 \cos(\theta_2)$$

$$F_y = F_1 \sin(\theta_1) + F_2 \sin(\theta_2)$$

$$F_r = \sqrt{F_x^2 + F_y^2}$$

$$\theta_r = \operatorname{atan}(F_y / F_x)$$

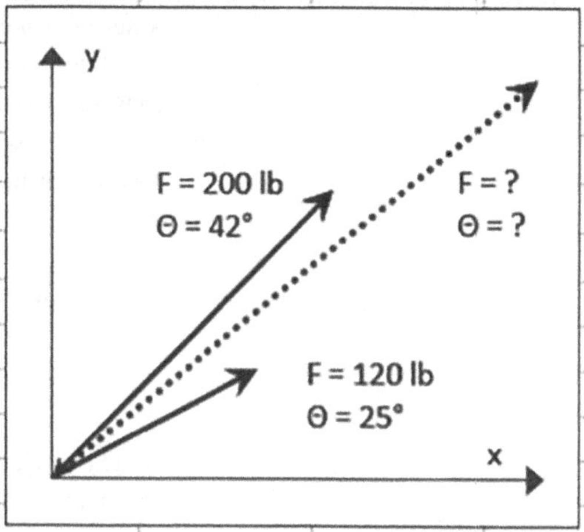

In the diagram: y-axis and x-axis shown. F = 200 lb, Θ = 42°. F = 120 lb, Θ = 25°. F = ?, Θ = ?

■ FIG. 9.11

▲	A	B	C
1	Scientific Numbers		
2			
3	Number	Significand	Power
4	45.678	4.5678	1
5	123.99	1.2399	2
6	293456	2.93456	5
7	4.342	4.342	0
8	1010.29	1.01029	3

■ FIG. 9.12

6. *A range (vertical or horizontal) in a worksheet contains the magnitude of some vectors. Write a UDF that accepts the range and returns the magnitude of the resultant vector using

$$F = \sqrt{\Sigma f_i^2}$$

7. See Problem 9 of Chapter 2; write a UDF with the header `Function SolExt(mass, Kd, Vw, Vs, n)` to return the mass of solute in the water after n extractions. The arguments are: *mass* is the amount of solute in the water phase before extraction, Kd is the distribution constant, Vw is the volume of water, and Vs is the volume of solvent.

8. Developing series expansions for π seems to have been a favorite pastime for mathematicians of old. Two of historic interest are as follows.

$$\text{Gregory} - \text{Leibniz}: \frac{\pi}{4} = 1 - \frac{1}{3} + \frac{1}{5} - \frac{1}{7} + \frac{1}{9} = \sum_{n=0}^{\infty} \frac{(-1)^n}{2n+1}$$

$$\text{Wallis}: \frac{\pi}{2} = \frac{2 \cdot 2}{1 \cdot 3} \cdot \frac{4 \cdot 4}{3 \cdot 5} \cdot \frac{6 \cdot 6}{5 \cdot 7} = \sum_{n=1}^{\infty} \frac{(2n)^2}{(2n-1)(2n+1)}$$

Write two UDFs and compare the results of the Gregory-Leibniz series with that of Wallis for $n = 1000$ and $n = 1 \times 10^6$. Do not try n greater than one million; to suggest 10 years ago that one lets even these small pieces of code loop a million times would have raised eyebrows! Neither series is of practical use, but they make excellent programming challenges. These functions may slow down your workbook if you leave n with a large value.

9. The Antoine equation is $\log_{10}(p^*) = A - \frac{B}{T+C}$ where p^* is in mmHg and T is the temperature in degrees Celsius. Write a UDF with the header `Antoine(A, B, C, P)`, which will return the boiling point (rounded to one decimal) of the liquid when the external pressure is P. Do not use any worksheet functions in your code. Data to test your formula: for benzene A, B, and C, are 6.90565, 1211.033, and 220.790, respectively, and its normal boiling point is 80.1°C.

10. You are planning to write code to perform matrix algebra on the numbers in a range. Your code will require that the range be square, that is, that the number of rows must equal the number of columns. In preparation for this write a UDF with header `Function IsSquare (myRange)` that will return TRUE or FALSE.

11. Write a UDF with the header `Function SumOfDiagonal (myRange)`, which returns the sum of the diagonal elements in `myRange` if it is square, or a suitable error message otherwise. This UDF should call the one above in a statement equivalent to `Test = IsSquare(myRange)`. The sum of the diagonal elements is often called the *trace* of the matrix.

12. Write an array UDF with the header `Function NormalArray (myRange)` which will return a new array equal to the input array normalized by dividing each element by the sum of the squares of all elements. In the code, declare a 100 by 100 array to hold the calculated

values. Do not bother to redimension this. It will be simpler if you use `Option Base 1`. For your first attempt, you may use a worksheet function but then challenge yourself to make it work without the Excel function.

13. Write a UDF that takes in a two-column array (up to 100 by 2) and returns an array such that, in each row, the left column has the larger value and the right column has the smaller.

14. A ball is thrown at 50 mph at an angle of 30 degrees to the horizontal. How high will it be when it has traveled 50 ft horizontally? Write a UDF to solve this using the following equation. Also, use it to make a plot of the ball's path. Do your results show that the maximum distance is achieved with a 45 degrees angle?

$$y = \frac{\sin\theta}{\cos\theta}x - \frac{g}{2v^2\cos^2\theta}x^2$$

15. A cylinder of radius r has a cone attached to its base. The sides of the cone make an angle θ to the axis of the cylinder. The object has a volume V. The surface area of the object is given by

$$S = \frac{2V}{r} + \pi r^2 \left(\csc\theta - \frac{2}{3}\cot\theta \right)$$

Write a UDF with the header `Function SurfaceArea(r, V, theta)` to compute the value of S. You might show by calculus that S is minimized when $theta = \cos^{-1}(2/3)$ or about 48.2 degrees. Use the function to plot S vs theta and confirm this value. Think of a way in which you could perform a check on your function.

APPENDIX: SUPPLEMENTARY MATERIAL

Supplementary material related to this chapter can be found on the accompanying CD or online at https://doi.org/10.1016/B978-0-12-818249-9.00009-1.

VBA Subroutines

CHAPTER OUTLINE

Early in the history of personal computers, software developers added scripting languages to their applications. This allowed users to write *macros*, also known as *subroutines*. The term macro is short for macro-instruction; so, while the command *Run MyMacro* is considered a single instruction it probably invokes a large number of instructions. This makes macros extremely useful for repetitive operations, but in the case of VBA in Excel, we can extend the use to other purposes.

Macros can be either recorded or coded. A recorded macro can be fine-tuned by editing. Alternatively, one can record a short macro to learn the syntax of a particular operation and paste the result into a larger subroutine.

The words *macro* and *subroutine* are almost synonymous. If you record a macro called *Alpha*, VBA will create a subroutine with the same name. If you use the command *View Macros* you will see *Alpha* within the macro list. Perhaps the subroutine *Alpha* calls subroutines *Beta* and *Gamma*, and function *Kappa*. Then the entire group constitutes the macro *Alpha*. Generally, these would be kept on a single module sheet.

Liengme's Guide to Excel 2016 for Scientists and Engineers. https://doi.org/10.1016/B978-0-12-818249-9.00010-8

There is a special class of macros called *event macros* that are associated with either a specific sheet or with the entire workbook. They are automatically executed whenever a specific event occurs. For example, there could be a before-close macro, which saves the workbook whenever the user attempts to close it. We shall not be investigating these further.

EXERCISE 1: RECORDING A MACRO

For this exercise, our scenario is as follows: Each day you receive a file similar to Fig. 10.1. You wish to copy and paste this to another workbook and perform some operations on it. For the exercise, we shall add a sorting macro, which can be called with Ctrl + ⇧ Shift + Q. This is a very simple operation, but it will demonstrate the procedures needed to record a macro.

(a) Download the file Chap10Data.xlsx from the companion website. Copy and paste everything to Sheet1 of a new workbook and save it as a macro-enabled workbook named Chap10.xlsm.

(b) Begin the macro recorder using *Developer / Code / Macro Recorder* (*Developer / Record Macro* on a mac). Complete the record macro dialog as shown in Fig. 10.2 and click OK. The shortcut key label will change from *Ctrl +* to *Ctrl + Shift +* as you type in the box (on a Mac it is *Option + Cmd +* for a macro).

Observe in the *Code* group that the *Record Macro* command has changed to *Stop Recording* with a blue rectangular icon and a similar icon is displayed next to *Ready* in the status bar. Only actions are recorded, so there is no merit in rushing the process. To have a macro recording button always visible on the status bar, on a PC right-click (on a Mac hold the Ctrl key and click) the status bar and click the appropriate item to give it a check mark.

> The macros here have been kept fairly simple, so the reader is advised to have only the Chap10.xlsm workbook open when running them. Otherwise, you may have unexplained results.
>
> There is no room to explain every feature of the VBA used, but the reader is encouraged to read through the code and work out what each statement does.

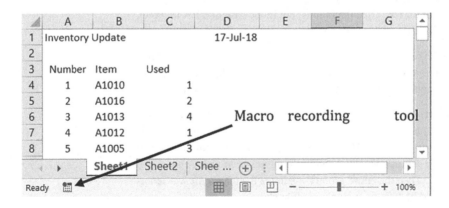

■ FIG. 10.1

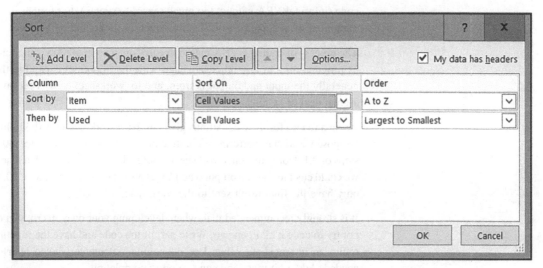

Record Macro ? X

Macro name:

SortData

Shortcut key:

Ctrl+Shift+ Q

Store macro in:

This Workbook ▼

Description:

Sort incoming inventory data

OK   Cancel

■ FIG. 10.2

Sort ? X

⁺ₐↃ Add Level   ✕ Delete Level   🖺 Copy Level   ▲ ▼   Options...   ☑ My data has headers

Column			Sort On		Order	
Sort by	Item	∨	Cell Values	∨	A to Z	∨
Then by	Used	∨	Cell Values	∨	Largest to Smallest	∨

OK   Cancel

■ FIG. 10.3

(c) Click anywhere within the data (for example cell A6) and use the command Data / Sort & Filtering / Sort (Data / Sort on a Mac). Fill in the sort dialog as shown in Fig. 10.3 and click OK. We sort first by the field Item, then by the field Used.

(d) If you look at the data, you will see that it is now sorted. We can turn the recording off either by opening the *Developer* tab to use the *Stop Recording* command or by clicking the recording icon in the status bar.

(e) Use *Developer / Code / Visual Basic (Developer / Visual Basic* on a Mac) and open the module for your file. We will not investigate every line in this code, but the meaning of some of it should be fairly obvious. Now we will show how the macro can be used in production.

(f) Delete all the data on Sheet1 and recopy the original data from Chap10Data.xlsx.

(g) Using either Ctrl + ⇧ Shift + Q or *View / Macros / View Macros (View / View Macros* on a Mac) run the macro *DataSort*. The data gets sorted.

(h) Save the workbook as Chap10.xlsm using the command *File / Save As* and choose the option *Excel Macro-Enabled*.

SUBROUTINES THAT COMPUTE

Before coding a calculation macro you should spend time planning its algorithm—a specific set of instructions for solving a problem. Some programmers like to do this with flow charts, while others prefer writing pseudo code (informal VBA without too much regard to syntax). For further information on these topics, consult an introductory programming book.

The process of coding a VBA macro to carry out a computation is the same as programming in any other language with the important exception that generally the input and output is from, and to, worksheet cells. This saves a great deal of coding.

Using a macro for the calculations gives us extra work but adds flexibility. Suppose we wish to perform an iterative process varying h from 1 to 100 in steps of 0.1. Doing this on a worksheet would take 1000 cells. With a macro we could elect to have an output when h was an integer or a multiple of 10, or only have the final result sent to the worksheet.

It is strongly recommended that, when developing your own macro, you do not try to code it all in one go. Write part of the code and have the function return an intermediate value. Use debugging statements (shown later) if needed. Add a bit more code and so on to completion.

NOTES ON THE VB EDITOR

1. If you run code with an error that causes the program to "hang," the offending line will be highlighted in the module. You will not be able to do anything on the worksheet while this condition lasts. The module can be reset within VBE with the command *Run/ Reset* or with the reset tool.

Try to familiarize yourself with the important commands on the VBE standard toolbar as shown in the following table.

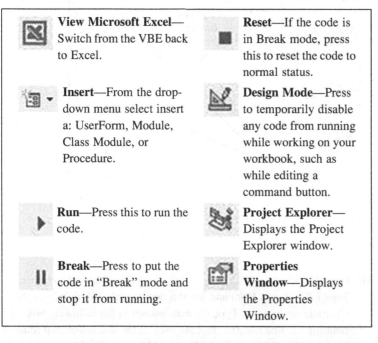

View Microsoft Excel—Switch from the VBE back to Excel.	**Reset**—If the code is in Break mode, press this to reset the code to normal status.
Insert—From the drop-down menu select insert a: UserForm, Module, Class Module, or Procedure.	**Design Mode**—Press to temporarily disable any code from running while working on your workbook, such as while editing a command button.
Run—Press this to run the code.	**Project Explorer**—Displays the Project Explorer window.
Break—Press to put the code in "Break" mode and stop it from running.	**Properties Window**—Displays the Properties Window.

2. If your subroutine goes into an infinite loop, use [Ctrl] + [Break] or [Esc] to terminate it. You can do this in Excel or in VBE. There is also a tool on the VBE toolbar to break a macro.

3. From the Excel window, a macro may be run using either *Developer / Code / Macros (Developer / Macros* on a Mac*)* or *View / Macros / View Macros (View / View Macros* on a Mac*)* to open a dialog box that lists all the available macros. You may also start macros from within VBE by (i) Using the *Run Sub* command in the *Run* menu, (ii) pressing [F5], or (iii) using one of the tools in the table before. You can also tap [F8] and run a macro one line at a time. Always ensure the current worksheet is active before using these tools.

EXERCISE 2: A COMPUTING MACRO

For our demonstration, we shall re-do the calculation of Problem 3 in Chapter 9—finding the resultant of two force vectors. There is no real advantage in using a macro here, but we want to start with an example that is not too complex. The simple algorithm is shown in the sidebar.

(a) Prepare a worksheet on Sheet2 of Chap10.xlsm as in Fig. 10.4, leaving B6 and C6 blank. The chart is optional.

> Simple algorithm for this macro:
> Get input data for worksheet.
> Compute Fx and Fy.
> Compute Fr and θr.
> Put values on worksheet.

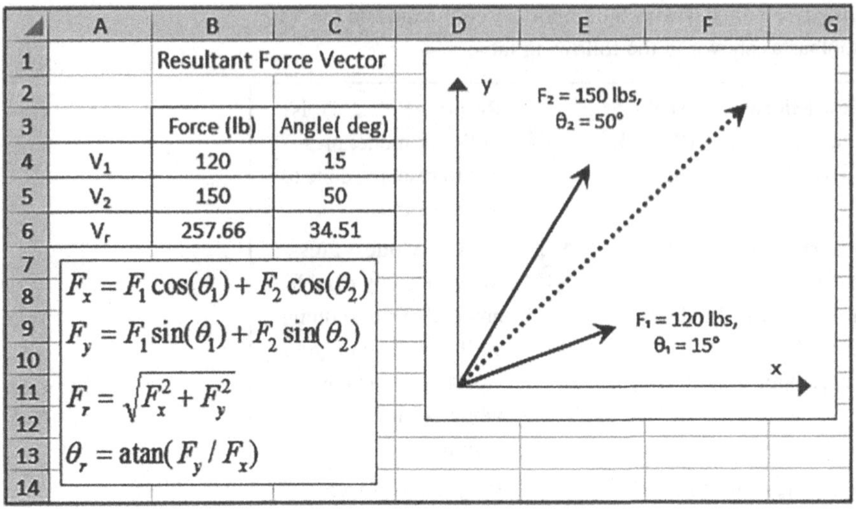
Resultant Force Vector
Force (lb)
Angle(deg)
V1
120
15
V2
150
50
Vr
257.66
34.51
$F_x = F_1 \cos(\theta_1) + F_2 \cos(\theta_2)$
$F_y = F_1 \sin(\theta_1) + F_2 \sin(\theta_2)$
$F_r = \sqrt{F_x^2 + F_y^2}$
$\theta_r = \text{atan}(F_y / F_x)$
y
F_2 = 150 lbs,
θ_2 = 50°
F_1 = 120 lbs,
θ_1 = 15°
x

■ FIG. 10.4

(b) Open the VBE using *Developer / Code / Visual Basic* (*Developer / Visual Basic* on a Mac) and insert a new module (Module2) on the Chap10.xlsm project. Type the code shown in the following box omitting the line numbers that are used in the discussion that follows.

(c) Return to Excel. Use *View / Macros / Macros / ViewMacros* (*View / ViewMacros* on a Mac) and run the macro *ResultantForceVector*. If you do not get the correct results, lines 17 and 22 give examples of how to get debugging information; just remove the leading apostrophe so they are no longer comments. Save the workbook.

```
1.  Option Explicit
2.  'To compute the resultant of two vectors
3.  Public Sub ResultantForceVector()
4.  Dim Answer, Pi
5.  Dim Force1, Force2, Theta1, Theta2, ForceX, ForceY, ForceR, ThetaR
6.  Range("B6").Select
7.  Const MyQuestion = "Is B6 where the resultant forces go?"
8.  Answer = MsgBox(MyQuestion, vbYesNo + vbQuestion, "Vectors")
9.   If Answer <> 6 Then Exit Sub
10.
11.     Pi = 4 * Atn(1)          'VBA calculation for Pi, 1 is in radians
12.
13.     Force1 = Range("B4").Value
14.     Theta1 = Range("C4").Value * Pi / 180 'convert degrees to radians
15.     Force2 = Range("B5").Value
```

```
16.        Theta2 = Range("C5").Value * Pi / 180
17.        'Debug.Print Force1; Theta1; Force2; Theta2
18.
19.        'calculate ForceX and ForceY
20.        ForceX = Force1 * Cos(Theta1) + Force2 * Cos(Theta2)
21.        ForceY = Force1 * Sin(Theta1) + Force2 * Sin(Theta2)
22.        'MsgBox "ForceX " & ForceX & " ForceY" & ForceY
23.
24.        'calculate R
25.        ForceR = Sqr(ForceX ^ 2 + ForceY ^ 2)
26.
27.        'calculate Theta
28.        ThetaR = Atn(ForceY / ForceX) * 180 / Pi 'convert radians to degrees
29.
30.        'place results in cells
31.        Cells(6, 2).Value = Round(ForceR, 2)
32.        Cells(6, 3).Value = Round(ThetaR, 2)
33.    End Sub
```

Comments on the code:

i. Lines 1–5: These are discussed just below the exercise.
ii. Lines 6–9: We have made no provisions to ensure the macro is run only from Sheet2. This code gives the user a chance to back out of the macro.
iii. Line 11: Defining π this way is superior to entering a value.
iv. Lines 13–17: This is where the program gets its input values from the worksheet. Note the use of meaningful names. Line 17 (with the apostrophe removed) is an example of debugging code.
v. Lines 19–22: Here we do the intermediate calculations. Line 22 is an interim output and can be used for debugging.
vi. Lines 24 and 28: We compute the final results.
vii. All the .Value phrases are optional, so code such as Force1=Range ("B4") would work. However, most VBA experts recommend the use of Value.
viii. IMPORTANT: If the values in B4:C5 are changed, the resultant information is not updated until the macro is rerun. This will be a good example of where a "change event" macro could be used to display a message whenever the input data was altered.

In a hurry? As an alternative to typing, download the file *Chap10Vector.bas* from the companion website and, in the VBE, use the command *File / Import* to load the macro.

Public or Private? The header of our macro begins with the word *Public*. This means the macro can be seen from any sheet in any open workbook.

The word is optional since it is the default setting. If *Public* is replaced by *Private*, the macro is no longer available through the *View* tab, but it may be run from a control placed on the worksheet. In Exercise 5 we see how this may be accomplished.

NAME THAT VARIABLE

We have used names like *Force1*, *ForceX* rather than *F1* and *FX*. The additional typing will be well invested. The meaningful names let you think about the algorithm. If you or someone else revisits the code months from now, the long names will aid in understanding what is going on. You may not find this argument convincing with such a simple program, but with longer programs, it becomes more important.

Simple variables like *j, k, n* for counters are fine (avoid the letters i, I, and l) as they can be mistaken for the digit 1. The names *x* and *y* when working with Cartesian coordinates are also acceptable.

We have used *Option Explicit* and hence need the Dim statements. This was discussed in the previous chapter. As the code gets longer, typo errors become more difficult to spot, so while this feature was not essential with UDFs, with subroutines it is indispensable.

Remember from the previous chapter that some words have special meanings in VBA. These keywords cannot be used for naming variables. So, although *Count* is not actually a keyword, the cautious programmer should use *Kount* or *MyCount*.

EXERCISE 3: BOLT HOLE POSITIONS

A milling machine is set up to drill holes in a circular pattern. The variable *Bcount* gives the number of holes while *Diam* is the diameter of the drilled circle. The first hole is drilled at an angle *Alpha* from a reference mark on the circle. Our task is to compute for each hole, the *x, y* coordinate values where 0, 0 are the coordinates of the circle's center. These values will be fed into the milling machine. The relevant equations are as follows:

$$\theta = n \cdot \frac{360}{Bcount} + Alpha$$

$$x = \frac{1}{2}Diam\cos\theta; y = \frac{1}{2}Diam\sin\theta$$

(a) On Sheet 3 of Chap10.xlsm, copy the entries from Fig. 10.5 but leave rows 8 to 15 blank.

(b) The flow chart for our macro is shown as follows. Open VBE, add a third module, and enter the subroutine shown. It is also available on the companion website as Chap10Bolt.bas.

(c) Run the macro a few times with various settings of the three input parameters.

(d) Save the workbook.

In line 8 we take care to open the correct worksheet. Then we delete any existing output. Recording a macro helped the author recall the syntax for lines 10 and 11. Input values are taken from the worksheet in lines 14–17. Then we have a FOR loop that iterates *Bcount* times calculating and outputting values of x and y to the worksheet. Note the use of Cells to place data on varying rows. The syntax is *Cells(row, column)*.

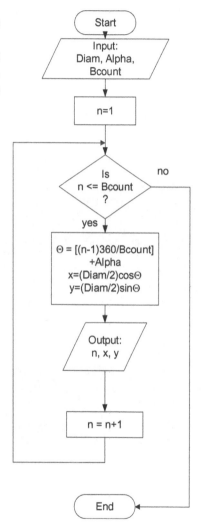

▲	A	B	C	D
1	Bolt circle hole calculation			
2				
3	Bolt hole diameter	6		inch
4	Offset angle	22		degree
5	Number of bolt holes	7		
6				
7	Bolt hole #	x [in.]	y [in.]	
8	1	2.78	1.12	
9	2	0.86	2.88	
10	3	-1.71	2.46	
11	4	-2.99	0.19	
12	5	-2.02	-2.22	
13	6	0.48	-2.96	
14	7	2.61	-1.47	
15				

■ FIG. 10.5

```
1    Option Explicit
2    Public Sub BoltHoleCircle()
3    Dim Pi, Diam, Alpha, Bcount, Theta, x, y, n
4    'declare constants
5       Pi = 4 * Atn(1)
6
7    'open Sheet3 and clear old data
8       Sheets("Sheet3").Select
9       Range("A8:C8").Select
10      Range(Selection, Selection.End(xlDown)).Select
11      Selection.ClearContents
12      Range("A8").Select
13
14   'input dia, alpha, bcount
15      Diam = Range("C3").Value
16      Alpha = Range("C4").Value
17      Bcount = Range("C5").Value
18
19   'Do calculations
20      For n = 1 To Bcount
21         Theta = ((n - 1) * 360 / Bcount) + Alpha
22         Theta = Theta * Pi / 180 'convert to radians
23         x = Round(Diam / 2 * Cos(Theta), 2)
24         y = Round(Diam / 2 * Sin(Theta), 2)
25      'output n, x, y
26         Cells(7 + n, 1).Value = n
27         Cells(7 + n, 2).Value = x
28         Cells(7 + n, 3).Value = y
29      Next n
30   End Sub
```

EXERCISE 4: FINDING ROOTS BY BISECTION

We next demonstrate a VBA implementation of the bisection method. The main subroutine *Bisection* calls a UDF named *MyFunction* to evaluate the function to be solved at specified x values. This will give us the flexibility of being able to recode only *MyFunction* when we wish to solve another equation.

A brief explanation of how the bisection method works: The equation to be solved is $f(x) = 0$. By trial and error, we have found that values of $f(a)$ and $f(b)$ have opposite signs. Clearly, the solution to our problem, $f(x) = 0$, lies in the interval $a < 0 < b$. We now find the midpoint ("bisect") of this interval defined by $m = (a + b)/2$ and evaluate $f(m)$. If $f(m)$ has the same sign as $f(a)$, then the root lies in $m < x < b$ as illustrated on the left of Fig. 10.6.

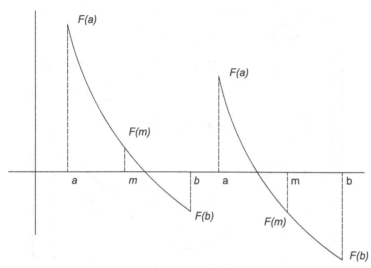

■ FIG. 10.6

Alternatively, if $f(m)$ and $f(b)$ have the same sign, the root lies in $m < x < b$ (see the right side of the figure). We can now bisect either m and b or m and a. In this way, we get a smaller interval. We may repeat this until we are close enough to zero to satisfy our precision needs.

This allows us to develop an algorithm for finding a root of $f(x)$:

    Start with values of a and b such that $f(a)$ and $f(b)$ have opposite signs
    Loop until the required accuracy is achieved
        Find the midpoint $m = (a + b)/2$
        If $f(m)$ and $f(b)$ have opposite signs
                Give a the value of m
        Else
                Give b the value of m
        End if
    End loop.

We will solve the transcendental equation $\exp(x) - \sin(x) = 0$. The graph of this (Fig. 10.7) shows there is one root between 0 and 1, and another between 2.5 and 3.5. With starting values of $a = 0$, $b = 1.0$ we might expect the algorithm to converge to the root near 0.5, likewise starting values of $a = -0.5$, $b = 4.0$ should give the root near 3.

(a) On Sheet 4 of Chap10.xlsm, set up a worksheet similar to that in Fig. 10.7 behind the chart. You may wish to experiment using worksheet functions to implement the Bisection algorithm before moving on to the

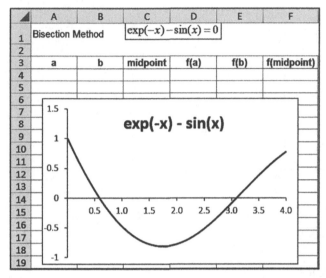

	A	B	C	D	E	F
1	Bisection Method		$\exp(-x) - \sin(x) = 0$			
2						
3	a	b	midpoint	f(a)	f(b)	f(midpoint)
4						
5						

■ FIG. 10.7

VBA coding. The formula =SIGN(A1) returns the value -1, 0, or $+1$ depending on the value in A1 (negative, zero, or positive).

(b) On the Chap10.xlsm project in the VBE, add Module 4. Enter the code shown as follows. A statement ending with a space followed by an underscore is continued on the next line. It may be entered as shown or as a single line without the underscore. The code is available on the companion website as Chap10Bisect.bas.

Experiment with various input values. Then modify the function *Kappa* to solve

$$f(x) = 4^x + 5^x - 100,$$

which has a root between 0 and 3. Make a plot to confirm this. Remember to save the workbook.

```
Option Explicit

Function Kappa(x) As Single
    Kappa = Exp(-x) - Sin(x)
End Function

Public Sub BisectionMethod()
    'declare variable types
    Dim a As Single, b As Single, midpoint As Single
    Dim f_a As Single, f_b As Single, f_m As Single
    Dim Kount As Integer, MyRow As Integer
```

```
' Select worksheet and clear old data
    Sheets("Sheet4").Select: Range("A4:F4").Select
    Range(Selection, Selection.End(xlDown)).Select
    Selection.ClearContents
    Range("A4").Select

 ' get starting values a and b
    a = InputBox("For f(a) give value of {a}", _
                           "Bisection Method")
    b = InputBox("For f(b) give value of {b}", _
                           "Bisection Method")
    f_m = Kappa((a + b) / 2)

' test that f(a) and F(b) has opposite signs
    If Sgn(Kappa(a)) = Sgn(Kappa(b)) Then
      MsgBox "f(a) and f(b) have the same sign"
      Exit Sub
    End If

'initialize Kount
    Kount = 1
```
```
'the algorithm, precision set to ± 1.0E-6
  Do While Abs(f_m) > 0.000001
    midpoint = (a + b) / 2
    f_a = Kappa(a)
    f_b = Kappa(b)
    f_m = Kappa(midpoint)

'output a, b, midpoint, f_a, f_b, f_m
    Cells(Kount + 3, 1) = a
    Cells(Kount + 3, 2) = b
    Cells(Kount + 3, 3) = midpoint
    Cells(Kount + 3, 4) = f_a
    Cells(Kount + 3, 5) = f_b
    Cells(Kount + 3, 6) = f_m

 'Compare sign of a with sign of f(b)
    If Sgn(f_m) <> Sgn(f_b) Then
      a = midpoint
    Else
      b = midpoint
    End If

    Kount = Kount + 1
 'Check that number of interations
```

```
    If Kount >= 50 Then
      MsgBox "Not converging"
      Exit Sub
    End If
Loop
MsgBox "Solution at x = " & _
    midpoint & " when f(x) = " & f_m
End Sub
```

EXERCISE 5: USING ARRAYS

Variable arrays can be used to organize the way input/output data is stored. The worksheet in Fig. 10.9 represents a table of soil contamination levels of carbon tetrachloride. In this exercise, the variable array for each soil contamination level is named *SCL*. We declare this as SCL(20) using the Dim command—see the following code. The number 20 indicates the size of the array. We can refer to SCL(1) through SCL(20). An integer variable *j* will be used in a Do Loop to input all the SCL values until it lands on a blank cell. We have chosen to have the variable array *Exceed* store the index (position) of SCL values greater than 0.01 rather than the actual values. This is a useful technique with large arrays since the integer value of the index takes little storage room.

```
Option Explicit
Public Sub SoilContamination()
'declare all variables
Dim SCL(20), Exceed(10)
Dim j, h, high, Answer

'initialize counting variables
    j = 1; h = 0

'start Do Loop; test for a value in the cell
    Sheets("Sheet5").Select
    Range("B3").Select
    Do While ActiveCell.Value > 0
    'input Soil Contamination Level
        SCL(j) = ActiveCell.Value
        If SCL(j) > 0.01 Then ' test for level
            h = h + 1
            Exceed(h) = j
            ActiveCell.Font.Bold = True
        End If
```

```
                'move down 1 row
                ActiveCell.Offset(1, 0).Activate
                j = j + 1
        Loop
'output h (number of occurrences _exceeding 0.01)
        Range("D3").Select
        ActiveCell.Offset(-1, 1).Value = _
         "There are " & h & " samples _
         with CTC > 0.01 ppm"
        For j = 1 To h
                'output SCL's that exceed 0.01
                high = Exceed(j)
                ActiveCell.Value = SCL(high)
                ActiveCell.Offset(0, 1) = "ppm"
                'move down 1 row
                ActiveCell.Offset(1, 0).Activate
        Next j

        Const MyQuest = _
                "Ready to reset the worksheet?"
        Const MyText = "Soil Contamination"
        Answer = MsgBox(MyQuest, vbYesNo _
                + vbQuestion, MyText)
    If Answer = 6 Then Clearworksheet
End Sub

Sub Clearworksheet()
        Range("B3").Select
        Do While ActiveCell.Value > 0
            ActiveCell.Font.Bold = False
            ActiveCell.Offset(1, 0).Activate
    Loop
    Range("E2").Clear
    Range("D3").Select
    Do While ActiveCell.Value > 0
        ActiveCell.Clear
        ActiveCell.Offset(0, 1).Clear
        ActiveCell.Offset(1, 0).Activate
    Loop
End Sub
```

◢	A	B	C	D	E	F	G	H
1	Soil Contamination Levels							
2		Data		Results	There are 3 samples with CTC > 0.01 ppm			
3		**0.021**		0.021	ppm			
4		0.010		0.020	ppm	Run Macro		
5		0.001		0.015	ppm			
6		0.005						
7		0.010						
8		**0.020**						
9		0.002						
10		0.001						
11		**0.015**						
12		0.010						

■ FIG. 10.8

(a) Open Chap10.xlsm and on Sheet5 construct the worksheet shown in Fig. 10.8. Ignore the bold format and everything after column C except the text in D2. The other entries come from the macro, and the box will be discussed in the next exercise.

(b) On Module 5 in VBE, code the macro consisting of the two subroutines *SoilContamination* and Clearworksheet shown before.

(c) Run *SoilContamination* and experiment with different data.

Suppose you have an array defined by MyArray(4,5) and the array elements get values through some calculation. Now you want to display these on the worksheet. This code will serve that purpose:

```
Range("A10").Select
ActiveCell.Resize(4,5).Value=MyArray
```

EXERCISE 6: ADDING A CONTROL

We have been running macros using from Excel with *Developer / Code / Macros* (*Developer / Macros* on a Mac) or with *View / Macros / View Macros* (*View / View Macros* on a Mac), or from VBE with either the Run command or using a tool.

Another way is to add a control or a shape to the worksheet. This is very convenient and ensures you are running the macro that matches the active sheet. It is also an excellent way to have another user, perhaps one less familiar with Excel, run a macro in a workbook you have developed.

■ FIG. 10.9

(a) Open Sheet5 of Chap10.xlsm. Use *Developer / Insert* and click the first tool in the *Forms* group; see Fig. 10.9. (Use *Developer / Button* on a

Mac) The mouse pointer turns to a cross (+); drag the mouse to outline a small rectangle about one column wide by two rows deep.

(b) When you release the mouse button, the Assign Macro dialog opens, enabling you to select which macro the control will run; SoilContamination is the appropriate one here.

(c) Right-click the button and you are presented with various options. Use *Edit Text* to customize the control.

(d) Now when you click the control, the associated macro will run. Save the workbook.

The alternative way is to use *Insert / Illustrations / Shapes (Insert / Shape / ... on a Mac)* to add the workbook, on a PC right-click (on a Mac hold the (Ctrl) key and click) on the shape and select *Assign Macro* and select the appropriate subroutine for the resulting dialog box.

EXERCISE 7: USER FORMS (NOT AVAILABLE ON MAC)

User forms provide a data input or output interface. They can also be used to launch programs that input and output on the worksheet. A surveyor takes three measurements: distance (d) from the building, the angle of elevation (θ) of theodolite, and height of theodolite (h_1) above ground level; see Fig. 10.10. The height of the building is given by $h = d\tan(\theta) + h_1$. The user form in Fig. 10.11 is associated with a subroutine that calculates the height of a building.

(a) We begin the exercise by designing the User Form. Learning to make a User Form is like learning to drive a car; someone must get you started, telling you what all the knobs are about, but you have to practice and learn how to use them smoothly. What follows tells you the very basic information; refer to Figs. 10.12 and 10.13 as you read.

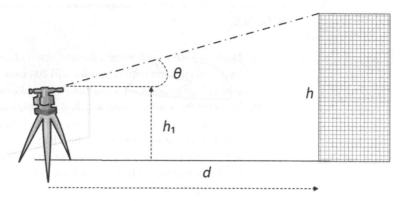

■ FIG. 10.10

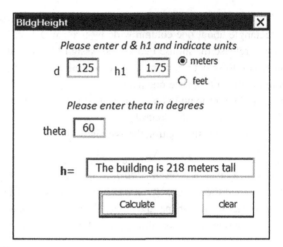

■ FIG. 10.11

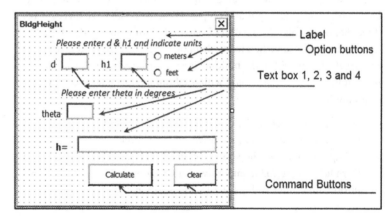

■ FIG. 10.12

i. Begin in VBE by using the command *Insert / UserForm*. Now you have a blank form onto which we will add controls such as Text labels, Text boxes, Command Buttons, and Option Buttons.

ii. We need the Toolbox from which to select the controls: have this displayed by using View / Toolbox or by clicking the crossed hammer-&-wrench icon next to the Help icon. If the mouse is allowed to hover over a tool, a screen tip displays its purpose. Do not worry that the controls do not look quite right as you add them; we fix that in step (iv).

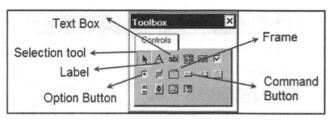

■ FIG. 10.13

    iii. Begin by dragging the Label tool (A) onto the form and typing the text `Please enter d & h1 and indicate units`. The Selection tool (arrow) lets you resize and position controls. Add the second label.

    iv. With a group of options buttons, we want to be able to select and have the others automatically be deselected. To do this we need first to add a Frame to the form and then place the Option Buttons inside it.

    v. Finally, add the Text Boxes using the tool labeled ab.

    vi. Now you may want to change the appearance of some controls. Click each one (just once!) in turn and look in the Properties box; see Fig. 10.14. We use this to alter such things as labels (e.g., take the frame label off), size, color, font, and so on. Make sure you click the User Form area, and in Properties change its name to `BldgHeight`.

(b) Now we need to write some code for the command buttons telling VBA what to do when either one is pressed. Double click each command button in turn and add the subroutines shown as follows. These go together in one module. Command Button 1 is the Calculate button.

(c) In VBE, add Module 6 and this simple subroutine:

```
Sub get_UserFormBldgHeight()
  BldgHeight.Show
End Sub
```

(d) Finally, following the instruction in Exercise 5, add a command button to Sheet 6 of Chap10.xlm and associate it with the subroutine `get_UserFormBldgHeight`. Now you are ready to experiment with the project we have just completed.

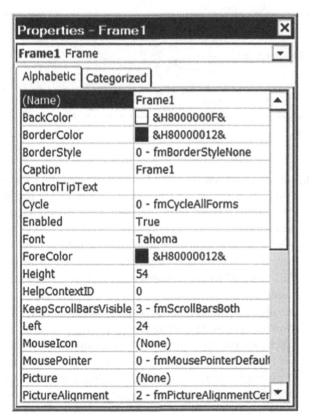

■ **FIG. 10.14**

```
Option Explicit

Public Sub CommandButton1_Click()

'declare variables
  Dim d As Single, theta As Single
  Dim h1 As Single, h As Single
  Dim Pi As Single, h_sigfig As Single
  Dim units As String
' declare constant
  Pi = 4 * Atn(1)
'test for units
  If OptionButton1 = True Then
    units = " meters"
  Else
    If OptionButton2 = True Then
      units = " feet"
```

```
        Else
              units = "please enter d, h1 units"
        End If
   End If

'input d, theta, & h1
   d = Val(TextBox1.Value)
   theta = Val(TextBox2.Value)
   h1 = Val(TextBox3.Value)
'calculate building height
   h = d * Tan(theta * Pi / 180) + h1

   'round off h
   h_sigfig = Round(h, 0)
'output h
   If OptionButton1 = True Or _
      OptionButton2 = True Then
      TextBox4.Value = "The building is " & _
                   h_sigfig & units & " tall"
   Else
       'indicate error
      TextBox4.Value = units
   End If

End Sub

Private Sub CommandButton2_Click()
'Clear the contents of the text boxes
   TextBox1.Text = " "
   TextBox2.Text = " "
   TextBox3.Text = " "
   TextBox4.Text = " "
   OptionButton1 = False
   OptionButton2 = False
End Sub
```

PROBLEMS

1. Use the program in Exercise 2 and make the code work so that the resultant angle will display properly in each of the four quadrants (360 degrees). Extend the program to find the resultant of three vectors.
2. *Write a computer program to calculate the force of drag of fluid flowing over a cylindrically shaped object using the following formula

$$Fd = \frac{\frac{1}{2} \cdot Cd \cdot \rho \cdot v^2 \cdot A}{gc}$$

where Fd is the force of drag, Cd is the drag coefficient, ρ is the density of the fluid, v is the velocity, A is the frontal area of the object, and gc is the gravity constant. The program should work for air flowing over an object shaped as a convex $Cd = 1.2$ or concave $Cd = 2.3$.

The program should be a VBA subroutine with input and output in the Excel worksheet. The program should work in US units. Use as input; velocity in mph, shape of the end (concave or convex), length in ft, and radius in inches, with the output in lbf. Incorporate a *Select Case* statement to handle the shape (concave or convex).

3. *Use the program in Exercise 5 so that it works for four columns of input in the worksheet.

Student1	Student2	Student3	Student4
45	80	60	100
75	58	37	85
90	59.5		50
80			

4. Write a computer program that grades "pass" or" fail" for the following numeric grades:
   Where 60 and above is passing. Replace the number with either *Pass* or *Fail*. Your program should work with any reasonable number of columns or rows.

5. Review Problem 2 in Chapter 2. Write or record a macro that will clear all the input cells on the worksheet when the user clicks a control button.

6. You have a worksheet with number in a range starting in A2. The size of the range is between 10 and 500. Each cell holds a value that is between 10 and 99, inclusive. So we might have 50, 45, 65, 60, 70, 75. You want them sorted in an odd way: all the numbers ending with 0 first, then all those ending with 1, and so on. The numbers are to be sorted in increasing order within each group. So the example would give 50, 60, 70, 45, 65, 75. Write a subroutine that takes the numbers in column A and generates the sorted list in column B. There are many sorting algorithms; pick the one that is easiest to code.

7. Review Problem 8 of Chapter 5. Construct a worksheet named `Recycle` with rows 1 to 8 as in Fig. 10.15. Add a button. Using the following code as a start, write a subroutine to generate the data in row 9 and down.

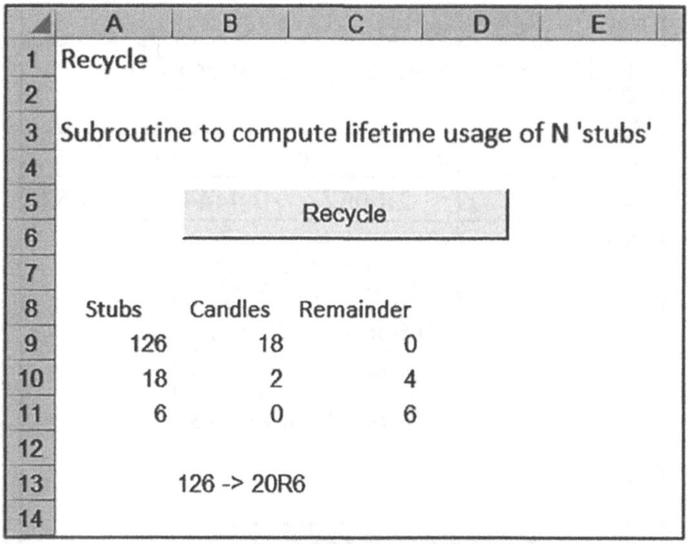

	A	B	C	D	E
1	Recycle				
2					
3	Subroutine to compute lifetime usage of N 'stubs'				
4					
5		Recycle			
6					
7					
8	Stubs	Candles	Remainder		
9	126	18	0		
10	18	2	4		
11	6	0	6		
12					
13		126 -> 20R6			
14					

■ FIG. 10.15

```
Sub Recycler()
 Worksheets("Recycle").Select
 Cells(9, 1).Select
 Range(Selection, Selection.End(xlDown)).Select
 Range(Selection, Selection.End(xlToRight)).Select
 Selection.ClearContents
 Cells(9, 1).Select
 stubs = InputBox(Prompt:="How many stubs?")
 If stubs = "" Then Exit Sub
   stubs = Val(stubs) 'convert text to number
 .........
End Sub
```

8. Modify the subroutine in Problem 7 to display a user-form that requests both the number of stubs to be recycled and how many are needed to make a new candle.

9. An engineer needs to solve cubic equations hourly. Make a worksheet to help with this—see Fig. 10.16. Use Newton's method to get the roots. She will enter the coefficients into A4:D4, click on the shape and have a subroutine enter the real roots in B7:B9. But wait! We need to give the

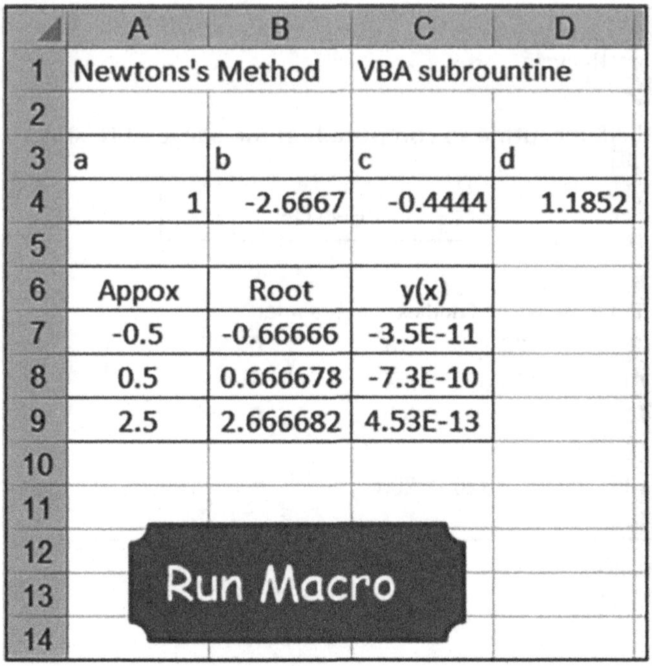

◢	A	B	C	D
1	Newtons's Method		VBA subrountine	
2				
3	a	b	c	d
4	1	-2.6667	-0.4444	1.1852
5				
6	Appox	Root	y(x)	
7	-0.5	-0.66666	-3.5E-11	
8	0.5	0.666678	-7.3E-10	
9	2.5	2.666682	4.53E-13	
10				
11				
12				
13		Run Macro		
14				

■ **FIG. 10.16**

subroutine some starting values. Approximate roots can be read off a plot made with a graphing calculator.[1] These are entered into A7:A9 before running the macro. The values in C7:C9 let us know how good the roots are.

APPENDIX: SUPPLEMENTARY MATERIAL

Supplementary material related to this chapter can be found on the accompanying CD or online at https://doi.org/10.1016/B978-0-12-818249-9.00010-8.

---

[1]Or a software simulation of a graphing calculator such as Microsoft Mathematics 4.0 available at no cost from http://www.microsoft.com/en-ca/download/details.aspx?id=15702.

Chapter 11

Modeling I

CHAPTER OUTLINE

In the preceding chapters, we have concentrated on Excel features. In this chapter, we are more concerned with getting answers to problems. It is very likely that some of the topics will be outside your interest areas, but please read every exercise since a major objective of the chapter is to show how to lay out a worksheet. This is something with which many new users have difficulty.

EXERCISE 1: POPULATION MODEL

An ecological niche contains two species: the prey and the predator. A theoretical analysis of the problem has yielded equations for the successive population of the two species:

$$N_{t+1} = (1.0 - b(N_t - 100))N_t - kN_tP_t$$
$$P_{t+1} = qN_tP_t$$

where:

N_t = the population of prey in generation t

b = the net birth-rate factor

k = the kill-rate factor

Liengme's Guide to Excel 2016 for Scientists and Engineers. https://doi.org/10.1016/B978-0-12-818249-9.00011-X

269

P_t = the population of predators in generation t

q = efficiency in use of prey.

We wish to observe how this model predicts the changing populations and to examine the sensitivity of the model to the values of the parameters. Fig. 11.1 shows the worksheet we will make. The values for the five parameters were taken from an ecology textbook. The N and P values are the populations in a square kilometer.

(a) Open a new workbook. On Sheet 1 enter the data shown in rows 1 through 9 of Fig. 11.1. Format the worksheet as shown. It is sometimes useful to document the worksheet as we have in rows 3 and 4.

(b) Select A6:E7 and using *Formulas / Defines Names / Create from Selection (Formulas / Create from Selection* on a Mac) name the cells in row 7.

(c) Fill the range A10 to A110 with the series 0 through 100. One method is to type the first two numbers, select A10:A11 and drag the fill handle down to A110. Another is to enter 0 in A10 and then with A10 select use *Home / Editing / Fill (Home / box under Σ* on a Mac) and select Series. Fill the dialog box shown in Fig. 11.2.

(d) In B10 and C10 enter =No and =Po, respectively.

(e) The formulas in B11 and C11 for the next generation are =(1-b*(B10-100))*B10-k*B10*C10 and =q*B10*C10. Copy these down to row 110 by selecting them both and clicking C11's fill handle.

(f) Construct a chart similar to that shown in Fig. 11.1. This shows just 40 generations, but more should be included to observe the effect of altering the parameters.

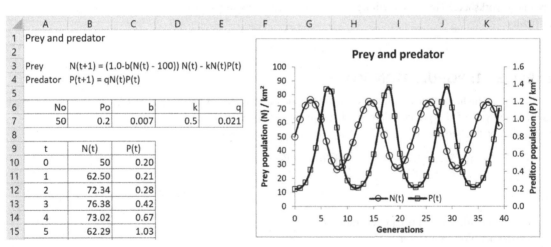

	A	B	C	D	E
1	Prey and predator				
2					
3	Prey	N(t+1) = (1.0-b(N(t) - 100)) N(t) - kN(t)P(t)			
4	Predator	P(t+1) = qN(t)P(t)			
5					
6		No	Po	b	k	q
7		50	0.2	0.007	0.5	0.021
8					
9	t	N(t)	P(t)		
10	0	50	0.20		
11	1	62.50	0.21		
12	2	72.34	0.28		
13	3	76.38	0.42		
14	4	73.02	0.67		
15	5	62.29	1.03		

■ **FIG. 11.1**

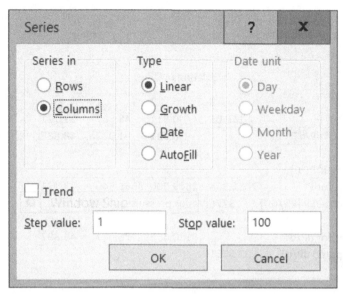

■ FIG. 11.2

(g) We can now experiment with the parameters. After observing the effect of changing one parameter, reenter its original value before changing the next.

  (i) Change the value of No to 60, 70, ...

 (ii) Change the value of Po to 0.25, 0.3, ...

(iii) Change the value of b to 0.0055, 0.0006, 0.00065, ...

(iv) Change the value of k to 0.25, 0.3, 0.19, 0.18, ...

Which parameters may be changed slightly while maintaining a stable state? Can you find one parameter for which a small change results in either a population explosion or an extinction?

(h) Save the workbook as Chap11.xlsx.

EXERCISE 2: VAPOR PRESSURE OF AMMONIA

We have a table[1] (A3:B12 in Fig. 11.3) showing the measured vapor pressure of ammonia at temperatures in the range 20 to 60°C. Our task is to estimate the vapor pressure at 75°C and to calculate the density of ammonia at

---

[1]This problem is from *Mathematical Methods in Chemical Engineering* by Jenson and Jeffreys, Academic Press (2000). The authors use the method of finite differences to get values: vapor pressure. = 3880 kN/m^2 and density = 29.0 kg/m^3.

	A	B	C	D	E	F	G	H
1	Ammonia: vapor pressure and density							
2								
3	Temp (°C)	Pressure (kN/m²)		Calculations				
4	20	805		Power	3	2	1	0
5	25	985		Coeff of P	7.14E-03	-0.511	48.887	-27.121
6	30	1170		Coeff of dP/dT		0.021	-1.022	48.887
7	35	1365						
8	40	1570		Specified T (°C)	75			
9	45	1790		Terms for P	3011.364	-2875.000	3666.540	-27.121
10	50	2030		Computed P (kN/m²)	3776	Vapor pressure at 75°C		
11	55	2300						
12	60	2610		Terms for dP/dT		120.455	-76.667	48.887
13				Computed dP/dT	92.6751			
14								
15				ΔH	1265	kJ/kg		
16				density ρ	25.5	kg/m³		

■ FIG. 11.3

this temperature from the Clausius-Clapeyron equation, given that the heat of vaporization (ΔH) is 1265 kJ/kg.

If we plot this data, we find that a third-order polynomial has an R^2 value of 1.00.

(a) On Sheet2 of Chap2.xlsx, enter all the text seen in Fig. 11.3 and the values in A4:B12.

(b) The values in E4:H4 are the powers for each LINEST term; enter these for use later. Select E5:H5 and enter =LINEST(B4:B12,A4:A13^{1,2,3}) as an array formula. Ignore F6:H6 for now.

In rows 8 through 10 of the Calculations area, we estimate the vapor pressure of ammonia at 75°C.

(c) Enter the specified temperature in E8. We now use the polynomial terms to compute each term in $aT^3 + bT^2 + cT + d$. The formula in E9 is =E5*$E$8^E4 and this is copied across to H9. In E10 we obtain the required vapor pressure by summing the terms with =ROUND(SUM(E9:H9),0). Note: this could be calculated directly using the trend function by =ROUND(TREND(B4:B12,A4:A12^{1,2,3},E8^{1,2,3}),0).

The Clausius-Clapeyron equation is shown below, where ΔH is the latent heat in kJ/kg, T is in Kelvins and V_L and V_V are the volumes in cubic meters of one kilogram of the liquid and vapor, respectively.

$$\frac{dP}{dT} = \frac{\Delta H}{T(V_V - V_L)}$$

We make the approximation that the volume of the liquid V_L is negligible compared to the volume of the vapor V_V. Then we rearrange the equation to give:

$$\rho = \frac{1}{V_V} = \frac{T}{\Delta H}\left(\frac{dP}{dT}\right)$$

We have a polynomial expression for P, which we may differentiate to get dP/dT. The first term in P is aT^3 giving $3aT^2$ as the first term in dP/dT, and so on. So the coefficient for the first term is $3a$; for the second it is $2b$, and so on.

(d) In F6 enter the first coefficient for dP/dT using =E4*E5. Copy this across to H6 to get the other terms.
(e) Now that we have the coefficients for dP/dT, we may compute each term in $3aT^2 + 2bT + c$. In F12 enter =F6*$F$8^F4 and copy across to H12. Sum these in E13 using =SUM(F12:H12) to give dP/dT.
(f) Enter the given value for ΔH in F15 and compute the density in E16 using the formula =((E8+273.15)/E15)*E13. Note the use of 273.15 to convert from degrees Celsius to Kelvin.
(g) Save the workbook.

EXERCISE 3: STRUCTURE MEMBER FORCE ANALYSIS

This Exercise will demonstrate a practical use of matrix algebra to solve a system of linear equations. Consider the structure represented by Fig. 11.4. We will assume the members of the structure are massless and are freely jointed one to another. Our task is to find the force in each member. For equilibrium, at each joint, the sum of the horizontal and the sum of the vertical components of all forces must be zero. From this, we develop the equations shown in the table below the diagram. Next, we need to translate this into an Excel worksheet.

We will assume every member is in tension so the forces act away from each joint. In the solution, a negative value will indicate compression.

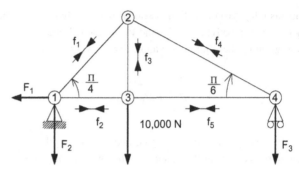

■ **FIG. 11.4**

Joint	Horizontal	Vertical
1	$-F_1 + \frac{\sqrt{2}}{2}f_1 + f_2 = 0$	$\frac{\sqrt{2}}{2}f_1 - F_2 = 0$
2	$-\frac{\sqrt{2}}{2}f_1 + \frac{\sqrt{3}}{2}f_4 = 0$	$-\frac{\sqrt{2}}{2}f_1 - f_3 - \frac{1}{2}f_4 = 0$
3	$-f_2 + f_5 = 0$	$f_3 = 10,000$
4	$-\frac{\sqrt{3}}{2}f_4 - f_5 = 0$	$\frac{1}{2}f_4 - F_3 = 0$

(a) On Sheet3 of Chap11.xlsx enter the text shown in Fig. 11.5. Do not bother with the explanatory text boxes.

(b) In A4:H11 enter the coefficient for the equation as shown in the table using the formulas =SQRT(2)/2, =SQRT(3)/2, and so on, as appropriate.

(c) In J15:J22 enter the constants for each equation; all but one are zero.

(d) In A15:H22, use MINVERSE to compute the inverse of the coefficient matrix. And in K4:K11 use MMULT to multiply the inverse matrix and the constants matrix to give the solutions to the problem.

This looks like a simple problem, but the book from which it is adapted contained a simple mistake. The author of that book had the wrong sign before ½ in the equation for the vertical components of joint 2. The matrix algebra still gave an answer—the wrong answer! Always look to see if there is another piece of information that can be used to double check the result. In this case, we know that the sum of the external forces F_1 and F_2 must balance the 10,000 N mass.

(e) Write a formula in D26 that checks this. Use conditional formatting to color A26:E26 red if there is an imbalance; otherwise, have them colored green. Note we have merged cells in A26:C26 and D26:E26.

(f) Save the workbook.

◢	A	B	C	D	E	F	G	H	I	J	K
1	Structural Analysis										
2	F1	F2	F3	f1	f2	f3	f4	f5			
3			Matrix of equation coefficients							forces	Solution
4	-1	0	0	0.707	1	0	0	0		F1	0.00
5	0	-1	0	0.707	0	0	0	0		F2	-6339.75
6	0	0	0	-0.707	0	0	0.866	0		F3	-3660.25
7	0	0	0	-0.707	0	-1	-0.5	0		f1	-8965.75
8	0	0	0	0	-1	0	0	1		f2	6339.75
9	0	0	0	0	0	1	0	0		f3	10000.00
10	0	0	-1	0	0	0	0.5	0		f4	-7320.51
11	0	0	0	0	0	0	-0.866	-1		f5	6339.75
12											
13											
14				Inverse of Matrix						Constants	
15	-1.00	0.00	-1.00	0.00	-1.00	0.00	0.00	-1.00		0	
16	0.00	-1.00	-0.37	-0.63	0.00	-0.63	0.00	0.00		0	
17	0.00	0.00	0.37	-0.37	0.00	-0.37	-1.00	0.00		0	
18	0.00	0.00	-0.52	-0.90	0.00	-0.90	0.00	0.00	×	0	
19	0.00	0.00	-0.63	0.63	-1.00	0.63	0.00	-1.00		0	
20	0.00	0.00	0.00	0.00	0.00	1.00	0.00	0.00		10000	
21	0.00	0.00	0.73	-0.73	0.00	-0.73	0.00	0.00		0	
22	0.00	0.00	-0.63	0.63	0.00	0.63	0.00	-1.00		0	
23											
24		{=MINVERSE(A4:H11)}						{=MMULT(A15:H22,J15:J22)}			
25											
26		Double check		0.000000							

■ FIG. 11.5

EXERCISE 4: CIRCUIT ANALYSIS

In this exercise, we again use matrix algebra. Within Fig. 11.6 we have a circuit with three "meshes." We are asked to find the current in each mesh given that $V_s = 120\,V$, $R_1 = 2\Omega$, $R_2 = 5\Omega$, $R_3 = 2\Omega$, $R_4 = 4\Omega$, $R_5 = 15\Omega$, and $R_6 = 5\Omega$. We will assume the current in each mesh flows clockwise.

We shall apply Kirchhoff's voltage law, which states the algebraic sum of the voltages around a closed loop in a circuit is zero. This gives us three equations:

$$V_s - I_1 R_1 - (I_1 - I_2)R_2 - I_1 R_3 = 0$$
$$-(I_2 - I_3)R_4 - (I_2 - I_1)R_2 = 0$$
$$-I_3 R_5 - I_3 R_6 - (I_3 - I_2)R_4 = 0$$

The spreadsheet and circuit diagram for Kirchhoff's Law:

	A	B	C	D	E	F
1	Kirchhoff's Law					
2						
3	Vs	120 V				
4	R1	R2	R3	R4	R5	R6
5	2 Ω	5 Ω	2 Ω	4 Ω	15 Ω	5 Ω
6						

	I1	I2	I3	Constants	Solutions	
7		Coefficients				
8	I1	I2	I3			
9	9	-5	0	120	I1	20
10	-5	9	-4	0	I2	12
11	0	-4	24	0	I3	2

■ **FIG. 11.6**

Rearranging the equations in preparation for the matrix solution, we write:

$$(R_1 + R_2 + R_3)I_1 - R_2I_2 = V_s$$

$$-R_2I_1 + (R_2 + R_4)I_2 - R_4I_3 = 0$$

$$-R_4I_2 + (R_4 + R_5 + R_6)I_3 = 0$$

Does something appear to be missing? Where is the range holding the inverse matrix of the coefficients? Read on.

(a) Open Chap11.xlsx and on Sheet 4 begin the worksheet by copying from Fig. 11.6 all the text entries and all the values in columns A through F. The resistors have a custom format of 0 W and are in Symbol font to make the W into the Ω symbol. The voltage value has a custom format of 0 V to get the unit.

(b) Label the resistors using the labels above them. Use the resistance values to calculate the coefficients.

(c) Select F9:F11 and enter this array formula: =MMULT(MINVERSE(A9: C11),D9:D11). So that is where the matrix gets hidden—inside the multiplication! The author is not a great fan of nesting complex formulas when it is unnecessary; after all, there is lots of space on a worksheet. Also, nesting can be a source of error, but it is fairly safe in this example. Save the workbook.

EXERCISE 5: LADDER DOWN THE MINE

The purpose of this exercise is to show that some problems can be solved graphically, especially where limited precision is required. The ladder in the mine is a problem often presented in computing textbooks. Generally, this problem is used as an example of a maximization/minimization problem. We will solve it graphically.

Our ladder is to be taken around a 123 degrees corner where a 9 ft passage joins a 7 ft passage. We assume the "ladder" is actually a piece of machinery that cannot be tilted. It can be shown (see, e.g., *Applied Numerical Analysis, Gerald and Wheatley, Addison-Wesley*) that the maximum length of the ladder is found by minimizing the expression:

$$L = \frac{w_1}{\sin(\pi - a - c)} + \frac{w_2}{\sin(c)}$$

where the variables are as shown in Fig. 11.7. Note that a is an angle.

For a change of pace, the reader is asked to develop the worksheet on Sheet 5 of Chap11.xlsx and to save the workbook upon completion. Some items to note:

(a) The cells A5:C5 have been named as a, $w1\_$, and $w2\_$, respectively. The underscore was applied by Excel in the naming process since W1 and W2 are valid cell references.

(b) The cells A5 and B5 were given a custom format so as to display *deg* and *ft*. The *Format* Painter on the *Home / Clipboard* (*Home* on a Mac) group was used to give other cells these formats. This is quicker than reformatting every cell.

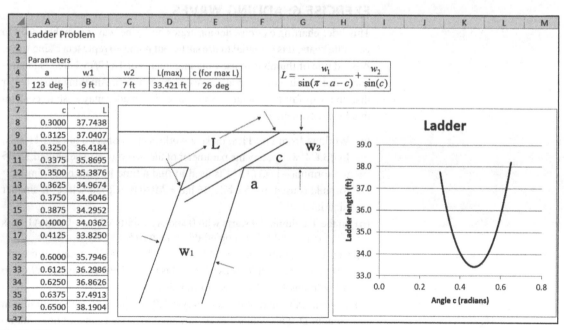

■ FIG. 11.7

(c) The formula in B8 is =w1_/SIN(PI()-RADIANS(a)-A8) + w2_/SIN (A8) and is copied to B36. Since the value in the cell named *a* is in degrees, we need the RADIANS function here.

(d) A quick chart was first made with values of *c* from 0.1 to 0.9; this showed the minimum lies in $0.4 < c < 0.6$. So values from 0.3 to 0.65 in 0.0125 increments were used in the final chart.

(e) On the assumption that the developer of the worksheet might wish to print it for documentation purposes (to show the boss how he got the result): (i) the equation was added with *Insert / Symbols / Equation* (same on a Mac), (ii) Microsoft Visio was used to make the diagram, and (iii) some rows were hidden to give a smaller print out. This last feature means that when making the chart we need to right-click it and open the Select Data dialog and using the *Hidden and Empty Cells* button, specify that hidden rows are to be plotted.

(f) The formula in D5 is =MIN(B8:B36). To find the value of *c* that generated the minimum *L* value, we use in E5 =DEGREES(INDEX(A8:A36, MATCH(D5,B8:B36,0))). The MATCH function finds the row where the minimum is found; INDEX takes the cell in that row from the *c* column; DEGREES converts the value from radians to degrees.

EXERCISE 6: ADDING WAVES

This brief charting exercise demonstrates how sine waves may be added to generate beats. It is essential to use sufficient points to represent a sine wave. A good rule of thumb is to use time increments of $1/(12f)$ where *f* is the frequency of the sine function as expressed by $\text{Sin}(2\pi f)$. Anything much larger than this will fail to generate a sine wave. The reader may wish to confirm this by experimentation.

(a) Working from Fig. 11.8, make a worksheet on Sheet 6 of Chap11.xlsx. In B3:C4, we specify the parameter of the two waves to be plotted. In B5 the formula =1/(12*B3) is used to find a time increment. A similar formula is used in C5 while B6 has =MIN(B5:C5) to select the smaller time interval.

(b) In the T column A9 starts with 0 and A10 starts with $t = 0$ and A10 has =A9+$B$6, and this is copied down to A207.

(c) Cell B9's formula is =B$4*SIN(2*PI()*B$3*$A9). A careful use of mixed references allows us to copy this to C9. Then we can use B9:C9 to fill columns B and C down to row 207.

(d) Make an XY chart of the data in A9:C207 before adding the formulas in column D. This data series is formatted to show a smooth line with no markers. The two axes are formatted with None for the Major tick

	A	B	C	D
1	Adding Sine Waves			
2		Wave A	Wave B	
3	Frequency	440	500	
4	Amplitude	1	1.5	
5	delta	0.000189	0.000167	
6	Δt	0.000166667		
7				
8	Time	Wave A	Wave B	Wave C
9	0.00000	0.00000	0.00000	0.00000
10	0.00017	0.44464	0.75000	1.19464
11	0.00033	0.79653	1.29904	2.09557
12	0.00050	0.98229	1.50000	2.48229
13	0.00067	0.96316	1.29904	2.26220
14	0.00083	0.74314	0.75000	1.49314
15	0.00100	0.36812	0.00000	0.36812
16	0.00117	-0.08368	-0.75000	-0.83368
17	0.00133	-0.51803	-1.29904	-1.81707
18	0.00150	-0.84433	-1.50000	-2.34433
19	0.00167	-0.99452	-1.29904	-2.29356

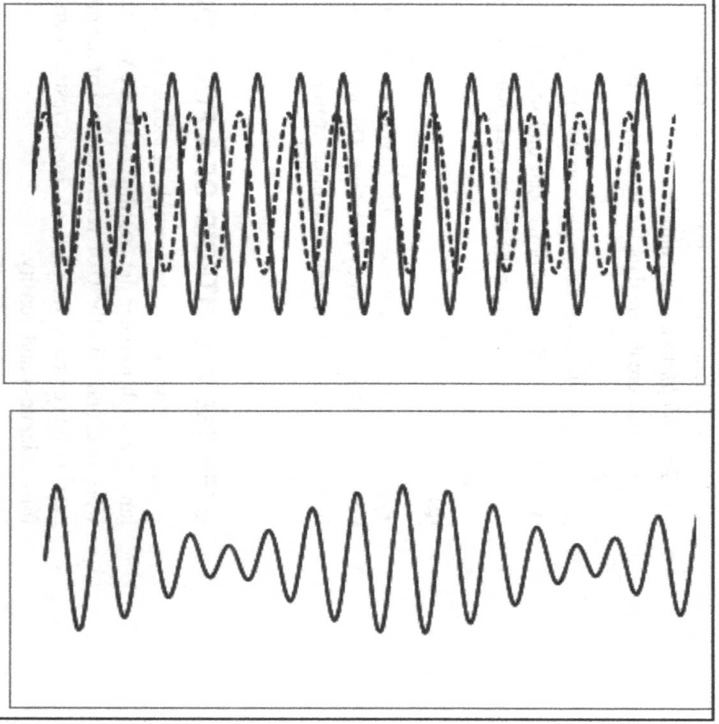

■ FIG. 11.8

marks, Minor tick marks, and Tick mark labels. With a chart containing so much data, it can be useful to format the data series setting the *Line Style Width* to a small number.

We next create the data for the superimposition of the two waves and plot it.

(e) The formula in D9 is simply =B9+C9. This is copied down the column.

(f) Select A9:A207, hold down (Ctrl), and select D9:D207. Use the *Insert / Chart* command to make the second chart; this is also a smooth XY chart.

(g) Save the workbook.

The lower chart clearly shows the development of beats when two sound waves of similar frequency interfere. The techniques of this exercise may be expanded to experiment with three or more waves.

EXERCISE 7: CENTROID OF A POLYGON

In this exercise, we see an example of a UDF that solves a real-world problem. We wish to enter the coordinates of the vertices of a polygon into a worksheet and have the UDF return the area of the polygon and the coordinates of its centroid. We shall, of course, restrict ourselves to plates of uniform thickness and density.

Let the polygon have N sides (and N vertices). The formulas we shall use are as follows:

$$A = \frac{1}{2}\sum_{i=0}^{N-1}(x_i y_{i+1} - x_{i+1} y_i)$$

$$C_x = \frac{1}{6A}\sum_{i=0}^{N-1}(x_i + x_{i+1})(x_i y_{i+1} - x_{i+1} y_i)$$

$$C_y = \frac{1}{6A}\sum_{i=0}^{N-1}(y_i + y_{i+1})(x_i y_{i+1} - x_{i+1} y_i)$$

These equations are applicable for nonself-intersecting polygons (no line in the polygon's drawing crosses another); this is not a serious constraint, as we are interested in polygons that can be made into physical objects.

The summations in our equations go from zero to $N-1$, so there will be N terms in each. This shows that we must include the first vertex at the start and at the end of our list of coordinates.

Going counterclockwise around the polygon, the formulas give positive values; otherwise, they are negative. We shall avoid this through the use of the *Abs* function in the macro.

(a) Open Chap11.xlsx. Review Chapter 9 if necessary and enter this function on a new module sheet.

```
Function Centroid(myrange)
  Dim temparray(3)
  Set mydata = myrange
  mylast = mydata.Count / 2
  area = 0 'area
  cx = 0 'centroid x-value
  cy = 0 'centroid y-value
  For j = 1 To mylast - 1
    xj = mydata(j, 1)
    xk = mydata(j + 1, 1)
    yj = mydata(j, 2)
    yk = mydata(j + 1, 2)
    term = (xj * yk - xk * yj)
    area = area + term
    cx = cx + (xj + xk) * term
    cy = cy + (yj + yk) * term
  Next j
  area = Abs(area / 2)
  cx = Abs(cx / (6 * area))
  cy = Abs(cy / (6 * area))
  temparray(0) = area
  temparray(1) = cx
  temparray(2) = cy
  Centroid = temparray
End Function
```

Some features to note are as follows: (i) The input range is passed to a VBA object with the *Set* statement allowing us to use subscripts as in $xj = mydata(j,1)$; (ii) the use of the *term* variable to save computing the same quantity three times; and (iii) the use of the array *temparray* so that we can return three values to the cells with our function. This also requires the function to be entered as an array function.

(b) Referring to Fig. 11.9, the range A5:B9 holds the coordinates for our first test polygon, which is a rectangle. A chart is to be made using these on Sheet7.

	A	B	C	D	E	F	G
1	Centroid of a Polygon						
2							
3	Rectangle						
4	x	y					
5	2	2					
6	10	2					
7	10	8					
8	2	8					
9	2	2					
10							
11	area	Cx	Cy				
12	48	6	5				
13							
14	Triangle						
15	x	y					
16	0	0					
17	16	0					
18	8	12					
19	0	0					
20							
21	area	Cx	Cy				
22	96	8	4				
23							
24							
25							
26	General						
27	x	y					
28	0	0					
29	10	0					
30	10	12					
31	7	12					
32	7	9					
33	3	9					
34	3	12					
35	0	12					
36	0	0					
37							
38	area	Cx	Cy				
39	108	5	5.5				

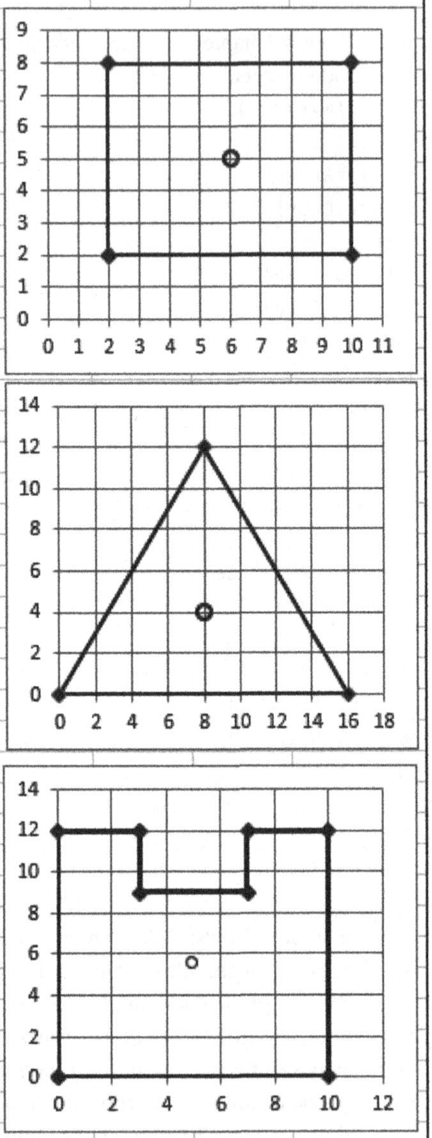

■ FIG. 11.9

(c) Select A12:C12, enter =Centroid(A5:B9) and commit the array formula with (Ctrl)+(⇧ Shift)+(Enter⏎). Note that the result is the expected one; this gives us the confidence to use the method for complex polygons.

(d) Select and copy B12:C12 then use the Paste Special method to add a new data series to the chart. This plots the centroid.

(e) Complete the worksheet for the next two polygons.

(f) Save the workbook. As it now contains a macro, it must be saved as a macro-enabled workbook, and it will be given the extension *.xlsm*.

You may wish to experiment by making changes to the coordinates for the last polygon or by extending the worksheet with further polygons.

EXERCISE 8: FINDING ROOTS BY ITERATION

There are many iterative methods of finding the roots of an equation. These include successive approximations, bisection, secant, and the Newton-Raphson methods. Powerful and instructive as these are we shall look at only one simple example. Chapter 12 introduces the Solver tool, which saves us a great deal of work in locating the roots of equations.

Scenario: It cost $P(w)$ dollars to produce w pounds of a chemical where $P(w) = 1000 + 2w + 3w^{2/3}$. The chemical sells for $4/lb. How much must be sold to break even?

Clearly, the break-even quantity is given by:

$$\text{Revenue} = \text{Expenditure}$$

$$4w = 1000 + 2w + 3w^{2/3}$$

This tells us we need to solve: $1000 - 2w + 3w^{2/3} = 0$

If we write this as $w = 500 + (3/2)w^{2/3}$ and assume w is small enough that $w^{2/3}$ is insignificant compared to w, then we get our first approximation of $w = 500$.

(a) On Sheet9 of Chap11.xlsm, enter the text seen in Fig. 11.10. There is no need to enter what is in columns E and F.

(b) In B4 enter the starting approximation of 500.
We then use this value of w (500) to compute the right-hand side of our equation $500 + (3/2)w^{2/3}$ as the next approximation.

(c) In B5 use the formula =500+(3/2)*B3^(2/3) to compute the next approximation. In D4 enter =B5-B4 so that we can observe the convergence more clearly.

(d) Copy B5:C5 down to next row to get the next approximation.

	A	B	C	D	E	F
1	Successive approximation method.					
2						
3	Iteration	w	diff		Cell	Formula
4	0	500			B4	Starting guess
5	1	594.4941	94.49408		B5	=500+(3/2)*B4^(2/3)
6	2	606.0530	11.55892		C5	=B5-B4
7	3	607.4233	1.370263			
8	4	607.5851	0.161859			
9	5	607.6042	0.019111			
10	6	607.6065	0.002256			
19	15	607.6068	1.01E-11			
20	16	607.6068	1.14E-12			
21	17	607.6068	0			
22	18	607.6068	0			
23	19	607.6068	0			
24	20	607.6068	0			

$$\text{Solving}: \quad 1000 - 2w + 3w^{\frac{2}{3}} = 0$$
$$\text{As}: \quad w = 500 + \frac{3}{2}w^{\frac{2}{3}}$$

■ FIG. 11.10

(e) Repeat until the required agreement between successive approximations is obtained. In our case, the answer is that 607 pounds must be sold to break even. We just need the successive approximations to have the same integer value so there was no real need to go down to row 20 but this does show how the method can converge to an exact answer. Note that rows 11–18 are hidden in the figure.

(f) Save the workbook.

PROBLEMS

1. *Using the method of successive approximations (see Exercise 8), find the molar volume V of CO_2 at 1 atm and 500K using the van der Waals equation

$$\left[P + a\left(\frac{1}{V}\right)^2\right](V - b) = RT$$

For CO_2, $a = 3.592$ $L^2 \cdot atm \cdot mole^{-1}$ and $b = 0.04267$ $L \cdot atm$. For your first approximation use the value given by the Ideal Gas Law

$$\frac{PV}{n} = RT$$

Let R have a value of 0.082057 $L \cdot atm \cdot K^{-1} \cdot mol^{-1}$.

2. *The relationship between the friction factor c_f and the Reynolds number Re for turbulent flow in a smooth pipe is given by $\sqrt{\frac{1}{c_f}} = -0.4 + 1.74 \text{Ln}\left(Re\sqrt{c_f}\right)$. Using the method of successive approximations, find c_f for $Re = 1 \times 10^4$. Hint: we expect $c_f \ll 1$.

3. *Use matrix math to find the forces in each member of the structure shown in Fig. 11.11. To simplify the formulas the authors found it useful to have two named cells holding the formulas `=SIN(Radians (30))` and `=COS(RADIANS(30))`.

4. Fig. 11.12 shows an electrical network. Assume currents $I_1, I_2,$ and I_3 are running clockwise around the three loops. Ohm's law relates the current to the voltage drop at each resistor. Use Kirchhoff's voltage law (sum of voltages in a closed loop is zero) to find the three currents. Then use

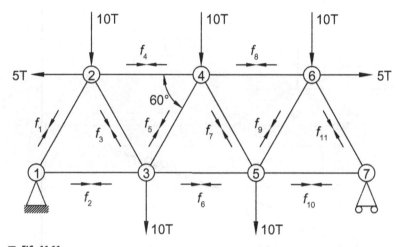

■ FIG. 11.11

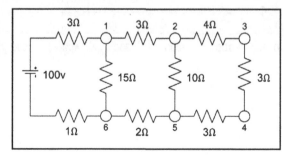

■ FIG. 11.12

Kirchhoff's Current Law (sum of currents flowing into a node is zero) to find the voltage at each node. How well do the two sets of results agree?

$$I_{ij} = \frac{V_i - V_j}{R_{ij}}$$

For each loop $\sum$ *(voltage drop each element)* $= 0$.

At each node $\sum$ *(current flowwing in)* $= 0$

5. A researcher is studying a system in which there are two competing processes. He wants to draw a trendline through the several points at the start of the data and another through several points toward the end; see Fig. 11.13. Your worksheet should allow him to specify the x-value

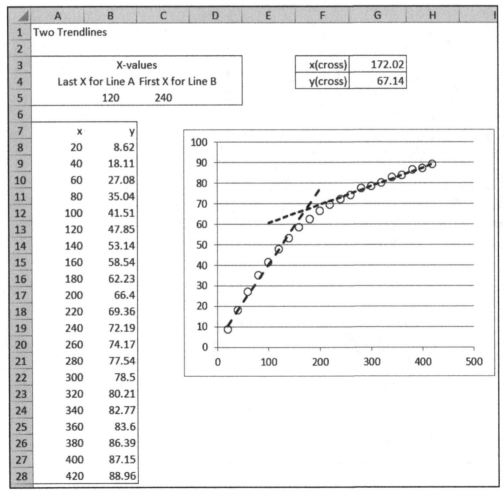

	A	B	C	D	E	F	G	H	I
1	Two Trendlines								
2									
3		X-values				x(cross)	172.02		
4	Last X for Line A	First X for Line B				y(cross)	67.14		
5	120	240							
6									
7	x	y							
8	20	8.62							
9	40	18.11							
10	60	27.08							
11	80	35.04							
12	100	41.51							
13	120	47.85							
14	140	53.14							
15	160	58.54							
16	180	62.23							
17	200	66.4							
18	220	69.36							
19	240	72.19							
20	260	74.17							
21	280	77.54							
22	300	78.5							
23	320	80.21							
24	340	82.77							
25	360	83.6							
26	380	86.39							
27	400	87.15							
28	420	88.96							

■ FIG. 11.13

cell to use for the last point in the first trendline and that for the first
x-value in the second trendline. (Hints: In Fig. 11.13 no trendlines were
inserted; the dotted lines are from two supplementary data series. Note
that slopes and intercepts were computed based on the *x*-values but
additional points were plotted. The functions MATCH and INDIRECT
will be helpful.)

6. The celebrated Newton (or Newton-Raphson) method for finding roots
is an iterative process where one starts with a guess and computes a new
approximation using $x_{n+1} = x_n - f(x_n)/f'(x_n)$. The process is repeated
until $f(x)$ is zero or sufficiently close to zero for one's practical purpose.
The method is readily implemented in Excel. Your task is to find the
three roots of $x^3 - 3x + 1 = 0$. Fig. 11.14 shows a possible layout.
The chart in the figure was produced using Microsoft Mathematics
4.0 which may be downloaded[2] at no cost. From the chart we can read
off approximate roots as -2, 1.5, and 0.5; use these for your first guess.

7. *The Beattie-Blackman equation of state is given in Problem 20 in
Chapter 2. It is readily shown that the appropriate Newton approximate
formula (see Problem 6) to compute V is

$$V_{k+1} = V_k + \frac{RTV_k^4 + \beta V_k^3 + \gamma V_k^2 + \delta V_k - PV_k^5}{RTV_k^3 + 2\beta V_k^2 + 3\gamma V_k + 4\delta}$$

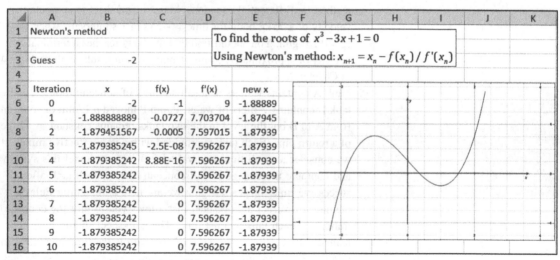

	A	B	C	D	E	F	G	H	I	J	K
1	Newton's method				To find the roots of $x^3 - 3x + 1 = 0$						
2											
3	Guess		-2		Using Newton's method: $x_{n+1} = x_n - f(x_n)/f'(x_n)$						
4											
5	Iteration	x	f(x)	f'(x)	new x						
6	0	-2	-1	9	-1.88889						
7	1	-1.888888889	-0.0727	7.703704	-1.87945						
8	2	-1.879451567	-0.0005	7.597015	-1.87939						
9	3	-1.879385245	-2.5E-08	7.596267	-1.87939						
10	4	-1.879385242	8.88E-16	7.596267	-1.87939						
11	5	-1.879385242	0	7.596267	-1.87939						
12	6	-1.879385242	0	7.596267	-1.87939						
13	7	-1.879385242	0	7.596267	-1.87939						
14	8	-1.879385242	0	7.596267	-1.87939						
15	9	-1.879385242	0	7.596267	-1.87939						
16	10	-1.879385242	0	7.596267	-1.87939						

■ FIG. 11.14

[2]Website http://www.microsoft.com/en-ca/download/details.aspx?id=15702.

Use the Newton method to find the volume of methane at 0°C and 200 atm; given $A_0 = 2.2769$, $B_0 = 0.05587$, $a = 0.01855$, $b = -0.0187$, and $c = 1.283 \times 10^5$. It may help to use separate columns for the numerator and denominator of the second term on the right.

8. Consider a body of mass m starting at rest and falling under the influence of gravity (g) while subject to a drag which is proportional to the square of the body's velocity (v). We may write

$$\frac{dv}{dt} = g - \frac{c_d}{m}v^2,$$

where c_d is the drag coefficient. Making the approximation

$$\frac{dv}{dt} \cong \frac{\Delta v}{\Delta t} = \frac{v(t_{i+1}) - v(t_i)}{t_{i+1} - t_i}$$

the velocity equation may be rewritten in the form

$$v_{i+1} = v_i + \left[g - \frac{c_d}{m}v^2\right](t_{i+1} - t_i).$$

This suggests that we may use the starting conditions ($v(0) = 0$) we may compute an approximate value for the velocity after a short time $\Delta t = (t_{i+1} - t_i)$. In Chapter 14 we find this is Euler's approximation method for differential equations. Make a worksheet to compute the velocity of a 75 kg body for time $= 0$–25 s in 1-s intervals, when $c_d = 0.25$ kg/m. You should find that the body reaches a terminal velocity of approximately 54 m/s. Compare your results with Problem 4 in Chapter 7.

9. Tank A contains V liters of a salt solution of concentration C_0 while tank B contains V liters of water. Pump P moved p L/min of solution from A to B but the level in the two tanks remains constant by virtue of a return pipe. Each tank is provided with a very effective mixer. If we consider a small time interval (*delta*) we can model this system and compute the change in salt concentration in each tank after successive time intervals. Again we are here anticipating Euler's method. Construct a worksheet that allows you to plot the concentrations in each tank against time. The worksheet should have four named cells C_0, V, p, and *delta* where *delta* is the time increment in minutes. For each time increment the change in concentration for tank A and B is

$\Delta C_A = \frac{p(C_B - C_A)delta}{V}$, $\Delta C_B = \frac{p(C_A - C_B)delta}{V}$ respectively. Show that with $V = 50\,\text{L}$, $p = 5\,\text{L/min}$, $C_0 = 10\,\text{kg/L}$ the difference in the concentration in A and B is less than 1×10^{-3} kg/L within about 40 min.

APPENDIX: SUPPLEMENTARY MATERIAL

Supplementary material related to this chapter can be found on the accompanying CD or online at https://doi.org/10.1016/B978-0-12-818249-9.00011-X.

Chapter

12

Using Solver

The Excel tools Goal Seek and Solver can save a great deal of time with complex mathematics. From a practical point of view, the simple tool Goal Seek is redundant. It has limited scope and is far outpaced by Solver. So why is it there? Simply because, being easier to use, it is less intimidating for the mathematically challenged; we shall spend a brief time on it.

Solver, which is leased by Microsoft from Frontline Systems Inc., was developed primarily for solving optimization (maximum and minimum) problems. However, it can also be used to solve equations, and that is where we shall start. You may wish to visit www.solver.com to learn more about this product and its variations. The site also has a tutorial for using the Excel Solver, but it concentrates on optimization problems; we will do more with Solver.

In this chapter we will see examples where Solver is used: (i) for equation solving, (ii) for curve fitting or regression analysis, and (iii) some simple optimization problems.

Liengme's Guide to Excel 2016 for Scientists and Engineers. https://doi.org/10.1016/B978-0-12-818249-9.00012-1
291

Solver needs to be installed on your computer and loaded into Excel. It is likely this happened when Excel was installed. To check, open the *Data* tab and look in the *Analysis* group for a *Solver* icon. If you do not see it, use *File / Options / Add-ins* and click the *Go...* the button next to *Manage Excel Add-ins* (on a Mac *Tools / Excel Add-ins...*) and then check the checkbox next to solver. The Excel Help has nothing more about Solver, as Solver has its own Help facility.

EXERCISE 1: GOAL SEEK

Suppose you have an equation such as $Exp(-x) - Sin(x) = 0$ and you know (perhaps from making a simple plot) that this has a root such that $0 <= x <= 1$. You could set up a worksheet similar to Fig. 12.1 (please ignore the screen captures of Goal Seek dialogs for now), and by altering the value in A5 and watching B5 you could find what value of x makes the function zero. Think for a moment of what strategy you would adopt. You could confirm that there was a root within (0,1) by making A5 first 0 then 1 and observing that $f(x)$ changes sign. Next, you might next try the midpoint of 0.5 and then 0.6. As the sign and magnitude of B5 changes, you would modify the direction and amount by which you altered A5 until B5 was nearly 0—or you got tired of the game! Well, Goal Seek works the same way. Let us see Goal Seek at work.

(a) On Sheet1 of a new workbook, enter what you see in Fig. 12.1 (without the screen capture images). The formula in B5 is =EXP(-A5)-SIN(A5).

(b) Use the command *Data / Forecast / What-If Analysis / Goal Seek* (*Data / What-If Analysis / Goal Seek* on a Mac) to open the Goal Seek dialog.

(c) In the *Set Cell* box type B5, in the *To Value* box type 0, and in the *By Changing Cell* enter A5. When these have been entered, click the OK button.

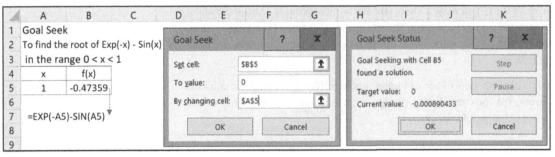

■ FIG. 12.1

Note that the *To Value* must be a number; it cannot be a cell reference. So if you want D5 to equal D6, then you will need a cell with =D5-D6, and this will be your *Set Cell* with 0 as the *To Value*.

(d) Goal Seek now displays its *Status* dialog giving you the option to either accept what it has found or cancel the operation. Click OK.

(e) Repeat steps (b) through (d) using different starting values in A5 (say 0, 1, and 0.5). Note how you have to reenter the problem each time you call up Goal Seek, and that the results vary slightly.

(f) If your starting value is 2, Goal Seek will find another root. Make a quick plot and see if you understand why.

(g) Save the workbook as Chap12.xlsx.

Goal Seek quits when it has made a certain number of trials (iterations), when a certain time period has passed, or when two answers are within a certain range of each other (convergence limit). Each of these settings may be changed by opening the *File / Options / Formulas (Excel / Excel Options / Calculation* on a Mac) dialog—see Fig. 12.2 (Fig. 12.2 shows the dialog on a Mac). Changing the *Maximum Change* from the default value of 0.001 to 0.0000001 has a significant effect on the final value in B5. The reader should experiment. Please remember that Formula settings will be retained until changed, so if you wish always to use the default values you must reset them before saving the file and exiting Excel.

EXERCISE 2: SOLVER AS A ROOT FINDER

As an introduction to Solver, we will use it to solve the same problem as in Exercise 1.

(a) Open Sheet1 of Chap12.xlsx. In A5 enter the value 1.

(b) Use the command *Data / Analyze / Solver (Data / Solver* on a Mac) to open the dialog shown in Fig. 12.3. Solver can be added if not visible by *File / Options* and selecting the *Add-ins* tab (on a Mac it is added using *Tools / Excel Add-ins…*), and then press the Go button next to the Manage Excel Add-ins selector. Check next to the solver add-in and press OK. You can see this is much more detailed than Goal Seek. For the time being, we will concern ourselves with just the top part of the dialog.

(c) For this problem: The *Set Target* is B5, with *Value Of* selected and set to 0, and *By Changing Cells* is A5. Click the *Solve* button.

(d) Note that this is the only time we shall use the Value Of setting; in the future, we will use constraints for this type of problem. This is explained in the next topic.

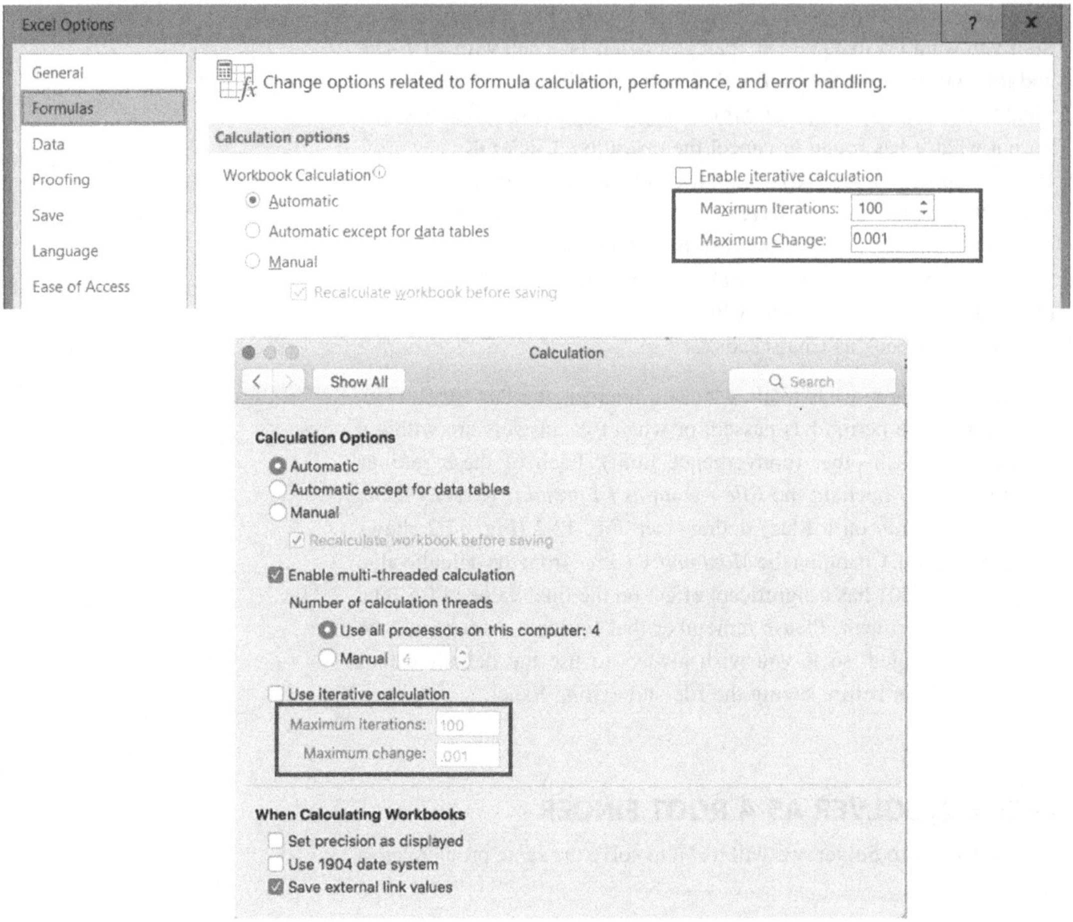

■ **FIG. 12.2**

(e) Solver finds an answer, and the Results dialog pops up—Fig. 12.4. Note that you can accept the answer or return to the original values. The reports are relevant only for optimization problems, not for a Value Of problem. Click OK.

(f) Set A5 to 0 and try again. Note (i) that Solver remembers the problem and (ii) the results are more consistent.

(g) Open Solver again but before you click Solve note that it has remembered the previously used setting. Now open the Options dialog (Fig. 12.5). We shall not make any adjustments for this problem, but you may wish to use Solver's Help to learn a little about some of the settings on the *All Methods* and the *GRG Nonlinear* tabs.

(h) Save the workbook.

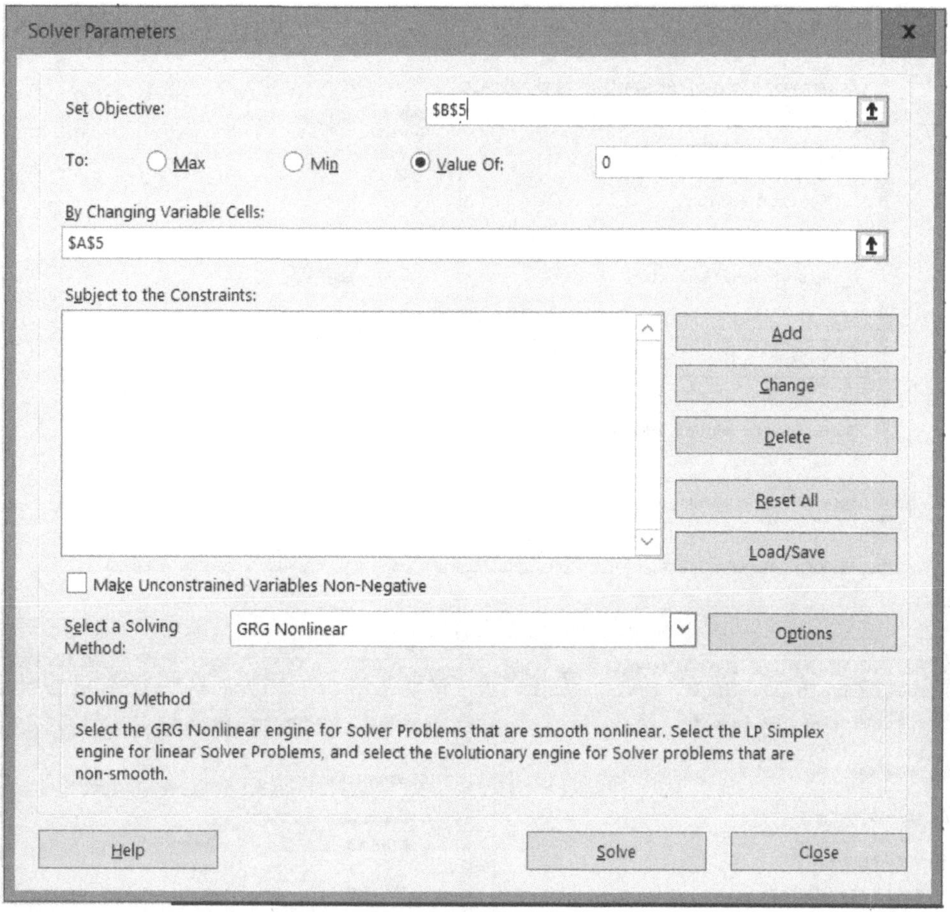

■ FIG. 12.3

SOLVING EQUATIONS WITH CONSTRAINTS

If in the last Exercise, you did look at Solver's Help and read about the Precision setting, you saw that it said: *Controls the precision of solutions by using the number you enter to determine whether the value of a constraint cell meets a target or satisfies a lower or upper bound.* This might sound irrelevant to the problem, but it is not.

With a starting value of 1 in A5 and the default value of Precision at 0.000001 (that's five zeroes after the decimal), Solver's answer made B5 a value of $-2.1E-08$ (your result could differ slightly). But when Precision is set to 0.00000000001 (ten zeroes after the decimal), the value was $1.8E-12$. Higher precision leads to an answer closer to zero.

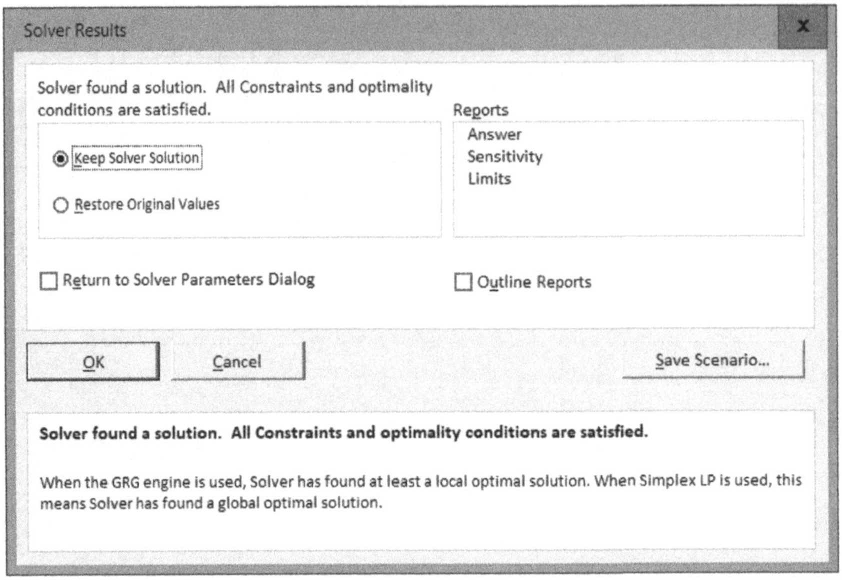

■ FIG. 12.4

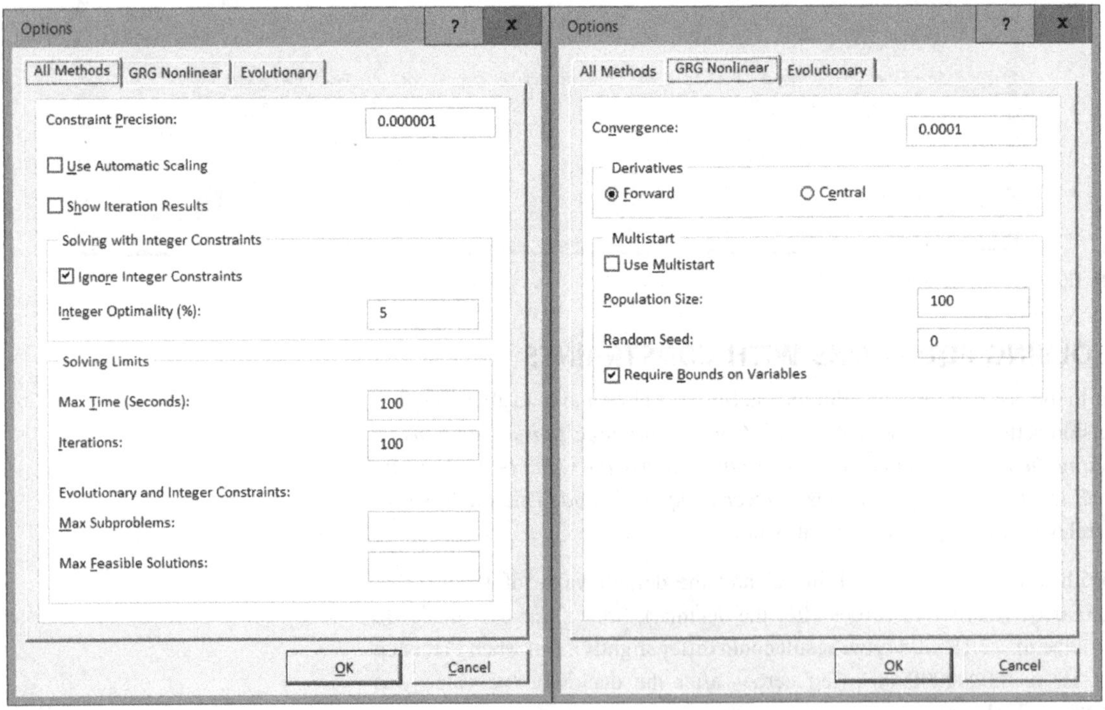

■ FIG. 12.5

This is because: **when the *Value Of* model is used in Solver, it is treated as a constraint problem**. Indeed, to solve the last problem, we could have cleared the *Set Target Cell* and entered a constraint in the form $B$5=0. This becomes very important when you want more than one cell to take on a certain value. Suppose your model requires all cells in D1:D10 to become zero. If we insist on using the *Value Of* method, we need a single cell, and so we might write =SUMSQ(D1:D10) in D11 and use it as the *Set Target Cell*. Of course, =SUM(D1:D10) would not work since cells with positive and negative values could sum to zero. A far better way is to use the constraint setting of D1:D10=0. This is the method we use in the next exercise.

EXERCISE 3: FINDING MULTIPLE ROOTS

In this exercise, we will find the roots of a cubic equation. We need to pause for a moment to look at Fig. 12.3. What do we make of the box *Make Unconstrained Variables Non-negative*? The term *variables* clearly applies to whatever we enter in the *By Changing* box and the meaning of *constrained* (and, hence, *unconstrained*) will become clear as we do this exercise. The bigger question is related to the *Select a Solving Method*. There are three options: GRG Nonlinear, Simplex LP, and Evolutionary. We do not have space to investigate these. The Evolutionary engine is far too complex to delve into. Of the other two, suffice to say that the GRG engine does work with linear problems while the Simplex engine does not work with non-linear. Conclusion: we will work with the GRG engine. The bold reader might wish to return to Exercise 2 and attempt to use the Simplex LP engine: he/she will be rewarded by a message that the model is nonlinear: indeed, we know $Exp(-x) - Sin(x) = 0$ is not a linear equation.

For the pedagogical reason, we set ourselves a problem whose answer can be readily found without using a spreadsheet. We will look for the roots of the simple cubic $x^3 + 8x^2 - 9x - 72 = 0$. Simple factorizing shows the roots to be 3, −3, and −8. Knowing the answer will help us compare using Solver with the *Value Of* setting compared to using constraints. We will see the constraints method gives superior results. Solver, like Goal Seek, homes in on the root that is closest to the initial value (sometimes called the *guess*) without passing through a minimum or maximum of the function. So we will be careful with our starting values. In more complex cases, one needs to experiment to find the multiple roots.

(a) On Sheet2 of Chap12.xlsx, copy from Fig. 12.6 the text and values in columns A and B. We have used "guesses" of 4, −4, and −10 in B5:B7.

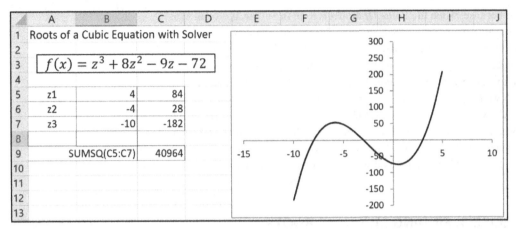

The spreadsheet shows:

	A	B	C
1	Roots of a Cubic Equation with Solver		
3	$f(x) = z^3 + 8z^2 - 9z - 72$		
5	z1	4	84
6	z2	-4	28
7	z3	-10	-182
9	SUMSQ(C5:C7)		40964

■ **FIG. 12.6**

(b) Select column B and label it x using the name box. The formula in C5 is `=x^3+8*x^2-9*x-72`, and this is copied down to C7. In C9 we have `=SUMSQ(C5:C7)`.

(c) We begin by using the "traditional" method with a cell using SUMSQ formula as our target cell.

(d) Use Solver as in Exercise 2 with the *Set Target* as C9, *Value Of* selected and set to 0, and *By Changing Cells* as B5:B7. Uncheck *Make Unconstrained Values Non-Negative*. Click the Solve button in the top right corner. We get results that are reasonably close to the known roots.

Now we will solve the same problem with no target cell but with a constraint.

(e) Reenter 4, −4, and −10 in B5:B7. Open Solver and clear the *Set Target* box.

(f) In the Subject to Constraints area, click the Add button to bring up the Add Constraints dialog; see Fig. 12.7. Enter the constraint \$C\$5:\$C\$7=0. You can type the range reference with or without the \$ symbols, or use the pointing method. Click the OK button. The Add button here is used to add additional constraints. Use the Solve button to have Solver seek a solution. Be careful to uncheck the *Make Unconstrained Variables Non-negative*; while there are occasions when we wish to restrict a variable to be nonnegative, it is pity that having this box checked is the default setting.

(g) Save the workbook.

The results are summarized in the following table. Did the Constraint method give exactly 3, −3, and 8? Not quite; the second value was −3.00000004030462 rather than integer 3 but the other two values are integers. Clearly, the constraint method gave superior results. So we will abandon the "traditional" method.

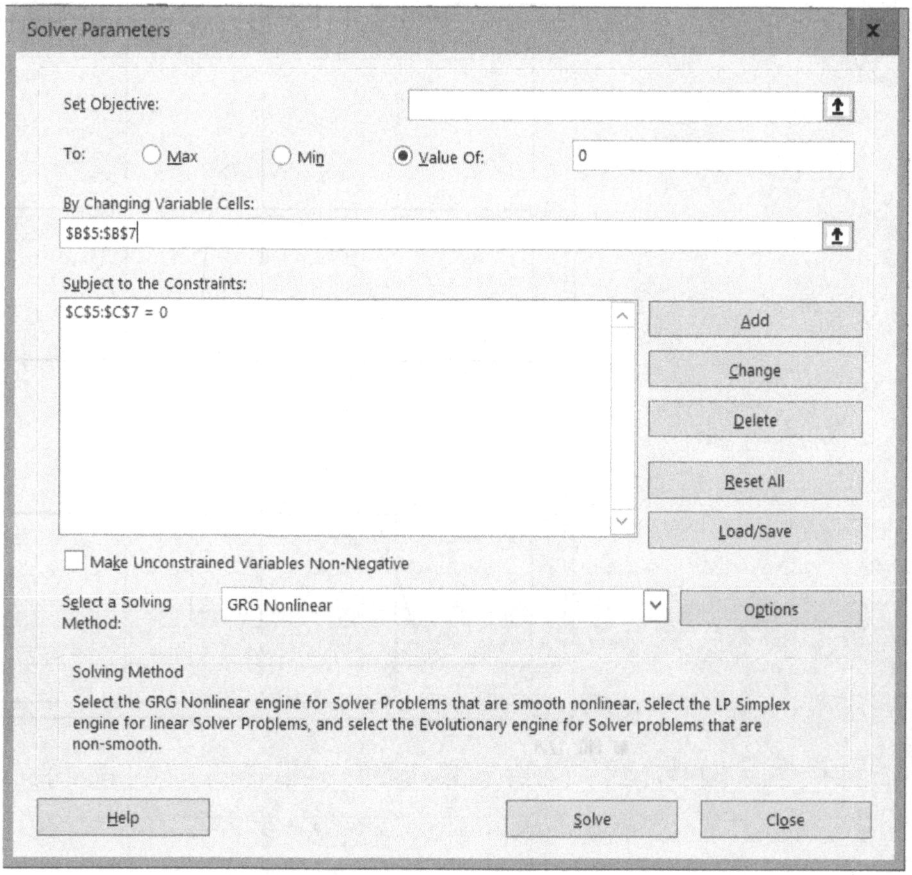

■ FIG. 12.7

Solver with SUMSQ cell			Solver with constraints		
Initial	Final	f(x)	Initial	Final	f(x)
4	3.0009101	0.0600832	4	3.0000000	0.0000000
−4	−3.0017323	0.0519645	−4	−3.0000000	0.0000012
−10	−8.0014418	−0.0793325	−10	−8.0000000	0.0000000

EXERCISE 4: SYSTEMS OF NONLINEAR EQUATIONS

In Chapter 4 we saw the use of Excel's matrix functions to solve systems of linear equations. In this exercise, we will solve the system of nonlinear equations

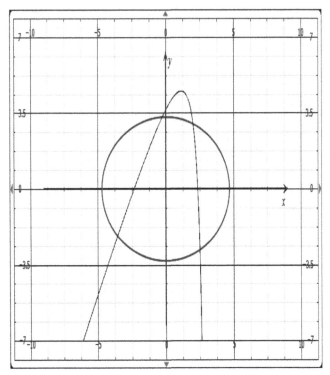

■ FIG. 12.8

$$x^2 + 2y^2 = 22$$

$$-2x^2 + xy - 3y = -11$$

Fig. 12.8 shows a plot[1] of the two functions and indicates that there are four sets of real roots. Although the plot has a few problems we can get rid of some starting values for Solver.

(a) Set up a worksheet on Sheet3 of Chap12.xlsx as shown in Fig. 12.9 but using the starting values of (2, 3.5), (0, 3.5), (−4, −2), and (2.5, −3) in A7:B10. The formulas in C7 and D7 (copied down to row 10) are C7: =A7^2+2*B7^2 and D7: =-2*A7^2+A7*B7 - 3*B7.

(b) Call up Solver. Clear the *Set Objective* box, for the *By Changing* use A7:B10, and enter the two constraints C7:C10=22 and D7:D10=-11.

---

[1]The plot was generated using Microsoft Mathematics 4, a software emulation of a graphics calculator. It is available free from http://www.microsoft.com/en-ca/download/details.aspx?id=15702. Rather uncharacteristically it has an *x*-axis with major steps of 0.5 divided unhelpfully into four.

	A	B	C	D
1	System of Non-linear Equations			
2				
3	$x^2 + 2y^2 = 22$			
4	$-2x^2 + xy - 3y = -11$			
5				
6	x	y	$f_1(x,y)$	$f_2(x,y)$
7	2.0000000000	3.0000000000	22.0000000000	-11.0000000000
8	-0.2762833099	3.3108660145	21.9999999997	-10.9999999994
9	-3.5496075201	-2.1679813714	22.0000000000	-11.0000000000
10	2.4925574968	-2.8095513098	22.0000000000	-11.0000000000
11				
12				
13	Starting values			
14	x	y		
15	2.0000000000	3.5000000000		
16	0.0000000000	3.5000000000		
17	-4.0000000000	-2.0000000000		
18	2.5000000000	-3.0000000000		

■ FIG. 12.9

Do remember to uncheck the *Make Unconstrained Variables Non-Negative* box. Open the *Options* dialog to set the *Constraint Precision* to 1×10^{-9}. Run Solver. Does running it a second time improve the results?

(c) Save the workbook.

EXERCISE 5: CURVE FITTING WITH SOLVER

In Chapter 7 we used various Excel functions (such as SLOPE, INTERCEPT, LINEST, and LOGEST) to fit experimental data to various mathematical models (linear, polynomial, exponential, etc.). We saw that the theory behind these fitting functions was based on the principle of minimizing the sum of the squares of the residuals. Solver was designed to perform maximization and minimization operations, and so lends itself to curve-fitting problems.

To demonstrate this method, we will do a simple linear fit with some test data taken from the NIST website (www.nist.gov). NIST offers many datasets, together with their fitting parameters, to enable others to test their regression programs.

Figs. 12.10 and 12.11 show the worksheet we need to make on Sheet4 of Chap12.xlsx. We shall fit the Norris dataset, shown in columns E and F of Fig. 12.11, to a linear equation $y = mx + b$.

	A	B	C
1	Least Squares fit using Solver		
2			
3		m	b
4	Solver	1.002116782479000	-0.262323072896842
5		SSR	26.61739854
6			
7	LINEST	1.002116818020450	-0.262323073773985
8	NIST	1.002116818020450	-0.262323073774029
9			
10	Solver error	-3.55414504494E-08	8.77187045223E-10
11	Linest error	4.21884749358E-15	4.44644321362E-14

■ FIG. 12.10

E	F	G
Norris Data Set		
x	y	yfit
0.2	0.1	-0.1
0.3	0.3	0.0
0.3	0.6	0.0
0.4	0.3	0.1
0.5	0.2	0.2
0.6	0.1	0.3
10.1	9.2	9.9
11.1	10.2	10.9
11.6	10.8	11.4
118.2	118.1	118.2
118.3	117.6	118.3
120.2	119.6	120.2
226.5	228.1	226.7
228.1	228.3	228.3
229.2	228.9	229.4
337.4	338.8	337.9
338.0	339.3	338.5
339.1	339.3	339.6
447.5	448.9	448.2
448.6	449.1	449.3
448.9	449.2	449.6
556.0	557.7	556.9
556.8	557.6	557.7
558.2	559.2	559.1
666.3	668.5	667.4
666.9	668.8	668.0
669.1	668.4	670.3
775.5	778.1	776.9
777.0	778.9	778.4
779.0	778.9	780.4
884.6	888	886.2
887.2	888	888.8
887.6	888.8	889.2
995.8	998	997.6
996.3	998.5	998.1
999.0	998.5	1000.9

■ FIG. 12.11

(a) Begin by entering the Norris data into E1:F38. For our starting "guesses," we will use values of 1 for both m and b in B4 and C4, respectively. Give B4 and C4 the names m and b, respectively.

(b) In G3 enter =m*x+b and copy down to G38. This is our fitted data.

(c) In C5 enter =SUMXMY2(F3:F38,G3:G38). This computes the sum of the squares of the residuals $\sum (y_i - \hat{y}_i)^2$ where the y-values are the experimental data and the $\hat{y}$-values are the fitted data. It is this quantity we wish to minimize—we are doing a *least-squares* fit.

(d) Call up Solver and configure it as shown in Fig. 12.12. After running Solver we have our fitting values for the two variables in B4 and C4 (m and b). The reader may wish to make an XY plot of E1:G38.

The remainder of the worksheet compares the Solver results with the Norris values (taken from the website) and the results from a LINEST formula. The errors are relative to the Norris values. LINEST clearly does a little better than Solver but for most practical purposes one seldom needs results with more than six decimal places and then there is no difference between the two methods.

EXERCISE 6: GAUSSIAN CURVE FIT

Having demonstrated that Solver provides a viable method of performing regression analysis, we will use it with a more challenging problem. Fig. 12.13 shows, in A10:B41, some experimental data that is to be fitted to a Gaussian curve. The function is given by:

$$y_i = h \exp\left(-\left(\frac{x_i - \mu}{\sigma}\right)^2\right) - b$$

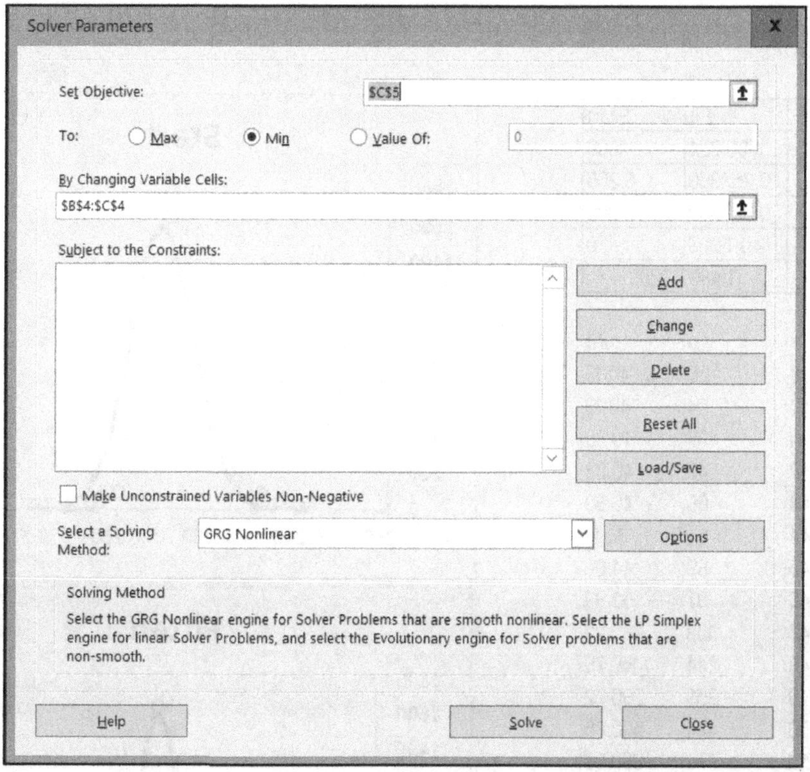

■ **FIG. 12.12**

where

y_i = the predicted value

h = the peak height above the baseline

x_i = the value of the independent variable

μ = the position of the maximum

σ = the standard deviation and

b = the baseline offset

(a) On Sheet5 of Chap12.xlsx, start a worksheet similar to that in Fig. 12.13. Begin by entering all the text and values except the values in C4.

(b) In B4:B7 use the same values as in A4:A7. Use A4:A7 to name the cells in B4:B7. We are going to vary the B4:B7 cells with Solver but will keep the C4:C7 values to remind us of our starting values.

(c) The formula in C11 is =h*EXP(-(((A11-mu)/sig)^2))+base. There may appear to be an extra pair of parentheses in this, but that is not the case; we need to allow for the fact that the negation operator has the highest priority.

▲	A	B	C	D	E	F	G	H	I	J
1	Gaussian Fit									
2										
3		Fit	Start							
4	h	1580.724	1600							
5	mu	0.253958	0.255							
6	sig	0.003653	0.005							
7	base	40.19186	0							
8	SSR	114867.7286								
9										
10	x	y	yfit							
11	0.239	25	40.19							
12	0.240	24	40.19							
13	0.241	39	40.20							
14	0.242	49	40.23							
15	0.243	56	40.39							
16	0.244	84	41.13							
17	0.245	66	44.06							
18	0.246	97	53.93							
19	0.247	158	82.20							
20	0.248	244	150.76							
21	0.249	353	290.72							
22	0.250	444	528.85							
23	0.251	773	860.69							
24	0.252	1196	1226.14							
25	0.253	1677	1515.80							
26	0.254	1654	1620.71							
27	0.255	1341	1497.48							
28	0.256	1173	1196.87							
29	0.257	933	830.50							
30	0.258	550	505.03							
31	0.259	220	275.55							
32	0.260	101	142.77							
33	0.261	97	78.68							
34	0.262	39	52.62							
35	0.263	26	43.65							
36	0.264	11	41.02							
37	0.265	16	40.36							
38	0.266	10	40.22							
39	0.267	13	40.20							
40	0.268	8	40.19							
41	0.269	5	40.19							

■ FIG. 12.13

(d) Construct a chart of the data in A10:C41. This will resemble the top chart in Fig. 12.13 where the markers are the *y*-values and the line the *yfit*-values.

You may have been wondering where the starting values for the *h, mu,* and *sig* parameters came from. The chart will answer this question. The height (*h*) appears to be about 1600; the midpoint seems to be in the range 0.25 and 0.26 so we use 0.255 for *mu*. The starting value for *sig* is found by experimentation. Try 1 in B6 and see the effect on *yfit*. Now try 0.5 and again see the effect on *yfit*. You will find that 0.005 gets *yfit* to more or less fit the *y*-values. The tails of the curve are not far from zero, so a starting value of 0 for *b* would be appropriate. So now we have reasonable starting parameters.

(e) To get ready for Solver we need a target cell holding the sum of the squares of the residuals. In B8 enter the formula =SUMXMY2(B11:B41,C11:C41).

(f) Use Solver to complete the task. The target cell is C8, which we wish to minimize by changing C4:C7. The resulting values are shown in Fig. 12.13 and the lower chart shows we have obtained a better fit.

(g) Save the workbook.

EXERCISE 7: SOLVER VS LINEARIZATION

Most often the purpose of fitting experimental data to an equation is to enable the experimenter to estimate one or more parameters. Traditionally, when the system's behavior was represented by a linear equation, the experimenter would plot the data points with the independent variable on the *x*-axis and the dependent variable on the *y*-axis; then slopes and intercepts would be read from the plot and the required parameters would be computed. When the equation was not linear, various methods were invoked to get a linear relationship.

In enzyme kinetics, the rate of a reaction (*v*) and the substrate concentration are governed by the nonlinear Michaelis-Menten equation. Biochemists have found two linearization methods to be helpful: the Lineweaver-Burk and the Eadie-Hofstee equations.

$$\text{Michaelis} - \text{Menten Eqn} \quad v = \frac{V_{max} S}{K + S}$$

$$\text{Lineweaver} - \text{Burk Eqn} \quad \frac{1}{v} = \frac{K}{V_{max}} \frac{1}{S} + \frac{1}{V_{max}}$$

$$\text{Eadie} - \text{Hofstee Eqn} \quad v = -K \frac{v}{S} + V_{max}$$

With the L-B equation, we see that a plot of $1/v$ against $1/S$ will have a slope of K/V_{max} and an intercept of $1/V_{max}$ while with the E-H equation one makes a plot of v against v/S which slope of K and an intercept of V_{max}. From the slopes and intercepts, one can then compute K and V_{max}.

(i) In Fig. 12.14 we have some experimental data in columns A and B.

(ii) The L-B model is used in columns D and E. The formulas in D5:E5 are $=1/A4$ and $=1/B4$, respectively, and these are copied down to row 14. The slope and intercept are found with $=SLOPE(E5:E14,D5:D14)$ in E15 and $=INTERCEPT(E5:E14,D5:D14)$ in E16. The estimated V_{max} value is computed in E17 with $=1/E16$ while K is found with $=E15*E17$.

(iii) The E-H model is used in columns F and G. The formulas in F5:G5 are $=B4/A4$ and $=B4$, respectively, and these are copied down to row 14. The slope and intercept are found with $=SLOPE(G5:G14,F5:F14)$ in G15 and $=INTERCEPT(G5:G14,F5:F14)$ in G16. The estimated V_{max} value is computed in E17 with $=G16$ while K is found with $=-G15$.

(iv) Columns H and I are used for the Solver method to do a direct fit to the M-M equation. The formulas in H5:I5 are $=B4$ and $=I\$17*A4/(\$I\$18+A4)$, respectively, and these are copied down to row 14. In I15 we use $=SUMXMY2(B4:B13,I5:I14)$ to compute the sum-of-the-squares-of-residuals (SSR). Cell I17 holds the V_{max} value and I18 the K value; we will start with 7 and 2, respectively. Then we call upon Solver to minimize the SSR value by changing the V_{max} and K values.

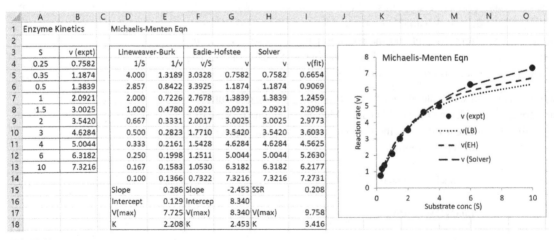

	A	B	C	D	E	F	G	H	I
1	Enzyme Kinetics			Michaelis-Menten Eqn					
2									
3	S	v (expt)		Lineweaver-Burk		Eadie-Hofstee		Solver	
4	0.25	0.7582		1/S	1/v	v/S	v	v	v(fit)
5	0.35	1.1874		4.000	1.3189	3.0328	0.7582	0.7582	0.6654
6	0.5	1.3839		2.857	0.8422	3.3925	1.1874	1.1874	0.9069
7	1	2.0921		2.000	0.7226	2.7678	1.3839	1.3839	1.2459
8	1.5	3.0025		1.000	0.4780	2.0921	2.0921	2.0921	2.2096
9	2	3.5420		0.667	0.3331	2.0017	3.0025	3.0025	2.9773
10	3	4.6284		0.500	0.2823	1.7710	3.5420	3.5420	3.6033
11	4	5.0044		0.333	0.2161	1.5428	4.6284	4.6284	4.5625
12	6	6.3182		0.250	0.1998	1.2511	5.0044	5.0044	5.2630
13	10	7.3216		0.167	0.1583	1.0530	6.3182	6.3182	6.2177
14				0.100	0.1366	0.7322	7.3216	7.3216	7.2731
15				Slope	0.286	Slope	-2.453	SSR	0.208
16				Intercept	0.129	Intercep	8.340		
17				V(max)	7.725	V(max)	8.340	V(max)	9.758
18				K	2.208	K	2.453	K	3.416

■ **FIG. 12.14**

Solver is more likely to be successful when the starting value ("guesses") is close to the actual answer. So we can use the results of either the L-B or the E-H for our initial values.

It is left to the reader to make the required data and generate the plot shown. Clearly, the Solver result is superior to those from the other two methods. This is not the place to discuss the statistical reasons but a plot showing the L-B predicted line and the raw data is informative.

EXERCISE 8: A MINIMIZATION PROBLEM

Scenario: An open-top tank is to be made from a sheet of metal by bending and welding (Fig. 12.15). The specifications are that the volume is to be $1.0\,m^3$ using the minimum sheet area. You are to find the dimensions a and b.

The worksheet to solve this problem is shown in Fig. 12.16 while Fig. 12.17 shows the Solver setup.

(a) Sheet 7 of Chap12.xlsx, enter the text shown in the figure. Enter the values in B4, B5, and H5.

(b) Select A4:B5, then while holding down Ctrl, select A8:B8 and A12: B12. Use *Formula / Defined Names / Create from Selection (Formulas / Create from Selection* on a Mac) to name the cells to the right of each text entry.

(c) Enter these formulas: in B8 $=(a\hat{}2)+(4*a*b)$ and in B13 $=a\hat{}2*b$. The parentheses in the first formula are just to improve its readability

(d) Setup and run Solver. Note that we have opted to use cell names rather than references; these must be typed in; the pointing method will result in cell references. Can you explain why the Simplex LP will not work with this problem? The results should be 1.26 for a and 0.63 for b. Save the workbook.

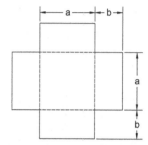

■ FIG. 12.15

	A	B	C	D
1	Minimization with Constraint			
2				
3	Variables			
4	a	2		
5	b	2		
6				
7	Target			
8	Area	20		
9	mimimize			
10				
11	Constraint			
12	Volume	8		
13	Required	1		

■ FIG. 12.16

EXERCISE 9: AN OPTIMIZATION PROBLEM

Sandbaggers Inc. processes sand to make semipure silica to sell to computer chip manufacturers at $50/ton. The company has Plant A and Plant B, in different locations. Plant A can process 450 tons/day at a cost of $25/ton, while Plant B does 550 tons/day for $20/ton.

There are three sand suppliers: Alpha, Beta, and Gamma. Today, Alpha has 200 tons of sand; they want $10/ton plus shipping of $2/ton to Plant A or $2.50/ton to Plant B. Beta's figures are 300 tons at $9 plus $1 or $1.50 while

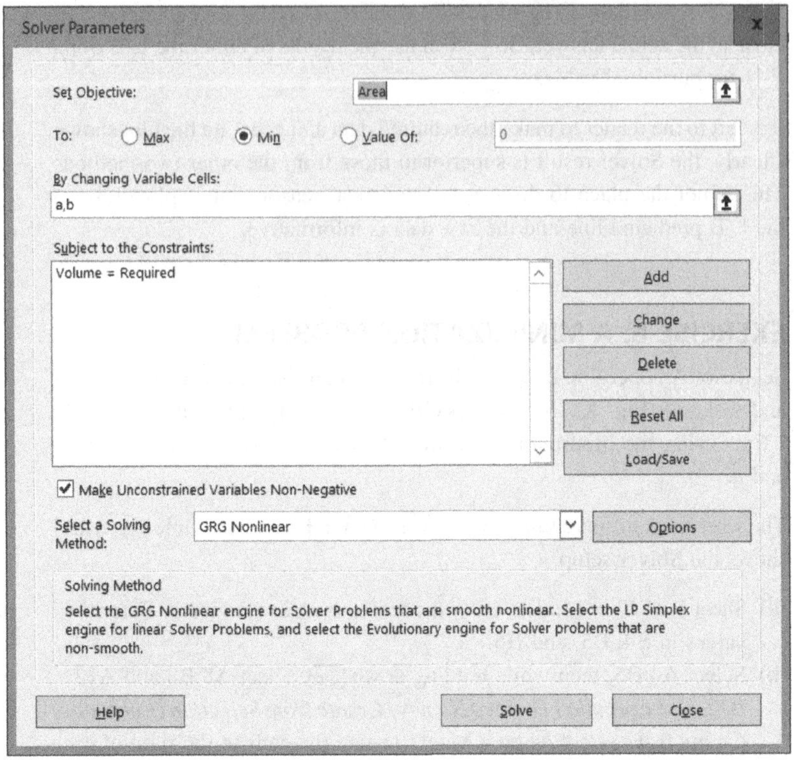

■ FIG. 12.17

Gamma has 400 tons at $8 plus $5 or $3 for shipping. Develop a business plan for Sandbagger's operation today.

This is a typical Solver optimization problem. There are three groups of data to be processed: (i) the constants; (ii) the independent variables (called the *decision variables*); and (iii) the dependent variables leading to a problem objective function, subject to some constraints. Our constants relate to the two plants and the three suppliers. The independent variables are how much sand from each supplier goes to each plant. The dependent variables are the expenses and income, with the profit being the objective function. The constraints are the finite amount each supplier has and the processing limit of each plant.

With this in mind, we plan a worksheet with different areas for the three groups of data. The constraints are placed in the Solver dialog. Fig. 12.18 shows our final worksheet.

▲	A	B	C	D	E	F	G	H	I
1	Sandbaggers, Inc								
2									
3		**Parameters**					**Market Plan**		
4	**Selling price**					Tons at each plant			
5	Sand	50	/ton				Plant A	Plant B	Total
6						Alpha	50	50	100
7	**Has two plants**					Beta	50	50	100
8		Plant A	Plant B			Gamma	50	50	100
9	Capacity	450	550	tons		Total	150	150	300
10	Op costs	25	20	$/ton					
11						**Expenses**			
12	**Can make these purchases**						Plant A	Plant B	Total
13		tons	cost			Alpha	$600	$625	$1,225
14	Alpha	200	$10			Beta	$500	$525	$1,025
15	Beta	300	$9			Gamma	$650	$550	$1,200
16	Gamma	400	$8			Operational	$3,750	$3,000	$6,750
17						Total	$5,500	$4,700	$10,200
18	**Shipping costs**								
19		Plant A	Plant B			Income	$15,000		
20	Alpha	$2	$3						
21	Beta	$1	$2			Profit	$4,800	To be maximized	
22	Gamma	$5	$3						

■ FIG. 12.18

(a) On Sheet 8 of Chap12.xlsx enter all the text shown in Fig. 12.18. Enter the values shown in columns B and C.

(b) Enter the values of 50 into G6:H8 as our starting values for Solver to work with. These are summed in row 9 and column I with formulas such as =SUM(G6:G8).

(c) In G13 enter =G6*($C14+B20) and copy this across and down to fill G13:H16. Sum these values in column I and row 16. Clearly, I17 gives the total of all expenses.

(d) Enter in G19 =I9*B5 (total income) and in G21 =G19-I17 (profit). This last item is our objective function.

(e) All that remains is to set up and run Solver.

(f) The *Set Objectives* should be G21; the *Max* option should be selected in the *To* line; the *By Changing* should be G6:H8.

(g) The *Constraints* are G9:H9<=B9:C9 (each plant has a specified capacity), and I6:I8<=B14:B16 (each supplier has a specified amount of sand for sale).

(**h**) On this occasion we will check the *Make Unconstrained Variables Non-negative* since variable are tons of sand—negative numbers would be meaningless.

(**i**) For a change, we will use the *Simplex LP* solving method. The reader may wish later to experiment with the *GRG Non-Linear* method.

The maximized profit comes out as $15,375 using all available sand. Save the workbook.

PROBLEMS

1. Download test data from www.nist.gov for a Gaussian fit. Use the method in Exercise 6 to fit the data. How do your values compare with the accepted values?

2. Redo Exercise 3 of Chapter 11 using Solver.

3. A uniform cable is hung between two towers of different heights. It assumes the shape of a catenary—see equation within the following figure. You are told that w is 10, the lowest point (y_0) is 5 m off the ground, and when $x = 50$, $y = 15$. Find the value of T.

Catenary Cable

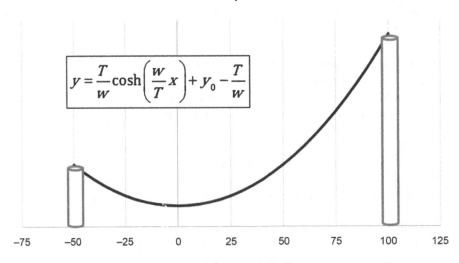

$$y = \frac{T}{w}\cosh\left(\frac{w}{T}x\right) + y_0 - \frac{T}{w}$$

4. The relationship between the friction factor c_f and the Reynolds number for turbulent flow in a smooth pipe is given by

$$\sqrt{\frac{1}{c_f}} = -0.4 + 1.74 \ln\left(Re\sqrt{c_f}\right)$$

Your task is to use Solver to find c_f for Re values of 1×10^4, 1×10^5, and 1×10^6. For starting values of c_f use 0.001.

5. *The vapor pressure (p° in Torr) of a pure liquid as a function of the absolute temperature can be expressed as $\log_{10}(p^\circ) = a - b/T$. The total vapor pressure of a three-component mixture is given by $P = x_1 p_1^\circ + x_2 p_2^\circ + x_3 p_3^\circ$ where x_i is the mole fraction of component i. Set up a worksheet to use with Solver to find the normal boiling point (the temperature at which $P = 760\,\text{Torr}$) of a three component mixture. For a working example, use the following data:

	a	b	x
Benzene	7.84125	1750	0.5
Toluene	8.08840	1985	0.3
Ethyl benzene	8.11404	2129	0.2

6. *A chemical plant[2] uses the following procedure: A volume V ft^3 of a solution of compound A having a concentration of a_o lb moles/ft^3 is allowed to react in a vat for t_r hours; the vat is then emptied, cleaned, and recharged for another cycle. It can be shown that the yield per unit time is given by:

$$yield = \frac{Va_0(1 - exp(-kt_r))}{t_r + t_c}$$

Make a worksheet using Solver to find the values of t_r that maximize the yield in each simulation shown in the following table. Test the statement that the value of t_r for maximum yield is the solution to the equation:

V	10	20	10	10	10
a_0	0.1	0.1	0.2	0.1	0.1
k	1	1	1	0.5	1
t_c	0.5	0.5	0.5	0.5	1

$$t_r - \ln(t_r k + t_c k + 1)/k = 0$$

---

[2]Carnahan and J. O. Wilkes, *Digital Computing and Numerical Methods*, Wiley, New York, 1973 (page 435).

7. *What will be the x, y-coordinates for the top right corner of the rectangle in the following figure left such that it touches the curve $3y = 18 - 2x^2$ and the area of the rectangle is maximized?

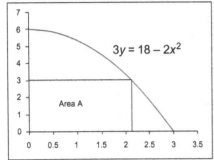

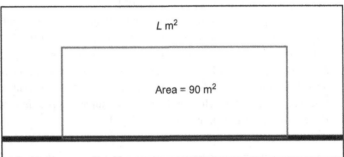

8. A rectangular enclosure (see figure to the right above) of $90\,\text{m}^2$ is to be built along a wall, any part of which may be used as one side of the enclosure. What value of L minimizes the amount of wire needed?

9. The volume V of liquid in a hollow horizontal cylinder of radius r and length L is given by the following equation where h is the depth of the water. Find h given $r = 2\,\text{m}$, $L = 5\,\text{m}$, and $V = 8\,\text{m}^3$.

$$V = \left[r^2 \cos^{-1}\left(\frac{r-h}{r}\right) - (r-h)\sqrt{2rh - h^2} \right] L$$

10. A metallurgist has four alloys with the composition shown in the following table. How many kilograms of each should be used to produce $600\,\text{kg}$ of a new alloy whose composition is 53.8% copper, 30.1% zinc, and 3.20% lead?

	Alloy1	Alloy2	Alloy3	Alloy4
Copper	52.0%	53.0%	54.0%	55.0%
Zinc	30.0%	38.0%	20.0%	38.0%
Lead	3.0%	2.0%	4.0%	3.0%

This problem can be cast as a system of four equations and solved with matrix mathematics, but it is interesting to use Solver. There are several possible approaches. Is your method linear or nonlinear? Can you explain why?

11. *A company has four sources of crude oil.[3] Crude from each source can produce specific amounts of various products. Thus from column B in the following figure we see that crude A converts to 60% gasoline, 20% heating oil, and so on. The company has a market for certain amounts of each product in a week (column G) and fixed supplies from each source (row 10). The profit per barrel is given in row 11. How many barrels of each type should be processed to maximize the profit?

	A	B	C	D	E	F	G
1	**Optimization Problem**						
2							
3			Source of Crude Oil				Product
		A	B	C	D	E	orders
4	Product						(bbls)
5	Gasoline	0.6	0.5	0.3	0.4	0.4	170,000
6	Heating Oil	0.2	0.2	0.3	0.3	0.1	85,000
7	Lube Oil	0.0	0.0	0.0	0.0	0.2	20,000
8	Jet Fuel	0.1	0.2	0.3	0.2	0.2	85,000
9	Loss	0.1	0.1	0.1	0.1	0.1	
10	Available (bbls)	100,000	100,000	100,000	100,000	100,000	
11	Profit/bbl	$100	$200	$70	$150	$250	

p	4.72	18.45	48.01	79.34	162.31	253.00
S	9.29	18.02	25.08	29.86	38.45	43.48

12. *For a change of pace, solve this magazine puzzle. Which three-digit number, when you divide it by the sum of its digits, gives you the sum of its digits plus one?

13. The Langmuir equation relates the amount of gas (S) absorbed on a surface to the pressure (p) of the gas $S = \dfrac{KS_{max}}{1 + Kp}$. Fit the data in the following table to find K and S_{max}.

---

[3]V. G. Jensen and G. V. Jeffreys, Mathematical Methods in Chemical Engineering 2nd ed., Academic Press, San Diego, 1977, (page 570).

14. In Problem 15 of Chapter 9, this equation was used to find the surface area of a cylinder with a conical base.

$$S = \frac{2V}{r} + \pi r^2 \left(\csc\theta - \frac{2}{3}\cot\theta \right)$$

We can show by calculus that S is a minimum when the angle of the cone is given by $\theta = \cos^{-1}(2/3)$. Use Solver to confirm that we did the differentiation correctly. How close is Solver's value to the expected one? Can you improve on this?

15. Sutherland's equations can be used to derive the dynamic viscosity of an ideal gas as a function of temperature:

$$\eta = \eta_0 \frac{T_0 + C}{T + C} \left(\frac{T}{T_0} \right)^{3/2}$$

t (°C)	10	20	30	40	50	60	70	80	90	100
η (expt)	17.9	18.4	18.9	19.3	19.8	20.3	20.8	21.2	21.7	22.10

where η is the viscosity at temperature T, η_0 is the viscosity, T is the input temperature in Kelvin, T_0 is the reference temperature, and C is Sutherland's constant for the specified gas. The following table lists some measured viscosity values for air. Given that for air, $\eta_0 = 18.27 \times 10^{-6}$ Pa s at 291.15 K, find C for air.

16. Refer to Problem 9 in Chapter 2. Make a new worksheet beginning with something similar to that in the following figure. Use Solver to find the n values that maximize the profit. Do not use the UDF you may have coded in Problem 7 of Chapter 9 but compute the m_1 values with an Excel formula.

	A	B	C	D	E	F	G	H	I	J	K
1	Extraction										
2											
3			Constants								
4	Labor cost		$10	each extraction				A	B	C	D
5	Solvent cost		$0.25	ml			Vs	50	50	50	50
6	Recovery revenue		$50	per g			n	2	0	0	0
7	m0		5	initial mass			m1	1.069	1.069	1.069	1.069
8	Kd		0.43	distribution const							
9	Vw		100	volume of water							

17. *Referring to the figure to the left below, an elastic cord[4] is stretched from A to C. When a vertical force P is applied at point B the cord deforms to the shape AB′C. The potential energy of the deformed system is given by

$$V = -Pv + \frac{k(a+b)}{2a}\delta_{AB}^2 + \frac{k(a+b)}{2b}\delta_{BC}^2$$

where δ_{AB} and δ_{BC} are the elongations of AB and BC and they are given by:

$$\delta_{AB} = \sqrt{(a+u)^2 + v^2} - a \text{ and } \delta_{BC} = \sqrt{(b-u)^2 + v^2} - b$$

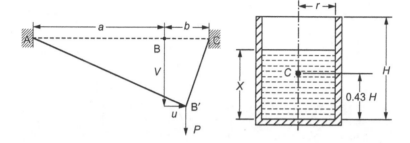

The principle of minimum energy predicts that the values of u and v will be such as to minimize V. Use Solver to find u and v. Use the values $a = 150$ mm, $b = 50$ mm, $k = 0.6$ N/mm, and $P = 5$ N.

18. *Referring to the diagram to the right above, a cylindrical vessel[5] with height H and radius r has a mass M. Its center of gravity is at point C. Water is added to a height x. The position of the center of gravity of two masses is given by

$$x_{cg} = \frac{m_1 x_1 + m_2 x_2}{m_1 + m_2}$$

where all distances are measured from a point on a line joining m_1 and m_2. Use Solver to determine the value of x such that the center of gravity of the vessel-water combination is as low as possible. Use $M = 115$ kg, $H = 0.8$ m, and $r = 0.25$ m.

---

[4]Numerical Methods in Engineering with MATLAB, Jaan Kiusalaas, Cambridge University Press 2005; page 406.

[5]*Numerical Methods in Engineering with MATLAB,* Jaan Kiusalaas, Cambridge University Press 2005; page 405.

19. The following table shows the weekly death rate during the 1905–06

Week	0	2	4	6	8	10	12	14	16	18	20	22	24	26	28	30
Deaths	3	9	20	40	99	150	380	600	800	830	670	350	180	85	30	5

Bombay epidemic. A study[6] by Kermack and McKendrick suggests this data should fit the curve $deaths = A\mathrm{sech}^2(bt - k)$ where t is the time in weeks and A, b, and k are constants. Estimate the three constants using Solver. Note that this is an example of a Solver problem where suitable initial values are crucial to finding a solution.

---

[6]Nicolas Bacaër, J. Math. Biology. (2012) 64:403–422.

Numerical Integration

Numerical integration (or *quadrature*) is used to evaluate a definite integral when there is no closed-form expression for the integral or when the explicit function is not known and the data is available in tabular form only. It consists of methods to find the approximate area under the graph of the function $f(x)$ between two x-values.

The simplest of these methods uses the *trapezoid rule*. If we divide the area under the curve into a sufficiently large number of parts, as shown in Fig. 13.1, then the area under the curve (the approximate integral) is given by:

$$I = \int_a^b f(x)dx \approx \sum_{i=1}^n A_i$$

We approximate the representative strip to a trapezoid. For a clearer drawing, only five strips are used. Obviously, more, smaller, strips are needed for a good approximation. Let there be n strips and hence $n+1$ data points. Using the fact that the area of a trapezoid is given by *average height × base*, we may write:

$$I \approx \Delta x \frac{y_1 + y_2}{2} + \Delta x \frac{y_2 + y_3}{2} + \cdots + \Delta x \frac{y_n + y_{n+1}}{2} \tag{13.1}$$

Liengme's Guide to Excel 2016 for Scientists and Engineers. https://doi.org/10.1016/B978-0-12-818249-9.00013-3

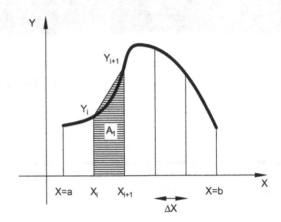

■ FIG. 13.1

With some simple mathematics, this can be rearranged as:

$$I \approx \frac{\Delta x}{2}\left(y_1 + 2\sum_{i=2}^{n} y_i + y_{n+1}\right) \tag{13.2}$$

Frequently it is Eq. (13.2) that is called the *trapezoid rule* but Eq. (13.1) can also be used in spreadsheets. We will use the trapezoid rule in form A in Exercise 1 to approximate an integral.

A better approximation to the integral is obtained using the *Simpson's ⅓ rule* in which the three points demarking two adjacent strips are joined with a parabola. The equation for this method is as follows:

$$I \approx \frac{1}{3}\sum_{i=1,3,5}^{n=2} (y_i + 4y_{i+1} + y_{i+2})\Delta x \tag{13.3}$$

The Simpson's ⅓ rule requires that there be an **even number of equally spaced** strips. Exercise 2 uses this approximation.

Approximating the curve through *four* adjacent points to a cubic equation gives the *Simpson's ⅜ rule*:

$$I \approx \frac{3}{8}\sum_{i=1,4,7}^{n=3} (y_i + 3y_{i+1} + 3y_{i+2} + y_{i+3})\Delta x \tag{13.4}$$

Surprisingly, the ⅜ rule is often less accurate than the ⅓ rule. However, unlike the ⅓ rule, it does not require an even number of strips. This is an advantage when the data is available only in tabular form (the explicit function being unknown) and there is an even number of data points—an odd number of strips.

Accuracy: If the interval a to b is divided into successively more strips, then, in principle, the accuracy should increase. This is not so in practice. When the number of strips becomes very large, the accumulated round-off error can become significant—see Exercise 4. Generally, the user knows the degree of accuracy required for a particular problem and can double the number of strips until two results differ by a value less than the requirement.

In some of the exercises and problems, we evaluate definite integrals with known values. There are two reasons for doing this: we can compare our approximations with the known values to check the accuracy, and it will give us the confidence to attack integrals with unknown values.

EXERCISE 1: THE TRAPEZOID RULE

We will use the trapezoid approximation to evaluate the integral $\int_{0}^{\pi} x \sin(x)dx$ and show that the approximation yields a result close to the known exact value which is π. We start this exercise with 10 strips, so we need 11 values of y. For this exercise, we will use the trapezoid rule in the form A. When you have completed this exercise, your worksheet should resemble that in Fig. 13.2.

(a) Open a new workbook and enter the text in A1:A6. Enter the following:

B3:	0	The lower limit of the integration
B4:	=PI()	The upper limit of the integration
B5:	10	The number of strips
B6:	=(upper-lower)/B5	

The last formula computes the value of Δx which we use to find the 11 x-values: *lower, lower* $+\Delta x$, *lower* $+2\Delta x$, and so on.

(b) Select A3:B6 and use *Formulas / Defined Names / Create from Selection (Formulas / Create from Selection* on a Mac) to give each cell a name from the text in its neighboring cell to the left. The shortcut for this is Ctrl + ⇧ Shift + F3 .

(c) Enter the text in row 8.

(d) Enter the following formulas:

A9:	=lower	The value of x_1.
A10:	=A9+delta	The value of x_2

◢	A	B	C	D
1	Trapezoid Rule			
2				
3	lower	0		
4	upper	3.141593	$\int_0^\pi x\sin(x)\,dx$	
5	n	10		
6	delta	0.314159		
7				
8	x	y	strip	
9	0	0	0.015249	
10	0.314159	0.097081	0.073261	
11	0.628319	0.369316	0.177782	
12	0.942478	0.762481	0.307501	
13	1.256637	1.195133	0.434471	
14	1.570796	1.570796	0.528337	
15	1.884956	1.792699	0.56106	
16	2.199115	1.779121	0.511512	
17	2.513274	1.477265	0.369293	
18	2.827433	0.873725	0.137244	
19	3.141593	3.85E-16		
20		Approx	3.115711	
21		Exact	3.141593	
22		Error	-0.8%	

■ **FIG. 13.2**

(e) Copy these down to row 19 to give the 11 *x*-values.

(f) In B9 enter the formula =A9*SIN(A9) to compute the value of *y1*. Copy this formula down to B19 giving the 11 *y*-values for the strips. Note how using the Eq. (13.1) form of the trapezoid rule allowed us simply to copy B9 to the other cells.

(g) In C9 enter the formula =delta*(B9+B10)/2 to compute the area of the first strip. Copy this formula down to C18 to get the areas of the other nine strips. Be careful NOT to copy it to C19.

(h) Enter the text in B20:B22. Enter the following formulas:

C20:	=SUM(C9:C18)	To sum the 10 trapezoid areas
C21:	=PI()	The exact result for the integral
C22:	=(C20-C21)/C21	Relative error

(i) Save the workbook as an Excel macro-enabled file called Chap13. xlsm. As we shall be adding modules later in the chapter, we may as well make it macro-enabled right away.

If all has gone well, you will see that the trapezoid rule approximates the integral to 3.115711, which is 0.8% low compared to the exact value. It is left as an exercise for the reader to modify the worksheet to show that doubling the number of stripes halves the relative error: with 20 strips the error is −0.2% and with 40 it is −0.05%.

EXERCISE 2: SIMPSON'S ⅓ RULE

This exercise uses Simpson's ⅓ rule (Eq. 13.3) to find the approximate value of $\int_{0}^{2} \exp(x^2)dx$, for which there is no simple analytical method for this integral. However, the known[1] value (to 12 decimal places) is 16.452627765507. We will use 20 strips (i.e., 21 x,y pairs) and generate a worksheet as in Fig. 13.3. Note that rows 17 to 24 have been hidden.

(a) Open the workbook Chap13.xlsm and on Sheet2 enter the text in A1:D8.
(b) Name the cells B3:B6 from the text to their left.
(c) Enter the following values or formulas:

B3:	0	The lower limit of the integration
B4:	2	The upper limit of the integration
B5:	20	The number of strips
B6:	=(upper-lower)/n	

The formula in B6 computes the value of Δx, which we use to find the 21 x-values: *lower, lower* $+\Delta x$, *lower* $+2\Delta x$, and so on.

(d) The numbers in A9:A29 are not essential but may help you understand how the Simpson's rule is implemented in the worksheet. Enter these using the Fill Series method you learned in an earlier chapter.
(e) Enter the initial x- and y-values in B9 and C9:

B9:	=lower	The value of x_1
C9:	=EXP(B9^2)	The value of y_1

---

[1]Wolframalpha.com gives the integral with either a 4-decimal place answer or a 101-decimal place one! Here it is to a mere 50 places: 16.45262776550723022473640445416 78875309084423771959967.

◢	A	B	C	D
1	Simpson's ⅓ Rule			
2				
3	lower	0		
4	upper	2		
5	n	20		
6	delta	0.1		
7				
8		x	y	term
9	1	0	1	6.08101
10	2	0.1	1.01005	
11	3	0.2	1.040811	6.59102
12	4	0.3	1.094174	
13	5	0.4	1.173511	7.74294
14	6	0.5	1.284025	
15	7	0.6	1.433329	9.85908
16	8	0.7	1.632316	
25	17	1.6	12.93582	110.44278
26	18	1.7	17.99331	
27	19	1.8	25.53372	227.99608
28	20	1.9	36.96605	
29	21	2	54.59815	
30			Approx	16.45521
31			Published	16.45263
32			Error	0.02%

$$I = \int_0^2 \exp(x^2)\,dx$$

■ FIG. 13.3

(f) Enter the second and subsequent values of x and y:

B10: =B9+delta   The value of x_2
C10: =EXP(B10^2)   The value of y_2

Copy B10:C10 down to row 29.
Since the steps (the Δx-values) are constant in the Simpson's method, we may write Eq. (13.3) as: $I = \frac{\Delta x}{3}\sum_{i=1,3,5}^{n-2}(y_i + 4y_{i+1} + y_{i+2})$. Thus we may compute each of the summation terms, find their total, and multiply the result by $\Delta x/3$ to approximate the integral.

(g) The first term in the summation is $(y_1 + 4y_2 + y_3)$ so in D9 enter the formula =C9+4*C10+C11. Check that your value agrees with that in Fig. 13.3.

(h) We need to copy this formula to each alternate cell in the range D11: D27. The easiest way to do this is select D9:D10 and drag the fill handle down to D27.

(i) Enter the text in C30. In D30 enter =ROUND(SUM(D9:D29)*delta/ 3,12) to find the integral. The rounding is so we can compare with the known value. Only five decimal places are displayed in the worksheet.

(j) Enter appropriate text and formulas in C31:D32 to show the published value (enter all 12 decimal places) and compute the percentage error.

(k) Save the workbook Chap13.xlsm.

The reader is encouraged to modify the worksheet to have 1000 steps (1001 data pairs) and observe how the error goes from 2.58×10^{-3} (or 0.02%) to a mere 4.26×10^{-6} (or 0.00003%).

EXERCISE 3: ADDING FLEXIBILITY

The worksheet in Exercise 2 evaluates a certain integral. For another function, we would need to edit the formula in C9 to reflect the new function to be integrated and copy this down to C29. Another way is to put the function in a module sheet and change the user-defined function each time we wish to evaluate a different integral.

We will begin by solving the same integral to confirm the integrity of the UDF.

(a) Open the workbook Chap13.xlsm. We wish to duplicate Sheet2. Hold down the Ctrl key and drag the Sheet2 tab to the right (you will see an icon of a sheet of paper overprinted with a+sign). Release the mouse button and tab labeled Sheet2 (2) will appear. Answer *yes* to various questions about Named cells; in this way cell B3 of the new sheet will have the name *lower*, and so on. On a PC right-click (on a Mac hold the Ctrl key and click) the tab and rename the worksheet as Sheet3.

(b) Use *Home / Editing / Clear (Home / Clear* on a Mac) (icon displays an eraser) on C31:D31 as we will be changing functions in this exercise.

(c) Open the Visual Basic Editor with the command *Developer / Code / Visual Basic (Developer / Visual Basic* on a Mac) or with the Alt+F11 shortcut. Insert a module sheet of the Chap13.xlsm project and enter this UDF

```
'Function to use with Simpson Rule worksheet
Function SimpFunc(x)
    SimpFunc = Exp(x^2)
End Function
```

Remember: whenever you edit a user-defined function, you must recalculate the worksheet by pressing F9 before any changes in the function will take effect.

(d) Return to Sheet3. Change the formula in C9 to =SimpFunc(B9) and copy this down to C29. Excel will ignore the uppercase letters, but in the Insert Function dialog in the User Defined category, our function will show as *SimpFunc*. The values should stay the same as before (see Fig. 13.3).

Now we will make a quick change to the UDF to evaluate

$$\int_{-1}^{1} \exp(-x^2)dx.$$

(e) Change the third line in the module function to read SimpFunc=Exp (-(x ^ 2)). Carefully note the position of the negation operator relative to the parentheses.

(f) Return to Sheet3. Change the values of lower and upper to 1 and 1, respectively. The value of the new function has been calculated. For this function, in the interval 1 to 1, the result[2] should be approximately 1.49365. Save the workbook.

EXERCISE 4: GOING MODULAR

In the last exercise, we used a UDF to facilitate computing the summation in Eq. (13.3). term by term on a worksheet. In this exercise, we use a UDF to both compute the terms and to perform the summation. To check our work more easily, we find the value of the same integral as in Exercise 1. In addition, we experiment with making the strip successively smaller and observe how the percentage error changes. By doing so, we demonstrate the power of Visual Basic. Our completed worksheet will resemble that in Fig. 13.4.

(a) Open the Visual Basic Editor with the Alt+F11 shortcut. On a new module sheet in Chap13.xlsm project, code the integrating function and the function to be integrated as shown following. Do not type the line numbers; they are for discussion purposes only. The statements in the *Integral* function are examined at the end of the exercise.

```
1.      'Simpson One-Third Rule Approximation
2.      Function Integral(a, b, n)
3.        Integral = 0#
4.        delta = (b - a) / n
5.        x = a
```

---

[2]How does your result compare to that of Wolframalpha: 1.493648265625 (rounded to 12 places)?

◢	A	B	C	D	E	F
1	Simpson's ⅓ Rule			With User-defined Integral Function		
2						
3						
4	lower	0				
5	upper	3.141593	π			
6	exact	3.141593	π			
7						
8	n	10	100	1,000	10,000	100,000
9	Approx	3.141765	3.141593	3.141593	3.141592654	3.141592654
10	Error	0.000172	1.7E-08	1.68E-12	-1.22569E-13	-3.45546E-12
11	% Error	0.0055%	0.0000%	0.0000%	0.0000%	0.0000%

The integral shown in the figure is:

$$\int_0^\pi x \sin(x)\,dx$$

■ FIG. 13.4

```
6.        For i = 1 To n Step 2
7.            Term = y(x) + 4 * y(x + delta) + y(x + 2 * delta)
8.            Integral = Integral + Term
9.            x = x + 2 * delta
10.       Next i
11.       Integral = Integral * delta / 3
12.   End Function
13.
14.   'The function to be integrated
15.   Function y(x)
16.       y = x * Sin(x)
17.   End Function
```

(b) On Sheet4 of the Chap13.xlsm enter the text and values shown in A1: A11 of Fig. 13.4.

(c) Name the cells B4:B6 with the text to the left of them.

(d) The formulas and values to be entered in B3:B10 are as follows:

B4:	0	The lower limit
B5:	=PI()	The upper limit
B6:	=PI()	The known value of the integral
B8:	10	The number of strips
B9:	=Integral(lower,upper,B7)	Value from Simpson's rule
B10:	=B8-exact	Error calculation
B11:	=B9/exact	Percentage error

(e) Format B11 to show a percentage value with four decimal places.

(f) Check that your results in B9:B11 agree with Fig. 13.4. If they do not, you may need to edit your module or your formulas.

(g) Enter the values in C8:F18 and copy B9:B11 across to column F.

(h) Select B7:B10 and drag the handle to column F. Change your *n* values to match those in the spreadsheet. Note that large numbers may be entered with a comma to make them more readable. Do not be surprised if your worksheet takes some time to respond. Microsoft Excel has to do a large number of calculations. Save the workbook.

Note how the absolute error progressively decreases up to $n = 10,000$ and then increases for larger *n* values. Here we are seeing the accumulated round-off errors beginning to creep in.

Notes on the Integral function:

Line	Comment
3	Type this as `Integral=0.0` to tell VBA that it is a real, not an integer, value. This initializes the value of the function to zero.
5	The lower limit of the integral is *a*.
6	We need to sum for odd values of *i* ($i = 1, 3, 5, ..., n$). The `step 2` phrase achieves this. It is equivalent to copying the formula in D9 of Exercise 2 to alternate rows.
7	This finds the $(y_i + 4y_{i+1} + y_{i+2})$ term for the area of a strip. We multiply by $\Delta x/3$ at the end of the calculation.
8	We keep a running total of the terms computed in line 7. We may read this as New Integral value = Old Integral value + Term value.
9	This statement computes the *x*-value for the next term.
11	The sum of the partial is multiplied by $\Delta x/3$.

EXERCISE 5: TABULAR DATA

There are times when the data to be integrated comes from an experiment and the implicit function is unknown. Which of the three rules should be used to evaluate the integral?

Trapezoid	may be used with any data but is the least accurate.
Simpson ⅓	requires an even number of equally spaced strips.
Simpson ⅜	requires equally spaced *x* values, may be less accurate than the ⅓ rule.

Suppose we have 63 strips (i.e., 64 data pairs). We may use the ⅜ rule for the first (or last) three and the ⅓ rule for the remaining 60 strips.

We have seen that increasing the number of strips (up to a point) improves the accuracy of these approximations. With tabular data, this option is not available. While we cannot increase the number of data points since we do not know the function, we can decrease the number by doubling the width of each strip. Essentially, this means we ignore every alternate data pair in the table. Obviously, our second value for the integral will be less accurate. This is where Romberg integration is useful. The Romberg integral is computed using: $I_R = I_h + \frac{I_h - I_{2h}}{2^n - 1}$ where I_h is the approximation with strip width h, I_{2h} the result with width $2h$, and I_R the improved result. Clearly, we may use Romberg integration only when the strips are evenly spaced.

In this exercise, we use the trapezoid rule to find an approximation to some tabulated data.

(a) On Sheet5 of the workbook Chap13.xlsm enter the text in A1:D14 as shown in Fig. 13.5.

(b) Enter the values in A4:B12. This is the experimental data that we wish to integrate.

(c) We will use the trapezoid rule in the form of Eq. (13.4) in this Exercise. Enter these formulas:

C4: =B4
C5: =2*B5   copy this down to C11by dragging
C12: =B12

	A	B	C	D	E
1	Tabular Data		Romberg Improvement		
2					
3	x	y	h=0.2	h=0.4	
4	1.8	6.050	6.0500	6.0500	
5	2.0	7.389	14.7780		
6	2.2	9.025	18.0500	18.0500	
7	2.4	11.023	22.0460		
8	2.6	13.464	26.9280	26.9280	
9	2.8	16.445	32.8900		
10	3.0	20.086	40.1720	40.1720	
11	3.2	24.533	49.0660		
12	3.4	29.964	29.9640	29.9640	
13					
14			h=0.2	h=0.4	Romberg
15	Integrals		23.99440	24.23280	23.91493
16	Exact		23.91445	23.91445	23.91445
17	Errors		0.334%	1.331%	0.002%

■ FIG. 13.5

Summing C4:C12 gives the bracketed part of Eq. (13.2). So the integral is completed by adding the entry:

C15: =0.2/2*SUM(C4:C12)

(d) In column D we will use strips of twice the width. Enter the formulas:

D4: =B4
D6: =2*B6 copy this to D8 and D10; do this cell by cell or by selecting D6:D7 and dragging down to D10.
D12: =B12
D15: =0.4/2*SUM(D4:D12)

(e) The Romberg integral is found with

E15: =C15+(C15−D15)/3.

(f) Save the workbook.

Is the Romberg value a better approximation? The data was actually generated using $y = \exp(x)$ with the values rounded to three decimal places. Therefore our result should approximate the integral

$$\int_{1.8}^{3.4} \exp(x)dx = \exp(3.4) - \exp(1.8) = 23.9145$$

In row 17, compute the percentage errors of the three values. Clearly, the Romberg value is the more accurate one.

EXERCISE 6: GAUSSIAN INTEGRATION

The Gaussian two-point integration formula, as derived in most elementary numerical analysis textbooks, has the wonderful simplicity of:

$$\int_{-1}^{1} f(t)dt = f\left(-\frac{1}{\sqrt{3}}\right) + f\left(+\frac{1}{\sqrt{3}}\right)$$

The four-point formula is only slightly more formidable:

$$\int_{-1}^{1} f(t)dt = \frac{5}{9}f\left(-\frac{\sqrt{3}}{5}\right) + \frac{8}{9}f(0) + \frac{5}{9}f\left(+\frac{\sqrt{3}}{5}\right)$$

The generalized formula for n points is as follows

$$\int_{-1}^{1} f(t)dt = \sum_{i=1}^{n} w_i f(t_i)$$

The accompanying table lists the values for the weights (w_i) and the points (t_i) for various numbers of points in the integration.

The degree of the polynomial function for which each integration formula is accurate is given by $2n - 1$. Thus the three-point formula is accurate for polynomials up to degree 5. When one is unsure of the number of points to use, successively use 2, 3 ... points until two results agree to the precision required.

The weights and point values in the table are for the limits of integration ± 1. To use them with other limits (a to b), we make the substitution:

$$x = \frac{(b-a)t + b + a}{2} \quad \therefore \quad dx = \left(\frac{b-a}{2}\right)dt \text{ giving } \int_a^b f(x)dt$$

$$= \frac{b-a}{2}\int_{-1}^1 f\left(\frac{(b-a)t + b + a}{2}\right)dt$$

n	$\pm t_i$	w_i
2	$1/\sqrt{3}$	1
3	0	8/9
	$\sqrt{3}/5$	5/9
4	0.33998 10435 84856	0.65214 51548 62546
	0.86113 63115 94052	0.34785 48451 37455
5	0	0.56888 88888 88889
	0.53846 93101 05727	0.47862 86704 99334
	0.90617 98459 38664	0.23692 68850 56189
6	0.23861 91860 83197	0.46791 39345 72691
	0.66120 93864 66264	0.36076 15730 48139
	0.93246 95142 03152	0.17132 44923 79170
8	0.18343 46424 95644	0.36268 37833 78363
	0.52553 24099 16329	0.31370 66458 77887
	0.79666 64774 13017	0.22238 10344 53966
	0.96028 98564 97439	0.10122 85362 90617
10	0.14887 43389 81631	0.29552 42247 14753
	0.43339 53941 29247	0.26926 67193 09997
	0.67940 95682 99032	0.21908 63625 15982
	0.86506 33666 88985	0.14945 13491 50581
	0.97390 65285 17188	0.06667 13443 08648

In this exercise, we use Gaussian integration to find the value of: $I = \int x^2 \cos(x)\,dx$. Since the definite integral evaluates to $(x^2-2)\sin(x) +2x\cos(x)+const$ our expected result is 0.4778267.

To make the worksheet more versatile, we will code a user-defined function for $x^2 \cos(x)$; in this way, we will be able to perform Gaussian integrations on other functions merely by editing the user-defined function. We will increase the number of terms until we have an approximate value to four decimal places.

  (a) Open Chap13.xlsm and invoke the VBE. Insert another module on the Chap13.xlsm project on which to code the following function:

```
Function gaussfunc(x)
    gaussfunc = x ^ 2 * Cos(x)
End Function
```

  (b) Return to the workbook and on Sheet6 enter the text shown in A1:E3 of Fig. 13.6.

  (c) Select the column headings B:E and use the command *Home / Cells / Format / Column Width* (*Format / Column Width* on a Mac) to set the width to 14. Or after selecting the headings, on a PC right-click

	A	B	C	D	E
1	Gaussian Integration				
2					
3	Terms	Weight	Point	Term	Approx
4	2	1	0.577350269	0.279303943	
5		1	-0.57735027	0.279303943	0.558607885
6					
7	3	0.555555556	0.774596669	0.238234398	
8		0.888888889	0	0	
9		0.555555556	-0.77459667	0.238234398	0.476468795
10					
11	4	0.652145155	0.339981044	0.071064922	
12		0.347854854	0.861136312	0.168076458	
13		0.347854854	-0.86113631	0.168076458	
14		0.652145155	-0.33998104	0.071064922	0.478282759
15					
16	5	0.478628671	0.538469231	0.119140142	
17		0.236926885	0.906179846	0.119993421	
18		0.568888889	0	0	
19		0.236926885	-0.90617985	0.119993421	
20		0.478628671	-0.53846923	0.119140142	0.478267125

■ FIG. 13.6

(on a Mac hold the $\boxed{\text{Ctrl}}$ key and click) and open the column width dialog from the shortcut menu.

(d) Enter the values shown in A4:A16 and in B4:B20. It is safer to use =5/9 in B7 and =8/9 in B8 rather than numerical values.

(e) Enter the values shown in C4:C20. For C4 you may wish to use =1/SQRT(3) and the negative of this in C5. Similarly for C7 use =SQRT(3/5) and the negative of this in C9.

(f) In D4 enter =B4*Gaussfunc(C4). Copy this down to D20. Delete D6, D10, and D15.

(g) Move to E5 and click on the Autosum button. Drag over D4:D5 to give the formula =SUM(D4:D5). Repeat this operation for the three other approximations.

(h) Save the workbook.

We can see that the four-point and the five-point approximations agree to four decimal places, so our task is complete.

EXERCISE 7: MONTE CARLO TECHNIQUES

There is no mathematical advantage to performing a numerical integration using a Monte Carlo technique. However, it does provide a simple way to illustrate a Monte Carlo calculation. A large worksheet would be needed to model a true stochastic process.

Consider a circle inscribed within a square with sides of l units. The radius (r) of the circle will be $l/2$. A large number (N) of darts are randomly thrown at the diagram, and the number (C) that fall within the circle is counted. If the throwing was truly random, then:

$$\frac{\text{Number of darts in circle}}{\text{Total number of darts}} = \frac{\text{Area of circle}}{\text{Area of square}} \text{ or } \frac{C}{N} = \frac{\pi r^2}{l^2} = \frac{\pi}{4}$$

Hence the value of π may be approximated from a simple dart-throwing experiment. We will use this Monte Carlo method to get an approximate value of the integral $I = \int_0^{10} (-x^3 + 10x^2 + 5x)dx$.

As in previous exercises, we have chosen an integral that can be solved analytically so as to be able to evaluate our method.

(a) On Sheet7 of Chap13.xlsm begin by constructing the table shown in H1:I12 of Fig. 13.7. The formula in I2 is =-H2^3+10*H2^2+5*H2. This is copied down to row 12. Make a chart of the data.

Our curve is enclosed by a 10 by 200 rectangle. We will use the RAND function to generate two random values (one between 0 and 10, the other between 0 and 200) from which we will find the position of the dart.

	A	B	C	D	E	F	G	H	I	J	K	L	M	N
1	Monte Carlo Integration													
2	x	y	in/out		Throws	1000								
3	8.00	101.25	1		Inside	562								
4	9.72	164.89	0		Area	1124.00								
5	4.80	184.59	0		Actual	1083.33								
6	8.86	80.85	1		%error	3.75%								
7	0.22	55.80	0											
8	4.71	151.43	0											
9	0.42	148.15	0											
10	9.06	95.84	1											
11	9.33	193.47	0											
12	5.00	48.24	1											
13	7.62	143.82	1											
14	4.33	77.67	1											
15	5.60	78.22	1											

■ FIG. 13.7

(b) In A3 enter =RAND()*10 and in B3 enter =RAND()*200. Copy these down to row 1002. Your values will not be the same as in the figure.

(c) The formula in C3 is =IF(B3>-A3^3+10*A3^2+5*A3,0,1). This returns 0 when the dart has fallen above the curve, and 1 otherwise. Copy it down to C1002.

(d) The formulas in column F are as follows:

F2:	=COUNT(A3:A1002)	Total darts thrown
F3:	=SUM(C3:C1002)	Darts inside the curve
F4:	=(200*10)*F3/F2	Area under the curve
F5:	=-(1/4)*10^4+(10/3)*10^3+(5/2)*10^2	Analytical result
F6:	=(F4-F5)/F5	Relative error

(e) Repeatedly press F9 to recalculate the worksheet. A new set of random numbers is generated each time leading to a new value in F5. The error seems to lie within a range of ±5% of the analytical value. Adding more random numbers would help. Alternatively, we could record the result for, say, 20 recalculations and take an average. This could be done automatically using a VBA subroutine. Save the worksheet.

PROBLEMS

1. Estimate $\int_{-1}^{0} \frac{1}{x+2} dx$ with $\Delta x = 0.2$ using the trapezoid rule. Try steps of 0.1 to see how much improvement there is.

2. Use Simpson's rule for the integral in Problem 1 with the same steps. Compare the four results with the value obtained by direct integration.
3. Write a UDF to evaluate $\int_{-1}^{0} \sqrt{2x+1}\,dx$. The header should include an argument n to specify how many strips to use.
4. *Find the area bounded by the two functions $f(x) = 4x^2$ and $g(x) = x^4$ in the range $0 <= x <= 2$.
5. *Calculate the definite integral $\int_{1}^{2} x \ln(x)\,dx$ using Simpson's rule with h values of 0.5, 0.25, and 0.125. The exact solution[3] may be found from $\int x \ln(x)\,dx = \left(\frac{x^2}{2}\right) \ln(x) - \frac{x^2}{4}$.
Show that the error $E(h)$ in Simpson's rule may be estimated from the inequality:

$$\frac{nh^5 m}{90} \le E(h) \le \frac{nh^5 M}{90}$$

where n times 2 is the number of intervals, m is $\min|f^{(4)}(x)|$, and M is $\max|f^{(4)}(x)|$ on $a < x < b$. The symbol $f^{(4)}(x)$ stands for the fourth differential of the function being integrated.
6. *With an odd number of data points in Fig. 13.8 we cannot use the Simpson's rule to find the area under the curve. However, we can use the

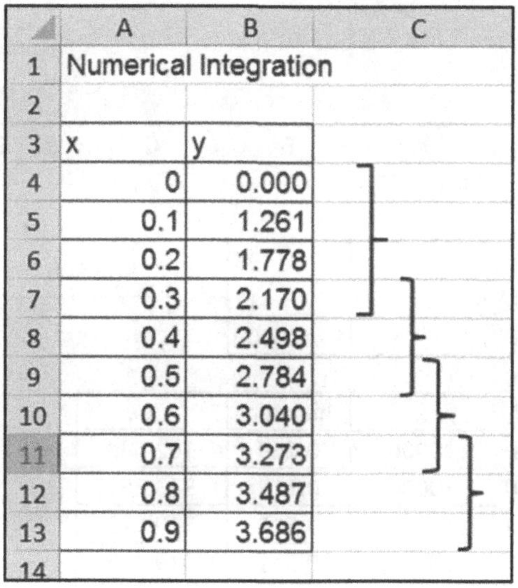

	A	B	C
1	Numerical Integration		
2			
3	x	y	
4	0	0.000	
5	0.1	1.261	
6	0.2	1.778	
7	0.3	2.170	
8	0.4	2.498	
9	0.5	2.784	
10	0.6	3.040	
11	0.7	3.273	
12	0.8	3.487	
13	0.9	3.686	
14			

■ FIG. 13.8

[3] A. Jeffrey, Mathematics for Engineers and Scientists, Chapman and Hall, New York, 1996 (page 770).

rule on the first four points, then the rule on groups of three for the rest of the data. What answer do you get?

7. Write a subroutine to find the area between the circumference of a circle of radius r and a horizontal line, which is a distance h from the circumference along a radius at right angles to the line—see diagram to left as follows. Your subroutine should read the r, h, and n values from the worksheet (Fig. 13.9), use a function to compute the area, and put data in A8:C14.

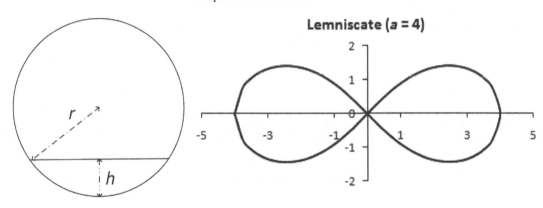

Lemniscate ($a = 4$)

	A	B	C	D
1	Monte Carlo			
2				
3	r	25		Run Sub
4	h	11		
5	n	10000		
6				
7	n	hit ratio	area	
8	10000	0.260	162.688	
9	10000	0.517	323.250	
10	10000	0.511	319.625	
11	10000	0.511	319.125	
12	10000	0.536	335.250	
13	10000	0.522	326.500	
14	Average	0.520	324.750	

■ FIG. 13.9

8. The equation of a lemniscate (see diagram to right above) is $\rho = a^2\cos(2\theta)$. The length (s) of an arc of its perimeter is given by $s = \int_0^\theta \frac{ad\theta}{\sqrt{\cos(2\theta)}}$.
   Make a worksheet that will find s for $0 <= \theta <= 90$. Make it general enough that it works for any a value.

9. *Using either the trapezoid or Simpson's method, evaluate the area under the curve represented by the data in the following table. Then plot the data and add a polynomial trendline. Use LINEST to get the coefficients of the polynomial and integrate the polynomial to find the area. How good is the agreement?

x	3.00	3.60	4.00	4.70	5.50	6.25	7.00	7.50
y	1.17	1.02	0.90	0.63	0.59	0.63	0.66	0.79
x	8.60	9.00	9.40	10.00	10.45	10.85	11.25	
y	1.20	1.45	1.65	2.05	2.26	2.62	3.00	

10. Use Simpson's rule and Gaussian integration with $n = 8$ to evaluate $\int_0^\pi \sin(x)dx$. Which method gives the better approximation?

11. The following table gives values for $f(x)$. Use the trapezoid rule to find the area under the curve with steps (h) values of 0.1, 02, and 0.4.

x	1	1.1	1.2	1.3	1.4	1.5	1.6	1.7	1.8
f(x)	1.543	1.669	1.811	1.971	2.151	2.352	2.577	2.828	3.107

▲	A	B	C	D	E
1	Simpson's Rule				
2					
3	Lower	0			
4	Upper	5			
5	Exact	2.339766			
6					
7	Delta	Integral	Error		
8	0.5	2.337191	-2.58E-03		
9	0.1	2.339766	-1.81E-07		
10	0.05	2.339766	-1.13E-08		
11	0.001	2.339766	0.00E+00		

$$\int_0^5 \frac{1}{1+(x-\pi)^2}dx$$

■ FIG. 13.10

12. *Compose a user-defined function to compute a Simpson's rule approximation for the integral shown in Fig. 13.10 for various step sizes. Visit Wolframalpha.com to find the formula for the exact value.

APPENDIX: SUPPLEMENTARY MATERIAL

Supplementary material related to this chapter can be found on the accompanying CD or online at https://doi.org/10.1016/B978-0-12-818249-9.00013-3.

Differential Equations

Differential equations occur in many physical problems. Let us look at some simple examples.

A body falling through the air is subjected to two forces: gravity acting downward and air resistance acting upward. The first force is constant, but the second is proportional to the body's velocity. This gives rise to a first-order differential equation.	$m\frac{dv}{dt} = g - kv^2$
Consider the chemical reaction $A + B \rightarrow C$ where the rate of reaction is proportional to the concentration of A and to the concentration of B. Let x be the amount of A and B reacted at time t, and let the initial concentration of A and B be a and b, respectively. These quantities will be related by the equation shown here.	$\frac{dx}{dt} = k_2(a-x)(b-x)$
The motion of a harmonic oscillator is the second-order differential equation shown here.	$\frac{d^2x}{dt^2} + \omega^2 x = 0$

Each of the examples earlier is readily solved. The analytical solution for other differential equations such as $dy/dx = (x+y)/(x-y)$ is more difficult

Liengme's Guide to Excel 2016 for Scientists and Engineers. https://doi.org/10.1016/B978-0-12-818249-9.00014-5

and for some, there is no analytical solution. In these cases, we may use numerical methods to find approximate solutions.

Consider the simple equation $dv/dt = g$ for a falling body when air resistance is ignored. This integrates to give $v = gt + c$ where c is the integration constant and g is a constant of known value. Thus we do not have a unique solution since any value of c will satisfy the differential equation. By inspection of the solution, we see that c is the value of v when t equals zero. We need to know this value in order to uniquely solve the equation. In general, to solve $dy/dx = f(x,y)$ over the x range $[a, b]$, we need to know the value of $y(a)$, which is called the *initial value*. Problems of this type are called *initial value* problems. With second-order differential equations, two integration constants arise. For an initial value problem, we need to know the initial value of the two values of the dependent variables. Alternatively, the problem may be defined by specifying some conditions at one value of x and others at another value of x. Such problems are called *boundary value* problems.

EXERCISE 1: EULER'S METHOD

Euler developed a method for finding the approximate solution to initial value problems. Let the differential equation to be solved have the form $dy/dx = y' = f(x,y)$ and let the initial value of y be y_0. Let the solution (i.e., the integral y') have the form $y = g(x,y)$—see Fig. 14.1.

Integrating y' from x_0 to x_1 yields $y_1 = y_0 + \int_{x_0}^{x_1} f(x, y)dx$. We may think of the second term as the area under the curve $f(x, y)$ between the two x-values.

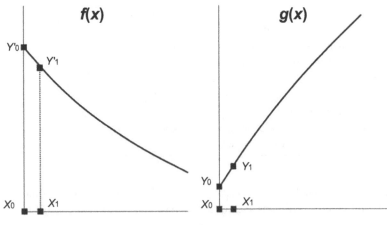

■ FIG. 14.1

Euler approximated this to the area of the rectangle defined by y'_0, y'_1, x_0, and x_1. The approximate value of y_1 is then given by $y_1 = y_0 + (x_1 - x_0)f(x_0, y_0) = y_0 + hf(x_0, y_0)$ when the x increment is represented by h.

Having found an approximation for y_1, we may now find an approximation for y_2 with $y_2 = y_1 + hf(x_1, y_1)$. In general, the value of the approximation at one point is found from the previous one using

$$y_{n+1} = y_n + hf(x_n, y_n) \qquad (14.1)$$

We are not generating $g(x, y)$ but numerical points which are approximations; we can improve the approximations by using smaller values of h but the further we move away from y_0, the more our approximations will deviate from $g(x, y)$. This exercise will demonstrate both facts.

In this exercise, Euler's method is used to find an approximate solution of the differential equation $dy/dx = xy$, with the initial value $y(0) = 1$. The approximation is compared to the analytical solution, namely, $y = \exp(x^2/2)$.

(a) Open a new workbook. On Sheet1 enter the text shown on A1:E5 of Fig. 14.2. Enter the values in A6:A11.

(b) In C2 enter 0.1 for the value of h. Name the cell as h.

(c) In B6 enter 0, the initial value of x. In C6 enter 1.0 for the initial value of y from the condition $y(0) = 1$; this corresponds to the first term on the right of the recursive formula when $n = 0$.

(d) In D6 enter =h*(B6*C6). This corresponds to the second term on the right of the recursive formula when $n = 0$. The parentheses are not essential here but are used to make it clear that we are computing the value $h*function$.

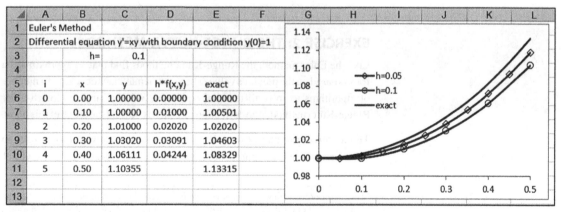

	A	B	C	D	E	F
1	Euler's Method					
2	Differential equation y'=xy with boundary condition y(0)=1					
3			h=	0.1		
4						
5	i	x	y	h*f(x,y)	exact	
6	0	0.00	1.00000	0.00000	1.00000	
7	1	0.10	1.00000	0.01000	1.00501	
8	2	0.20	1.01000	0.02020	1.02020	
9	3	0.30	1.03020	0.03091	1.04603	
10	4	0.40	1.06111	0.04244	1.08329	
11	5	0.50	1.10355		1.13315	
12						
13						

■ FIG. 14.2

(e) In B7 enter =B6+h to increment x. In C7 we compute the first approximation of y with =C6+D6. This corresponds to Eq. (14.1). Copy D6 down to D7.

(f) Copy the cells B7:D7 down to row 11. This computes the successive y approximations. Since we shall not be using the last value of $h \times f(x, y)$, delete D11.

(g) So that we can compare our approximations in the C column with the exact solution, in E6 enter =EXP(B6^2 / 2) and copy this down to E11. Save the workbook as Chap14.xlsm. We need the file to be macro-enabled for later exercises.

Clearly, our answer in the C column is not in very good agreement with the exact values in the E column. A better approximation may be obtained by reducing the size of h, the increment for the x-values as demonstrated in the chart in Fig. 14.2. This shows that (i) the deviation from the exact values increases with each iteration of the reclusion as expected, and (ii) decreasing the size of h significantly improves the solution values.

(h) Modify your worksheet to find the approximations of this differential equation for x-values from 0 to 0.5 with steps of 0.025. Save the workbook.

In this example, we have used the "crude" Euler method. In going from $y_1 = y_0 + \int_{x_0}^{x_1} f(x, y)dx$ to $y_1 = y_0 + hf(x_0, y_0)$ the integral term was approximated to the area of a rectangle. In the improved Euler method, it is approximated to a trapezoid. Compared to the original Euler method, this requires fewer calculations for comparable accuracy. We shall not examine the modified method. The next exercise uses a more modern method.

EXERCISE 2: THE RUNGE-KUTTA METHODS

Like the Euler method, the Runge-Kutta methods find that an approximation for y is based on the previous value. These mathematicians developed a number of algorithms to solve differential equations. We shall use the fourth-order Runge-Kutta method,[1] the derivation of which is beyond the scope of this book.

The iterative formula shown following may look somewhat formidable, so let us see how we can put it into a worksheet. We have seen in Exercise 1

---

[1]The fourth-order method is sometimes called the Kutta-Simpson formula since, when the right-hand side of the differential equation is a function of x alone, it reduces to Simpson's ⅓ rule.

how to evaluate the equivalent to k_1; this is the value of the differential function for various x- and y-values. The second parameter, k_2, is similar except that the x-value is incremented by h while the y-value is incremented by k_1. Each parameter increments y by a multiple of the parameter preceding it.

$$y_{n+1} = y_n + \tfrac{1}{6}(k_1 + 2k_2 + 2k_3 + k_4)$$
$$k_1 = hf(x_n, y_n)$$
$$k_2 = hf(x_n + \tfrac{1}{2}h, y_n + \tfrac{1}{2}k_1)$$
$$k_3 = hf(x_n + \tfrac{1}{2}h, y_n + \tfrac{1}{2}k_2)$$
$$k_4 = hf(x_n + h, y_n + k_3)$$

In the previous exercise we solved $dy/dx = xy$ with the initial value $y(0) = 1$ using Euler's method. Here we solve the same problem using the Runge-Kutta method so that we may compare the results.

(a) Open the workbook Chap14.xlsm and make Sheet2 active. Enter the text shown on A1:H5 of Fig. 14.3.
(b) Name the cell C4 as h.
(c) Enter the series of values in A6:A11 and B6:B11.
(d) In C6 enter the value 1.0. This is the initial condition $y(0) = 1$.

	A	B	C	D	E	F	G	H
1	Runge-Kutta Method							
2	Differential equation y' = x*y with boundary condition y(0)=1							
3								
4		h =	0.1					
5	i	x	y	k1	k2	k3	k4	Error
6	0	0.0	1.00000	0.00000	0.00500	0.00501	0.01005	0.000E+00
7	1	0.1	1.00501	0.01005	0.01515	0.01519	0.02040	-2.607E-11
8	2	0.2	1.02020	0.02040	0.02576	0.02583	0.03138	-2.684E-10
9	3	0.3	1.04603	0.03138	0.03716	0.03726	0.04333	-1.023E-09
10	4	0.4	1.08329	0.04333	0.04972	0.04987	0.05666	-2.854E-09
11	5	0.5	1.13315					-6.949E-09

■ FIG. 14.3

(e) Enter these formulas to compute the *k* parameters:

```
D6:  =h*(B6 * C6)
E6:  =h*((B6+h/2) * (C6+D6/2))
F6:  =h*((B6+h/2) * (C6+E6/2))
G6:  =h*((B6+h) * (C6+F6))
```

(f) In C7 enter =C6+(1/6)*(D6+2*E6+2*F6+G6) to compute the first approximation. Compare this formula with the first equation in the Runge-Kutta recursion formula earlier.

(g) Copy the cells D6:G6 to D7:G7. Then copy C7:G7 down to row 10 and copy C10 to C11. This computes the successive *y* approximations.

(h) To compute the error in our approximations enter in H6 =C6 -EXP (B6^2 / 2); after formatting to scientific, copy it down to H11. Save the workbook Chap14.xlsm.

We have shown that for the equation *dy/dx=xy*, the Runge-Kutta method is clearly far superior to Euler's method. It may be shown that this is true for all equations.

EXERCISE 3: SOLVING WITH A USER-DEFINED FUNCTION

In the previous exercise, we solved *dy/dx=xy*. We would need to make many edits to the worksheet to solve for another equation *dy/dx=f(x,y)*. Furthermore, the differential to be solved was typographically rather simple (just $x \times y$) but with more complex functions there is always the worry of mistyping the function somewhere. If we put the function *f(x, y)* in a module, we need edit only the module (and the initial value) to change our worksheet.

In this exercise we find the values of *y* that satisfy the equation *dy/dt=y−t²+1* with the initial value *y(0)=0.5*. We use *x*-values from 0 to 2.0 in increments of 0.25. We will compare the result with the exact solution of $y(t)=(t+1)^2 - 0.5\exp(t)$.

(a) Open the VBE and insert a module on the Chap14 project. Enter this function on the module sheet:
```
Function RKfunc(t, y)
    RKfunc = y-t^2+1
End Function
```

(b) On Sheet3 enter the text and values shown in A1:J7 of Fig. 14.5. Name the cells C4:C6 using the text in B4:B6.

(c) Enter the series of values in A8:A16.
Enter these formulas:

B8: =x0	The initial x value
C8: =y0	The initial y value
D8: =h*rkfunc(B8, C8)	The k parameters
E8: =h*rkfunc ((B8+h/2), (C8+D8/2))	
F8: =h*rkfunc ((B8+h/2), (C8+E8/2))	
G8: =h*rkfunc ((B8+h), (C8+F8))	
I8: =(B8+1)^2-0.5*EXP(B8)	The exact solution of $y(t)$
J8: =C8-I8	The error

(d) In B9 enter =B8+h to increment Δx.

(e) In C9 enter =C8+(1/6)*(D8+2*E8+2*F8+G8) to compute the first
approximation for y.

(f) Copy D8:J8 to line 16. We have no need of entries D16:G16 so these
can be deleted. Save the workbook.

If your values do not agree with Fig. 14.4 you need to check the
function in the module and the formulas on the worksheet. Remember
that formulas can be displayed with Ctrl+`. To check the function,

⊿	A	B	C	D	E	F	G	H	I	J
1	Runge-Kutta Method									
2	Solving dy/dx = f(x,y) with a user-defined function					Diff Eqn	dy/dt = y − t² + 1			
3						Exact	y(t) = (t+1)² − 0.5Exp(t)			
4	increment	h	0.25							
5	initial x value	x0	0							
6	initial y value	y0	0.5							
7	i	t	y	k1	k2	k3	k4		exact	error
8	0	0.00	0.5000000	0.37500	0.41797	0.42334	0.46521		0.5	0.00E+00
9	1	0.25	0.9204712	0.46449	0.50302	0.50784	0.54458		0.9204873	-1.61E-05
10	2	0.50	1.4256038	0.54390	0.57673	0.58084	0.61099		1.4256394	-3.56E-05
11	3	0.75	2.0039410	0.61036	0.63587	0.63906	0.66075		2.0040000	-5.90E-05
12	4	1.00	2.6407719	0.66019	0.67631	0.67833	0.68915		2.6408591	-8.71E-05
13	5	1.25	3.3172078	0.68868	0.69273	0.69324	0.69011		3.3173285	-1.21E-04
14	6	1.50	4.0089950	0.68975	0.67831	0.67688	0.65584		4.0091555	-1.60E-04
15	7	1.75	4.6849913	0.65562	0.62429	0.62038	0.57634		4.6851987	-2.07E-04
16	8	2.00	5.3052097						5.3054720	-2.62E-04

■ FIG. 14.4

move to a blank cell such as A20 and enter =rkfunc(3,1). This should return the value 0.25.

The reader is encouraged to experiment with smaller values of h and observe the changes in the final error. It will be necessary to extend the worksheet by dragging row 15 down as far as needed.

Now that we have solved one equation, let us see how readily we can solve another. We will make a copy of Sheet3 and make a few modifications to solve the equation $dy/dx = x^2 + y$ with the initial value $y(1) = 1$ to find the value of $y(1.5)$ using $h = 0.05$ and compare it with the exact solution $y(1.5) = 2.64232762$ (to eight places) found from $y(x) = -x^2 - 2x + 6\exp(x-1) - 2$.

(g) The first operation is to make a copy of Sheet3. We could, of course, use Copy and paste but we will explore another method. Click on the Sheet3 tab and while holding down the Ctrl key, drag the mouse pointer to the right to make a new sheet called Sheet3 (2). Note that all the named cells have their names preserved with this method.

(h) Open the VBA module and make a new function:

```
Function RKfunc2(x, y)
        RKfunc2 = X62 + y
End Function
```

(i) Edit the text cells G2, G3, and B7 to reflect the new function.

(j) Change the values in C4:C6 to 0.05, 1, and 1.

(k) Adjust the formula[2] in I8 to =-(B8ˆ2)-2*B8+6*EXP(B8-1)-2

(l) Use *Home / Editing / Find & Select / Replace (Edit / Find / Replace...* on a Mac) to change rkfunc to rekfunc2.

(m) Extend the worksheet so that the final x value is 1.5. Save the workbook.

Hopefully, your result for $y(1.5)$ agrees with that obtained from the analytical result to at least six decimal places. Experiment with a smaller h value to see how much improvement this results in your approximate value.

Not a great deal of work! But perhaps you are less than impressed by all this since we already know what the answers should be. In Problem 5 you will be asked to solve differential equations for which analytical solutions are hard to find.

---

[2]The parentheses around B8ˆ2 are needed because in Excel negation has a higher precedence that exponentiation; we want to square x before making the result negative.

SIMULTANEOUS AND SECOND-ORDER DIFFERENTIAL EQUATIONS

The Runge-Kutta method can be extended to solve systems of simultaneous differential equations and, by extension, differential equations of order higher than first. We will restrict ourselves to pairs of simultaneous differential equations and the second-order differential equations.

Consider a pair of simultaneous equations having the form:

$$y' = g(x, y, z)$$
$$u' = f(x, y, z) \tag{14.2}$$

The Runge-Kutta formulas for these equations are as follows:

$$y_{n+1} = y_n + \tfrac{1}{6}(k_1 + 2k_2 + 2k_3 + k_4)$$

$$u_{n+1} = u_n + \tfrac{1}{6}(q_1 + 2q_2 + 2q_3 + q_4)$$

$$k_1 = hg(x_n, y_n, u_n)$$

$$q_1 = hf(x_n, y_n, u_n)$$

$$k_2 = hg\left(x_n + \tfrac{1}{2}h, y_n + \tfrac{1}{2}k_1, u_n + \tfrac{1}{2}q_1\right)$$

$$q_2 = hf\left(x_n + \tfrac{1}{2}h, y_n + \tfrac{1}{2}k_1, u_n + \tfrac{1}{2}q_1\right) \tag{14.3}$$

$$k_3 = hg\left(x_n + \tfrac{1}{2}h, y_n + \tfrac{1}{2}k_2, u_n + \tfrac{1}{2}q_2\right)$$

$$q_3 = hf\left(x_n + \tfrac{1}{2}h, y_n + \tfrac{1}{2}k_2, u_n + \tfrac{1}{2}q_2\right)$$

$$k_4 = hg(x_n + h, y_n + k_3, u_n + q_3)$$

$$q_4 = hf(x_n + h, y_n + k_3, u_n + q_3)$$

Equations of second order and greater may be solved by transforming them into sets of simultaneous equations. For example, to solve $y'' = ay' + by + c$, we make the substitution $y' = u$. The introduction of the auxiliary variable u allows us to write the second-order equations as two simultaneous equations:

$$y' = u$$
$$u' = au + by + c \tag{14.4}$$

Combining these with the Runge-Kutta formulas for a pair of simultaneous ones we see that function g is now just a function of u. This reduces the k terms to:

$$k_1 = h(u_n)$$
$$k_2 = h\left(u_n + \tfrac{1}{2}q_1\right)$$
$$k_3 = h\left(u_n + \tfrac{1}{2}q_2\right) \tag{14.5}$$
$$k_4 = h(u_n + q_3)$$

EXERCISE 4: SOLVING A SECOND-ORDER EQUATION

In this Exercise we apply the equations developed before to solve $y'' = y' + y = \sin(x)$ with boundary conditions $y(0) = 0$ and $y'(0) = 0$. Our task is to obtain approximate values of y and y' when $x = 1$.

With the substitution $y' = u$, we get a pair of equations:

$$y' = u \qquad\qquad \text{initial value } y(0) = 0$$
$$u' = \sin(x) - y - u \qquad \text{initial value } u(0) = 0$$

Comparing these with Eq. (14.2) we see that $g = u$, so we will use the simplified k values of Eq. (14.5).

We also see that $f = \sin(x) - y - u$. The function f is referenced in each of the q terms, so it will be more convenient to use a module function. Furthermore, by changing the module you will be able to use the same worksheet for another function.

(a) With Chap14.xlsm open, go to the VBE and insert a new module. For this exercise, code the function:

```
Function f(x, y, u)
        f = Sin(x) - y - u
End Function
```

(b) Move to Sheet4 and enter the text and values shown in A1:K6 of Fig. 14.5.

(c) Select A3:D4 and name the cells in row 4.

(d) The formulas in row 7 are as follows.

	A	B	C	D	E	F	G	H	I	J	K
1	Second-order differential equation						$y'' + y' + y = \sin(x)$				
2							$y''(0) = y(0) = 0$				
3	xinit	yinit	uinit	h							
4	0	0	0	0.2							
5											
6	x	y	u	k1	q1	k2	q2	k3	q3	k4	q4
7	0.0	0.000	0.000	0.000	0.000	0.000	0.020	0.002	0.018	0.004	0.036
8	0.2	0.001	0.019	0.004	0.036	0.007	0.051	0.009	0.049	0.014	0.062
9	0.4	0.010	0.068	0.014	0.062	0.020	0.073	0.021	0.071	0.028	0.079

■ FIG. 14.5

A7:   =xinit

B7:   =yinit

C7:   =uinit

D7:   =h*C7

E7:   =h*f(A7,B7,C7)

F7:   =h*(C7+E7/2)

G7:   =h*f(A7+h/2,B7+D7/2,C7+E7/2)

H7:   =h*(C7+G7/2)

I7:   =h*f(A7+h/2,B7+F7/2,C7+G7/2)

J7:   =h*(C7+I7)

K7:   =h*f(A7+h,B7+H7,C7+I7).

(e) The formulas in row A8:C8 are shown here. Those in columns D to K may be copied from row 7.

A8:   =A7+h

B8:   =B7+(D7+2*F7+2*H7+J7)/6

C8:   =C7+(E7+2*G7+2*I7+K7)/6

D8:   =h*C8

(f) Copy row 8 down to row 12 to get a final value of $x=1.0$.
The RK approximations (to seven places) should be $y(1.0)=0.1193941$ and $y'(1.0)=0.3079599$. The analytical to seven decimals results are $y(1)=0.1193978$ and $y'(1)$ $i=0.3079638$.

(g) Try other values of h such as 0.1 and 0.05 to see if the approximations converge. You will need to expand the table to have $x=1.0$ in the final row. Save the workbook.

EXERCISE 5: THE SIMPLE PENDULUM

The equation of motion for a simple pendulum of length L is as follows

$$\frac{d^2\theta}{dt^2} + \frac{g}{L}\sin(\theta) = 0.$$

Most textbooks consider a pendulum that starts with a small displacement and uses the approximation $\sin(\theta) \approx \theta$. Our approximation will be to use the Runge-Kutta method to solve this second-order differential equation to show how the angle and angular velocity change with time. We will model a 0.75-m pendulum which is started with a displacement of 0.8 rad from the perpendicular.

As before, we start with the substitution $d\theta/dt = u$, giving:

$$\theta' = u \qquad\qquad \theta(t=0) = 0.8$$
$$u' = -\left(\frac{g}{L}\right)\sin(\theta) \qquad u(t=0) = 0$$

(a) On the same module used for Exercise 4, code the Pend function

```
Function Pend(L, angle )
    g = 9.8
    Pend = (-g/L) * Sin(angle)
End Function
```

The parentheses around g/L help in reading the formula.

(b) On Sheet5 enter the text and values shown in A1:K6 of Fig. 14.6.

(c) The formulas needed in row 7 start with:

A7:	=InitTime
B7:	=InitAngle
C7:	=InitVel
D7:	=h*C7
E7:	=h*Pend(Length, B7)
F7:	=h*(C7+E7/2)
G7:	=h*Pend(Length, B7+D7/2).

(d) Using what you learned in Exercise 4, complete the formulas in rows 7 and 8. Copy row 8 down to row 37.

(e) Make a chart showing how the angle and the velocity vary with time. Save the workbook.

Chapter 15 contains some further examples of the Runge-Kutta method.

	A	B	C	D	E	F	G	H	I	J	K
1	Pendulum										
2											
3	InitTime	InitAngle	InitVel	Length	h						
4	0	0.8	0	0.75	0.1						
5											
6	Time	Angle	Velocity	k1	q1	k2	q2	k3	q3	k4	q4
7	0.0	0.800	0.000	0.000	-0.937	-0.047	-0.937	-0.047	-0.916	-0.092	-0.894
8	0.1	0.753	-0.923	-0.092	-0.894	-0.137	-0.849	-0.135	-0.827	-0.175	-0.758
9	0.2	0.618	-1.757	-0.176	-0.757	-0.214	-0.661	-0.209	-0.640	-0.240	-0.520
10	0.3	0.408	-2.403	-0.240	-0.519	-0.266	-0.371	-0.259	-0.355	-0.276	-0.195

■ FIG. 14.6

PROBLEMS

1. Use Euler's method to solve the differential equation

$$\frac{dy}{dx} = x + y$$

   with the initial condition $x(0) = 0$, $y(0) = 0$. Use steps of 0.1 and 0.05 up to $x = 1$. How does your result compare with the analytical solution $y = e^x - x - 1$? How small must h be such that the $y(1)$ values with h and $h/2$ agree to within 1×10^{-6}?

2. *Use Euler's method with $h = 0.05$ to solve the differential equation $y' = -2xy$ with $y(0) = 1$ for $0 \leq x \leq 2$. The exact solution is $y = \exp(-x^2)$, but be careful how you make the Excel formula; remember, negation has higher priority than exponentiation.

3. Repeat Problem 2 using the Runge-Kutta method by copying and modifying the worksheet from Exercise 2. Graphically compare (Fig. 14.7) the errors in each method with $h = 0.05$.

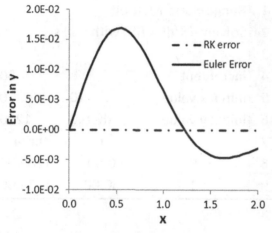

■ FIG. 14.7

4. With the instructions of Exercise 2 as a guide, use the Runge-Kutta method with $h = 0.1$ to solve the differential equation $y' = -2x - y$ with $y(0) = -1$ for $0 \leq x \leq 2$. Now modify your work to use a UDF that you have coded in VBA.

---

[3] R. L. Burden and J. D. Faires, *Numerical Methods (9th Ed)*, Brook/Cole, Boston, 2011 (page 292).

5. *With the instructions of Exercise 3 as a guide, use the Runge-Kutta method using $h=0.2$ to solve the differential equation $y' = 1/(x+y)$ with boundary condition $y(0)=2$ and find $y(1)$.

6. A ball at 1200 K is allowed to cool down in air at an ambient temperature of 300 K. Assuming heat is lost only due to radiation, the differential equation for the temperature of the ball is given by $\frac{d\theta}{dt} = -2.2067 \times 10^{-12} \left(\theta^4 - 81 \times 10^8\right)$ where θ is in K and t in seconds. Find the temperature at 1000 s using Runge-Kutta fourth-order method with step sizes $h=50$. Start by making a copy of the sheet used for Problem 5.

7. Your task is to redo Problem 6 dispensing with the four columns that compute the k parameters by calling a UDF—see Fig. 14.8. Thus in C9 you would use =RungeKutta(B8,C8,h). Of course, the RungeKutta function may itself call another function to compute $f(t, theta)$.

	A	B	C
1	Runge-Kutta Method		
2	Solving dθ/dt = f(θ) with UDFs		
3			
4	increment	h	50
5	initial x value	t0	0
6	initial y value	theta0	1200
7	i	t	theta
8	0	0.00	1200.00
9	1	50.00	1032.66

■ FIG. 14.8

8. *Water flows from an inverted conical tank with a circular orifice at the rate given[3] by

$$x'(t) = -0.6\pi r^2 \sqrt{2g} \frac{\sqrt{x}}{A(x)}$$

where r is the radius of the orifice, x is the height of the water level from the vertex of the cone, and $A(x)$ is the area of cross section of the tank x

---

[3]R. L. Burden and J. D. Faires, *Numerical Methods (9th Ed)*, Brook/Cole, Boston, 2011 (page 292).

units above the orifice. Let $r=0.1\,\text{ft}$, $g=32\,\text{ft/s}^2$. The tank was initially filled with $512\pi/3$ cubic feet of water to a level of 8 ft. Estimate (i) the water level after 10 min, and (ii) the time within, 1 min, taken to empty the tank.

9. *The circuit[4] shown in Fig. 14.9 contains a battery (E), an inductance (L), and a resistor (R) whose magnitude varies with its temperature and hence with the current passing through it. Its resistance can be expressed by $R=a+bi$, where a and b are constants and i is the current. The switch (S) is closed at time $t=0$ and the resulting current can be described by the differential equation: $\frac{di}{dt}=\frac{E}{L}-\frac{b}{L}i^3-\frac{a}{L}i$. Using Exercise 3 as a model, compute the current from $t=0$ to $t=0.8\,\text{s}$ in increments of 0.001 for the case $E=200\,\text{V}$, $L=3\text{H}$, $a=100\,\Omega$, and $b=50\,\Omega/\text{A}^2$. Since the independent variable t does not appear explicitly in the differential, the terms for the Runge-Kutta k's will involve only the current i.

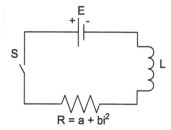

E

S

L

$R = a + bi^2$

■ FIG. 14.9

10. A 500-gal tank is filled with water with 20 lbs of dissolved salt. Fresh water flows in at 10 gal/min. How long will it take until there are just 5 lbs of salt in the tank? Assume perfect mixing. Solve[5] this without writing down a differential equation; see Fig. 14.10. Can you write a formula to compute t when $Salt=5$ (see cell H3 in figure)? As Δt gets smaller, your answer will converge to a more accurate value.

11. *Write a differential equation for Problem 10. Integrate this analytically. Make a plot of $Salt$ against time from the approximations in Problem 11 and compare its trendline to that predicted by your analytical solution; the exact values are in column H of Fig. 14.10 starting in row 9.

12. Suppose a ship moving at speed 6 m/s suffers a sudden loss of power. We will assume the distance s (meters) it moves in time t (seconds) is governed by the differential equation[6] $\frac{ds}{dt}=v_0\exp\left(-kt/m\right)$ with initial condition $s(0)=0$. Use the Runge-Kutta method to find how far it will move in the first 60 s if $k=44\times10^3\,\text{kg/s}$ and $m=2.55\times10^6\,\text{kg}$. For your first attempt, use steps of $h=10\,\text{s}$. Then repeat using steps of $h=5$ and $h=1$. Compare your approximations with the exact values computed with $s(t)=\frac{v_0m}{k}\left(1-\exp\left(-kt/m\right)\right)$.

---

[4]M. L. James et al., *Applied Numerical Methods for Digital Computation*, Harper & Row, New York, 1977 (page 406).

[5]Hint: Imagine the outflow can be paused. Let 10 gals flow in; what is the new concentration? Let the 10 gals escape in a flash. How much salt remains? Do this for successive minutes—columns A through C starting in row 9 of the figure. Can a single formula be used to find the amount after n minutes—columns E through G?

[6]J. R. Hanly, *Essential C++ for Engineers and Scientists*, Addison-Wesley, Reading, MA, 1977 (page 362).

	A	B	C	D	E	F	G	H
1	Tank Problem							
2								
3	Volume	500			t (for S= S_end)) =			72.73
4	S_start	20			when dt = 5			
5	S_end	5						
6	Flow	10						
7	dt	5						
8								
9	n	t	salt		n	t	salt	Exact
10	0	0	20		0	0	20	20
11	1	5	18.182		20	100	2.973	2.707
12	2	10	16.529		40	200	0.442	0.366

■ FIG. 14.10

13. Getting your information from a textbook and/or the Internet, repeat Problem 4 using (i) integration using the Taylor series method and (ii) the Runge-Kutta-Fehlberg method. The second method is used in programs such as MathLab and Maple for their OED routines.

14. For the initial value problem $y' = \cos^2\left(\frac{y}{x}\right) + \left(\frac{y}{x}\right), y(1) = 0$, using $h = 0.1$ determine an approximation of $y(2)$ using (a) the Euler method and (b) the RK4 method. Obtain an exact solution from Wolframalpha.com and compare the errors of the numerical methods.

15. For the initial value problem

$$y' = \left(\frac{y}{x}\right)(x^2 + 3 \ln y), y(1) = 1,$$

using $h = 0.1$ determine an approximation of $y(2)$ using the RK4 method. Obtain an exact solution from Wolframalpha.com.

16. A second-order differential equation problem. Consider an object of mass m falling under the influence of gravity g (e.g., a paratrooper) subject to a drag force which is proportional to the square of the object's velocity v. The equation of motion is as follows

$$m\frac{d^2y}{dt^2} = -mg + kv^2$$

where y is the object's distance from the ground. Remember that $v = dy/dt$. Following the technique shown in Exercise 4, set up a worksheet using the RK4 method with $h = 0.2$ to estimate the height (y) and velocity (v) for the first 20 s of the fall when $y(0) = 1200$ m. Use constants of $m = 80$ kg, $g = 9.81$ m s^{-2}, and $k = 0.25$ kg m^{-1}.

Make a plot of height and velocity against time. Compare your

results with the analytical solution[7] $v(t) = \sqrt{\frac{mg}{k}} \tanh\left(\sqrt{\frac{kg}{m}} \cdot t\right)$.

17. Using Exercise 3 as a guide, solve[8] the initial value problem

$$\frac{dx}{dt} = -x\ln(x); x(0) = 0.0001$$

using the RK4 method with $h = 0.2$ to show that the plot of x against t is sigmoid (S-shaped) with x asymptotically approaching 1.0. Wolframalpha.com give a solution with a small imaginary term (perhaps an error!) which if ignored should agree with your results.

---

[7]Dennis G. Zill, Differential Equations with Computer Experiments (page 86), PWS Publishing Company, Boston, 1995.

[8]This is the Gompertz Law which has been used in cancer studies. See J Aroesty et al., *Mathematical Bioscience* 17:243, 1973, and C M Newton, *Ann. Rev. Biophys. Bioeng.* 9:541, 1980.

Chapter

15

Modeling II

This chapter will give us the opportunity to model some practical problems using what we have learned in the last three chapters.

EXERCISE 1: THE FOUR-BAR CRANK

In this exercise, we examine an engineering mechanism used to generate a complex rotational motion from a simple one motion.

The four-bar mechanism (see Fig. 15.1) consists of three movable links (a, b, and c) and a fixed link d. The link a is rotated, causing link c to rotate. Our objective is to map the relationship between the angles θ and ϕ—see Fig. 15.2.

For the quadrilateral formed by the four links, the algebraic sum of the vertical component and the algebraic sum of the horizontal component must equate to zero. This gives the two equations:

$$a\sin\theta + b\sin\beta - c\sin\phi = 0 \quad \text{and} \quad a\cos\theta + b\cos\beta - c\cos\phi + d = 0.$$

Adding the squares of these gives the Freudenstein equation: $R_1 - R_2\cos\phi + R_3 - \cos(\theta - \phi) = 0$ where $R_1 = d/c$, $R_2 = d/a$, and $R_3 = (a^2 - b^2 + c^2 + d^2)/2ac$.

Liengme's Guide to Excel 2016 for Scientists and Engineers. https://doi.org/10.1016/B978-0-12-818249-9.00015-7

355

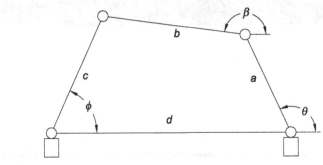

■ FIG. 15.1

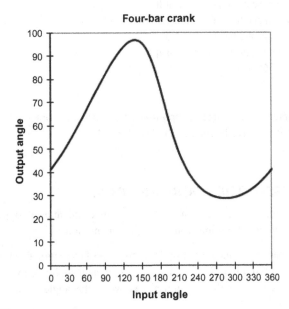

■ FIG. 15.2

We will use Microsoft Excel's Solver to find the output angle for input angles in the range of 0–360 degrees in 5 degrees steps. Our completed worksheet, prior to running Solver, will resemble Fig. 15.3.

(a) Start a new workbook and on Sheet1 enter all the text shown in Fig. 15.2. Cell A13 contains a formula, not text.

(b) Enter the specifications for the four-bar crank in B5:B8.

(c) Name the cells B5:B11 with the text to their left—it is OK to name empty cells!

	A	B	C	D	E	F	G	H
1	Four-bar Crank			Input Angle		Output Angle		Freudenstein's
2				Degrees	Radian	Degrees	Radian	Equation
3				0	0	57.3	1	0.6291
4	Crank lengths			5	0.0873	57.3	1	0.5540
5	a	1		10	0.1745	57.3	1	0.4760
6	b	2		15	0.2618	57.3	1	0.3956
7	c_	2		20	0.3491	57.3	1	0.3136
8	d	2		25	0.4363	57.3	1	0.2304
9	Ratio1	1		30	0.5236	57.3	1	0.1468
10	Ratio2	2		35	0.6109	57.3	1	0.0633
11	Ratio3	1.25		40	0.6981	57.3	1	-0.0193
12				45	0.7854	57.3	1	-0.1006
13	Unsolved			50	0.8727	57.3	1	-0.1797

■ FIG. 15.3

(d) Compute the *ratio* values with the formulas:

B9: =d/c
B10: =d/a
B11: =(a^2 -b^2+c_^2+d^2)/(2*a*c_)

(e) In D3 enter the value 0 and in D4 enter 5. Select these two cells and, by dragging the fill handle down to D75, make the series 0 to 360 in increments of 5.
(f) In E3 enter =RADIANS(D3) and double-click the fill handle to fill the formula down to F75.
(g) In F3 enter the formula =DEGREES(G3) and in G3 enter the value 1. Select both cells and double-click the fill handle to fill down to row 75.
(h) In H3 enter: =Ratio1*COS(E3) - Ratio2*COS(G3)+Ratio3 - COS(E3-G3). Refer to the Freudenstein equation before to ensure you have this correct.
(i) Give G3:G75 the name *Output* and H3:H75 the name *Equation*.
(j) In A13 enter: =IF(SUM(Equation)>0.00001, "Unsolved", "Solved"). *Merge and Center* this across A13:B13. This formula will display *Unsolved* until we invoke Solver and solve all 73 Freudenstein equations.
(k) Save the workbook as Chap15.xlsm as a precaution. Make it macro-enabled since we shall be adding modules later.
We are ready to have Solver make every cell in the *Equation* range equal to zero by changing the *Output* values.

(l) Call up Solver with *Data Analysis | Solver* (*Data | Solver* on a Mac). Clear the *Target* box; in the *By Changing Cells* box enter Output and add the Constraint Equation=0. Click the Solve button.

(m) After about 5–10 s (watch the Excel status bar as it displays messages like Trial Solution 42), Solver will have completed its task.

(n) Make a plot to show how the two angles are related as in Fig. 15.2, Save the workbook.

The companion website has a workbook called *FourBarCrank.xlsm*, which contains VBA code to make an animated diagram using the results of this exercise.

EXERCISE 2: TEMPERATURE PROFILE USING MATRIX ALGEBRA

Consider a thin metal sheet (Fig. 15.4) whose edges are maintained at specified temperatures and which is allowed to come to thermal equilibrium. Our task is to compute the approximate temperatures at various positions on the plate.

We need to make some assumptions. The first is that the two faces of the plate are thermally insulated. Thus there is no heat transfer perpendicular to the plate. The second assumption starts with the mean-value theory, which states: if P is a point on a plate at thermal equilibrium and C is a circle centered on P and completely on the plate, then the temperature at P is the average value of the temperature on the circle. The calculations required to use this theory are formidable, so we will use an

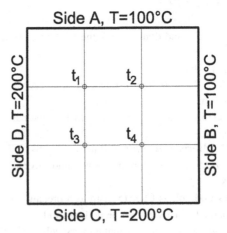

■ FIG. 15.4

approximation. We shall consider a finite number of equidistant points on the plate and use the discrete mean-value theory, which states that the temperature at point P is the average of the temperatures of P's nearest neighbors.

The most convenient way to arrive at the equidistant point is to divide the plate using equally spaced vertical and horizontal lines. In Fig. 15.4, two such lines have been drawn parallel to each axis. This gives four interior points for the calculation. With such a small number, the results will not be very accurate. However, the methodology is the same regardless of the number of points, and it is simpler to describe and test the method initially with four points.

Applying the averaging rule, the temperatures of the four interior points are given by:

$$t_1 = (100 + t_2 + t_3 + 200)/4$$
$$t_2 = (100 + 100 + t_4 + t_1)/4$$
$$t_3 = (t_1 + t_4 + 200 + 200)/4$$
$$t_4 = (t_2 + 100 + 200 + t_3)/4$$

These can be made more general by replacing the numbers with variables:

$$t_1 = (t_2 + t_3)/4 + (a + d)/4$$
$$t_2 = (t_4 + t_1)/4 + (a + b)/4$$
$$t_3 = (t_1 + t_4)/4 + (c + d)/4$$
$$t_4 = (t_2 + t_3)/4 + (b + c)/4$$

To facilitate the use of a matrix method, we will write each in a more systematic form:

$$t_1 = (0.00t_1 + 0.25t_2 + 0.25t_3 + 0.00t_4) + (a + d)/4$$
$$t_2 = (0.25t_1 + 0.00t_2 + 0.00t_3 + 0.25t_4) + (a + b)/4$$
$$t_3 = (0.25t_1 + 0.00t_2 + 0.00t_3 + 0.25t_4) + (c + d)/4$$
$$t_4 = (0.00t_1 + 0.25t_2 + 0.25t_3 + 0.00t_4) + (b + c)/4$$

This has given us a system of four equations in the form $T = MT + B$ where

$$T = \begin{bmatrix} t_1 \\ t_2 \\ t_3 \\ t_4 \end{bmatrix}, \quad M = \begin{bmatrix} 0 & 0.25 & 0.25 & 0 \\ 0.25 & 0 & 0 & 0.25 \\ 0.25 & 0 & 0 & 0.25 \\ 0 & 0.25 & 0.25 & 0 \end{bmatrix}, \quad B = \begin{bmatrix} (a+d)/4 \\ (a+b)/4 \\ (c+d)/4 \\ (b+c)/4 \end{bmatrix}$$

We can rearrange the equation in this way:

$$T - MT = B$$

$$(I - M)T = B$$

$$T = (I - M)^{-1} B$$

Matrix I is the identity matrix, a matrix in which diagonal elements have values of 1 and off-diagonal elements values of 0. If A is a matrix then we speak of A^{-1} as its inverse.

(a) On Sheet2 of Chap15.xlsm, enter all the text shown in Fig. 15.5.
(b) To define the problem, enter the temperature values in A4:D4. Name these cells with the text above them.
(c) Enter the values of the M matrix as shown in A7:D11.
(d) Enter the values for the Unit matrix in F8:I11 either by typing the numbers (it is OK to type just the 1s and leave blank cells for the zeros) or by selecting F8:I11, entering[1] the formula =MUNIT(4) and committing it with [Ctrl]+[⇧ Shift]+[Enter←].
(e) Now we make the $(I - M)$ matrix: in K8 enter the formula =F8 - A8. Copy this to K8:N11.
(f) To compute $[I - M]^{-1}$ select A15:D18, type the formula =MINVERS(K8:I11) and use [Ctrl]+[⇧ Shift]+[Enter←] to complete the array formula.

	A	B	C	D	E	F	G	H	I	J	K	L	M	N
1	Temperature Profile			Using Matrix Method										
2														
3	SideA	SideB	SideC	SideD										
4	100	100	200	200										
5														
6														
7		M matrix					I matrix					I - M		
8	0	0.25	0.25	0		1	0	0	0		1	-0.25	-0.25	0
9	0.25	0	0	0.25		0	1	0	0		-0.25	1	0	-0.25
10	0.25	0	0	0.25		0	0	1	0		-0.25	0	1	-0.25
11	0	0.25	0.25	0		0	0	0	1		0	-0.25	-0.25	1
12														
13														
14		Inverse of I - M				B matrix		T matix						
15	1.167	0.333	0.333	0.167		75		T1	150					
16	0.333	1.167	0.167	0.333		50		T2	125					
17	0.333	0.167	1.167	0.333		100		T3	175					
18	0.167	0.333	0.333	1.167		75		T4	150					

■ FIG. 15.5

---

[1]The function MUNIT was introduced with Office 2013 so typing is the only option for users of earlier versions.

(g) The formulas for the *B* matrix are as follows:

F14: =(SideA + SideD)/4
F15: =(SideA + SideB)/4
F16: =(SideC + SideD)/4
F17: =(SideB + SideC)/4

(h) All that remains is to multiply $[I - M]^{-1}$ by *B*. With I15:I18 selected, enter =MMULT(A15:D18, F15:F18) and commit the array formula with [Ctrl]+[⇧ Shift]+[Enter↵].

(i) There is an alternative method: Omit generating the $(I - M)^{-1}$ matrix and find the T matrix with the nested formula =MMULT(MINVERSE(K8: N11),F15:F18).

(j) Save the workbook.

EXERCISE 3: TEMPERATURE PROFILE USING ITERATION

In this exercise, we solve the same problem as in Exercise 2 but using a 5×5 grid instead of a 2×2 grid. For simplicity, Excel will use iteration that will be used to provide an approximate solution for the equations. In order to plot the temperature on a surface plot, it is assumed that the plate is 0.6×0.6 m divided into a 0.1 m mesh as seen in Fig. 15.6. The distance labels are in B3: H3 and A4:A10. The boundary conditions are in rows 4 and 10 and columns B and H. The values in B4, B10, H4, and H10 are not used in the central calculations but are set to the average of the adjacent boundary conditions to make the surface plot more readable.

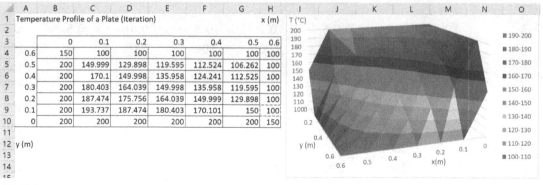

■ FIG. 15.6

(a) On Sheet3, enter the text in rows 1, 3, 4, and 10, and columns A, B, and H.

(b) The formula for C5 is the average of the adjacent temperatures, =AVERAGE(C4, D5, C6, B5)

(c) Copy this formula down to C9 by dragging the fill handle. If iterative calculations are not enabled in excel, you will get an error message like in Fig. 15.7. After clearing the error, your Excel will show arrows between C5 and C6, as seen in Fig. 15.8. This means that there is a circular reference in that the function in C5 is based on values in C6, while the function in C6 is based on the value in C5. This is bad accounting but does provide a way to solve simultaneous equations in Excel using iteration. Excel will calculate C5 based on the value in C6, and then calculate C5 based on what is in C6. This will be repeated until the values do not change much in a successive iteration.

(d) In order to enable Excel to perform iterative calculations, use *File | Options | Formulas* (*Excel | Preferences… | Calculation* on a Mac) and then check the box next to *Enable Iterative Calculation* (*Use iterative calculation* on a Mac). This will remove the recursive error indicator.

![Microsoft Excel dialog box]

Microsoft Excel   x

! There are one or more circular references where a formula refers to its own cell either directly or indirectly. This might cause them to calculate incorrectly.

Try removing or changing these references, or moving the formulas to different cells.

                                        OK   Help

■ **FIG. 15.7**

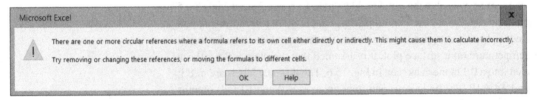

	A	B	C	D	E	F	G	H
1	Temperature Profile of a Plate (Iteration)							
2								
3		0	0.1	0.2	0.3	0.4	0.5	0.6
4	0.6	150	100	100	100	100	100	100
5	0.5	200	150					100
6	0.4	200	150					100
7	0.3	200	150					100
8	0.2	200	150					100
9	0.1	200	150					100
10	0	200	200	200	200	200	200	150

■ **FIG. 15.8**

(e) Copy the formulas through D5:G9. Note that there will be a small change in the values if you perform a recalculation using (F9).

(f) Select a cell in B4:H10. Insert a surface chart using *Insert | Charts | Waterfall | Surface (Insert | Area | Surface | 3-D Surface* on a Mac).

(g) On a PC right click (on a Mac hold the (Ctrl) key and click) on the chart. Adjust the *3D rotation* to have an *X rotation* of 70 degrees to provide a better viewing angle.

(h) Add axis titles to the Horizontal, Vertical, and Depth axes.

(i) Make the major units of the vertical axis 10 to get more resolution in the depth chart.

(j) Save your work.

EXERCISE 4: TEMPERATURE PROFILE USING SOLVER

In this exercise, we solve the same problem as in Exercise 3, but here we shall use Solver instead of iterations. Looking at the before (Fig. 15.9) and after (Fig. 15.10) screen captures of the worksheet will make it clearer for the reader what has to be done.

The *Model* range (C4:G8) and its borders are numeric values. The *Solution* range (C12:G16) has formulas calculating the average of each cell's four neighbors; for example, C12 has the formula =AVERAGE(C3,B4,D4,C5). We will have Solver change each of the *Model* cells until they equal their

	A	B	C	D	E	F	G	H
1	Temperature profile of metal plate				Unsolved			
2								
3	Model		100	100	100	100	100	
4		200	100.0	100.0	100.0	100.0	100.0	100
5		200	100.0	100.0	100.0	100.0	100.0	100
6		200	100.0	100.0	100.0	100.0	100.0	100
7		200	100.0	100.0	100.0	100.0	100.0	100
8		200	100.0	100.0	100.0	100.0	100.0	100
9			200	200	200	200	200	
10								
11	Solution		100	100	100	100	100	
12		200	125.0	100.0	100.0	100.0	100.0	100
13		200	125.0	100.0	100.0	100.0	100.0	100
14		200	125.0	100.0	100.0	100.0	100.0	100
15		200	125.0	100.0	100.0	100.0	100.0	100
16		200	150.0	125.0	125.0	125.0	125.0	100
17			200	200	200	200	200	

■ FIG. 15.9

	A	B	C	D	E	F	G	H
1	Temperature profile of metal plate					Solved		
2								
3	Model		100	100	100	100	100	
4		200	150.0	129.9	119.6	112.5	106.3	100
5		200	170.1	150.0	136.0	124.2	112.5	100
6		200	180.4	164.0	150.0	136.0	119.6	100
7		200	187.5	175.8	164.0	150.0	129.9	100
8		200	193.7	187.5	180.4	170.1	150.0	100
9			200	200	200	200	200	
10								
11	Solution		100	100	100	100	100	
12		200	150.0	129.9	119.6	112.5	106.3	100
13		200	170.1	150.0	136.0	124.2	112.5	100
14		200	180.4	164.0	150.0	136.0	119.6	100
15		200	187.5	175.8	164.0	150.0	129.9	100
16		200	193.7	187.5	180.4	170.1	150.0	100
17			200	200	200	200	200	

■ FIG. 15.10

corresponding *Solution* cell. The trick is that the *Solution* range contains formulas. This may sound a little like a circular reference, but it really is not.

(a) On Sheet3 of Chap15.xlsm enter all the text and numbers shown in rows 1 through 9 in Fig. 15.9, with the exception of F1.

(b) Select C4:G8, in the Name box, enter the word Model and press Enter↵ so as to name that range.

(c) In F1 enter =IF(C4=C12,"Solved","Unsolved"). This is our "flag" to tell us if Solver needs to be called should we alter the model.

(d) We want the border of the *Solution* range to match those of the *Model* range. In C11 enter =C3 and copy across to G3. Do the same for the other three borders.

(e) In C12 enter the formula =AVERAGE(C3,B4,D4,C5) and fill this down and across to G16. Note that the formula in the solution references the cells adjacent to its position in the model.

(f) Select C12:G16 and give it the name Solution. Enter the same text in C11.

(g) Now we call Solver. Clear the *Target* cell; in the *By Changing Cells* enter Model; and add the *Constraint* Solution=Model. Click the *Solve* button to generate the results shown in Fig. 15.10.

(h) Save the workbook.

The reader can experiment by changing the temperature setting for the four borders of the plate and rerunning Solver.

EXERCISE 5: EMPTYING THE TANK

Exercises 5 and 6 use the Runge-Kutta method for solving ordinary differential equations as was done in Chapter 14. In these exercises, we demonstrate the use of VBA functions to make these methods more convenient. In this exercise, we use one function for the differential equation and another to do the Runge-Kutta calculations.

Scenario: A cylindrical tank of diameter D has a short discharge pipe of diameter d at the bottom. The tank is initially filled with water to a height h. We wish to examine how changing the diameter of the pipe alters the rate of discharge of the tank. The problem chosen has an analytical solution. You may wish to find it and compare the results from it with those found using the Runge-Kutta approximation.

For a short pipe, we may assume the rate of change of h is as follows:

$$\frac{dh}{dt} = -\frac{d^2}{D^2}\sqrt{2gh}.$$

Working in metric units, we shall use $g = 9.8\,\text{ms}^{-2}$.

We begin by developing a user-defined function to compute dh/dt for any value of h. We would like a worksheet that lets us vary both the diameter of the pipe d and of the tank D. Clearly our equation could be rewritten as $dh/dt = -R^2\sqrt{2gh}$ where $R = d/D$ is the ratio of the two diameters. The required function is shown to the left below. Both Excel and VBA have a function for computing a square root (SQRT and Sqr, respectively). Whenever this is true, we are required to use the VBA function and not the Excel one in VBA macros.

The VBA function to perform the Runge-Kutta approximation is shown as follows. Compare the k expressions with those in Exercise 2 of Chapter 14. Since t does not appear to the right in the differential equation we are solving, there are no x terms in our k expressions. The y term becomes the *height* term. What was called h in Chapter 14, we call *incr* (short for increment) in our function. Since *height* is water height value, a variable called h might be confusing. The ratio term has been added so that it may be passed to the tank function. We must be careful in the worksheet to call this function with the *height*, *incr*, and ratio arguments in the correct order.

Note: In the interests of clarity, dimension statements have been omitted in all our VBA functions. The reader is strongly advised to set the VB Editor to require these. All numeric variables should be dimensioned as Double.

```
'The function of the differential equation
Function tank(height, ratio)
  Const g = 9.8
  tank = -(ratio ^ 2) * Sqr(2 * g * height)
End Function
```

```
'Function to compute Runge-Kutta approximation
Function RK(height, incr, ratio)
  k1 = incr * tank(height, ratio)
  k2 = incr * tank(height + k1 / 2, ratio)
  k3 = incr * tank(height + k2 / 2, ratio)
  k4 = incr * tank(height + k3, ratio)
  RK = height + (k1 + (2 * k2) + (2 * k3) + k4) / 6
End Function
```

(a) Open Chap15.xlsm and invoke the VBE. Insert a module on which to code the two functions shown before.

(b) On Sheet4, enter the text values shown in columns A through C of Fig. 15.11.

(c) Enter the values in B3:B6 and in B7 enter =C6*0.01/C5. The 0.01 converts the pipe diameter to meters.

(d) Enter 0 in A10 and =A10+B4 in A11. Copy this down to row 110. This gives us the time steps.

(e) In B10 enter =B3 to set the initial height.

(f) In B11 type the formula =RK(B10,$B$4,$B$7). Hopefully, your worksheet returns the value 0.999 in C11. An error value of #NAME! means that the name of the function in the cell does not match that in the module. If B11 shows #VALUE!, check (i) that the

	A	B	C	D	E
1	Tank Problem		Version 1		
2					
3	Initial height	1.00	meters		
4	Time increment	0.1	secs		
5	Tank diameter	1.00	meters		
6	Pipe diameter	5.00	cm		
7	Ratio of diameters	0.05			
8					
9	Time	Height		Summary	
10	0	1.000		time	height
11	0.1	0.999		0	1.00
12	0.2	0.998		5	0.95
13	0.3	0.997		10	0.89
14	0.4	0.996		15	0.84
15	0.5	0.994		20	0.79

■ FIG. 15.11

arguments in the formula point to the correct value and (ii) that the RK function is correctly coded.

(g) Copy B11 down to row 1110. This is when the trick of double-clicking the fill handle comes into its own.

Our data extends over more than 1000 rows and is too much to absorb. We need to make a summary.

(h) Enter the numbers 0 and 5 in D11 and D12, respectively. Select these two values and drag the fill handle down to row 33 to give the last value of 110.

(i) In E11 enter =VLOOKUP(D11,$A$10:$B$1110,2,TRUE) and copy this down to E33.

(j) Make a chart from either the Runge-Kutta data or the summary using a smooth XY chart with a line and no markers. Save the workbook.

EXERCISE 6: AN IMPROVED TANK EMPTYING MODEL

The worksheet in Exercise 5 has some faults: (i) it uses a great deal of space, and we had to make a summary table, and (ii) at any one time we can see data for only one pipe size. The faults are addressed in this exercise.

In the last worksheet (Fig. 15.11), the formula in B11 was =RK(B10,$B$4, $B$7). In B12 we compute a new height from the value in B11. This gets repeated. Why store every value on a worksheet? With VBA we could place data on the worksheet after so many (*n*) iterations. Consider the code to the left below.

```
'Function to make n calculations with RK
Function NewH(OldH, incr, ratio, n)
   For j = 1 to n
      NewH = RK(OldH, incr, ratio)
      OldH = NewH
   Next j
End Function
```

```
'Function to make n calculations with RK
Function NewH(OldH, incr, ratio, n)
   For j = 1 to n
      If OldH < 0.0001 Then
        NewH = 0
        Exit For
      Else
        NewH = RK(OldH, incr, ratio)
        OldH = NewH
      End If
   Next j
End Function
```

However, the Runge-Kutta method is not accurate when *OldH* gets small compared to *incr* and may result in negative values that are meaningless in this model. We, therefore, modify the iteration function so that it stops before completing the *n* iterations if *OldH* is small. The improved function is shown above to the right.

(a) Open Chap15.xlsm and on the existing VBA module enter the code for the function *NewH* as shown to the right above.
(b) Go to Sheet5 of the workbook and enter the text shown in Fig. 15.12.
(c) Enter the values in D3:D5 and name those cells with the text to the left.
(d) Enter the values in rows 7 through 9.
(e) In B12 enter the value 0. In B13 enter =B12+(incr*iter) and copy this down to row 32.
(f) In C12 enter =h0 and copy across to F12. In C13 enter =NewH(C12, incr, C$9,iter). Copy this across to F13. Copy C13:F13 down to row 32.
(g) Make a chart similar to that in Fig. 15.12. Save the workbook.

The use of the three UDFs has made it possible to generate a considerable about of useful data for the current problem.

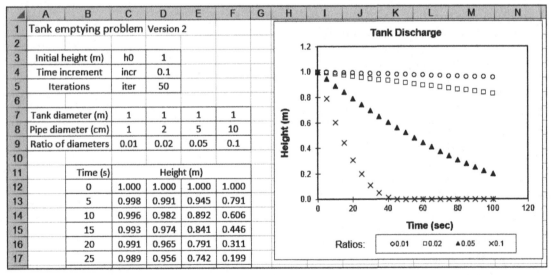

■ FIG. 15.12

PROBLEMS

1. In Exercises 5 and 6 we assumed the increment of 0.1 was sufficiently small for accurate results. Copy Sheet5 of Chap15.xlsm by dragging its tab to the right and modify the copied worksheet to use the same height but different *inc* values. To make the comparison easier, each column should use values of *iter* such that *inc* × *iter* = 5; this can be achieved with a formula in D5.

2. As an extension to Exercise 6, write a UDF that estimates the time for the tank to empty. It would be unwise to have the function loop until the height was zero (why?), so we will define "empty" as meaning the height is reduced by at least 99.95%.

3. Use Solver to find the molar volume in Problem 1 in Chapter 11.

4. Use Solver to find the forces in Problem 2 of Chapter 11.

5. Use Solver to find the nodal voltages in Problem 3 in Chapter 11.

6. Fig. 15.13 shows a network of water pipes.[2] Your task is to develop a Solver model to find the flow of water (Q_i) in each pipe. Two conditions must be satisfied: (i) the algebraic sum of the pressure drops around each closed loop is zero (use the UDF developed in Problem 2 of Chapter 9), and (ii) at any junction, the inflow equals the outflow.

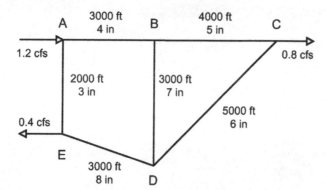

■ FIG. 15.13

[2]B. Maxfield, *Engineering with Mathcad*, Butterworth-Heinemann, Oxford, UK, 2006 (page 333).

Statistics for Experimenters

Microsoft Excel can be a powerful tool for statistical analysis. In this chapter, we look at a very small subset of these tools. The main focus is on the treatment of variability associated with data measurements. Some of the functions introduced are as follows:

AVERAGE	Calculates the arithmetic mean of the values in a dataset.
DEVSQ	Calculates the sum of the squares of the deviations of the values from their mean.
FREQUENCY	Calculates how often values in a dataset occur within a range of values in a bin.
STDEV	Calculates the sample standard deviation of the values in a dataset.
T.DIST (et al)	Calculates the probability for Student's *t*-distribution.
T.INV (et al)	Calculates the two-tailed t-value of Student's *t*-distribution.
TTEST	Calculates the probability associated with Student's *t*-test.

Liengme's Guide to Excel 2016 for Scientists and Engineers. https://doi.org/10.1016/B978-0-12-818249-9.00016-9

We shall also introduce some of the Data Analysis tools from the Analysis ToolPak.

EXERCISE 1: DESCRIPTIVE STATISTICS

An experimenter has collected 100 measurements and wishes to know some statistics of the data set, for example, the average or the sum. To save the task of entering 100 numbers we will have Excel generate some random numbers. The RAND or RANDBETWEEN functions are not appropriate here since they generate uniform distributions of values. So, we will use the Random Number generator tool found in the Data Analysis toolbox. To simulate the results from an experiment, we will request random numbers having a normal distribution with a mean (average) of 10 and a standard deviation of 0.5. The final worksheet will resemble Fig. 16.1.

(a) Open a new workbook. In A1 of Sheet1 enter the label data. Use the command *Data / Analyze / Data Analysis*[1] (*Data / Data Analysis* on a Mac) and select the item *Random Number Generation*. Complete the dialog box as shown in Fig. 16.2. Give A2:A101 the name data. Of course, the actual number will vary from user to user.

(b) To generate the statistics quickly, we will use another Data Analysis tool, namely, *Descriptive Statistics*. Complete the dialog box as shown in Fig. 16.3. Since we are working with random numbers readers will not get the same values as shown in column D.

	A	B	C	D	E	F	G	H
1	data		data				formulas	
2	10.20012							
3	8.954345		Mean	10.03594	10.03594		=AVERAGE(data)	
4	10.05579		Standard Error	0.05594	0.05594		=STDEV(data)/SQRT(COUNT(data))	
5	9.593495		Median	9.99994	9.99994		=MEDIAN(data)	
6	10.31294		Mode	#N/A	#N/A		=MODE.SNGL(data)	
7	9.838037		Standard Deviation	0.55939	0.55939		=STDEV(data)	
8	11.4453		Sample Variance	0.31291	0.31291		=VAR(data)	
9	9.899985		Kurtosis	-0.08594	-0.08594		=KURT(data)	
10	11.28357		Skewness	0.24594	0.24594		=SKEW(data)	
11	9.116269		Range	2.55930	2.55930		=MAX(data)-MIN(data)	
12	9.633214		Minimum	8.88600	8.88600		=MIN(data)	
13	9.755323		Maximum	11.44530	11.44530		=MAX(data)	
14	10.25377		Sum	1003.59432	1003.59432		=SUM(data)	
15	9.831372		Count	100	100		=COUNT(data)	
16	9.76027		Confidence Level(95.0%)	0.11099	0.0035077		=CONFIDENCE.NORM(0.95,E7,E15)	
17	10.38996							

■ FIG. 16.1

[1]If there is no Data Analysis item in the Analyze group, enable the add in using *File / Options / Add_ins* and click the *Go* button next to *Manage Excel Add-ins* (on a Mac *Tools / Excel Add-ins...*).

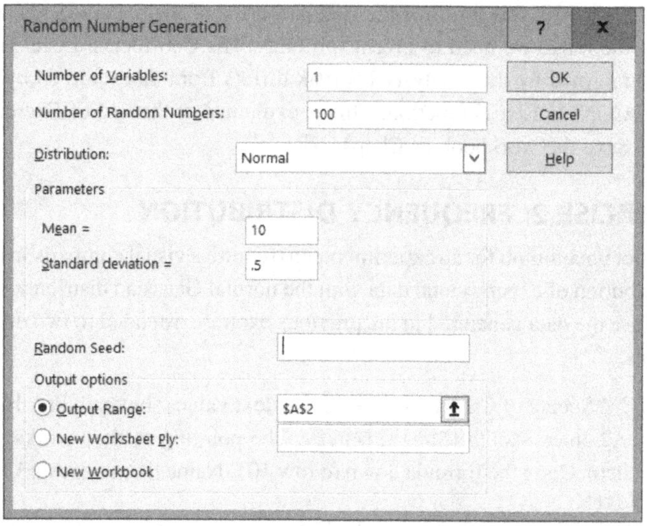

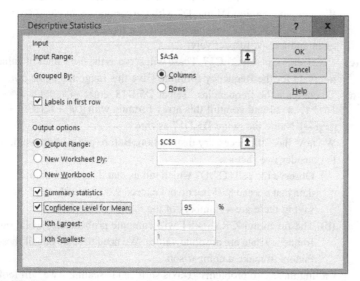

■ FIG. 16.2

■ FIG. 16.3

(c) For comparison we will generate the same statistics[2] using formulas—see columns E and F in Fig. 16.1. There is a worksheet function corresponding to all but two of the statistics generated by the

---

[2]The N/A result for the mode is not unexpected; it is most unlikely that with 100 random numbers (each with 15 decimal places) there would be any duplicates.

Data Analysis tool; for the standard error of the mean (SEM) and the range we need to invent formulas. The Confidence Level value returned by the Analysis ToolPak differs from that returned by the CONFIDENCE function. This is explained at the end of Exercise 3.

(d) Save the workbook as Chap16.xlsx.

EXERCISE 2: FREQUENCY DISTRIBUTION

It is not uncommon for an experimenter to require a visual comparison of the distribution of experimental data with the normal Gaussian distribution. We will use the data generated in the previous exercise, rounded to two decimal places.

(a) On Sheet2 of Chap16.xlsx, enter the text values shown in Fig. 16.4. In A2 enter =ROUND(Sheet1!A2,2). The pointing method works well here. Copy the formula down to row 101. Name the range A2:A101 as data.

(b) To compute the values to draw the normal curve we will need to know the mean (average) and the standard deviations of the data. We compute these in D1 and D2 with the formulas = AVERAGE(data) and = STDEV.P(data), respectively. Give these cells the names mean and stdev, respectively.

(c) Enter the values in C5:C17. These will serve as the *bin* values or value intervals for the frequency formula. Give this range the name *bin*.

(d) To compute the frequencies,[3] select D5:D18, enter =FREQUENCY (data, bin) and commit this array formula with Ctrl + ⇧ Shift + Enter ↵. Name the range D5:D17 as *freq*.

   We now have the necessary distribution. Before proceeding we need to consider two factors:

   (i) Observe the cells C7:D7 which tell us that there are 4 numbers in data that exceed 8.75 but do not exceed 9.00. Let's think of this as saying there are 4 numbers in the 8.875 ± 0.25 range.

   (ii) The function NORM.DIST will compute probabilities while our frequency data are absolute values. We need to scale one of these factors to make a comparison.

   Calling these two problems "*x*-axis shift" and "scaling," we will look at two alternative ways to achieve our goal. We will generate both sets of data and then make the charts.

   In Method 1 the *x*-axis shift is resolved by using the midpoint of the bin values. The normal distribution values get scaled as we shall see.

---

[3]We enter the FREQUENCY formula in a range with one cell more than the bin to ensure that there are no values larger than the last bin value. But note that we use only D5:D17 when naming the range.

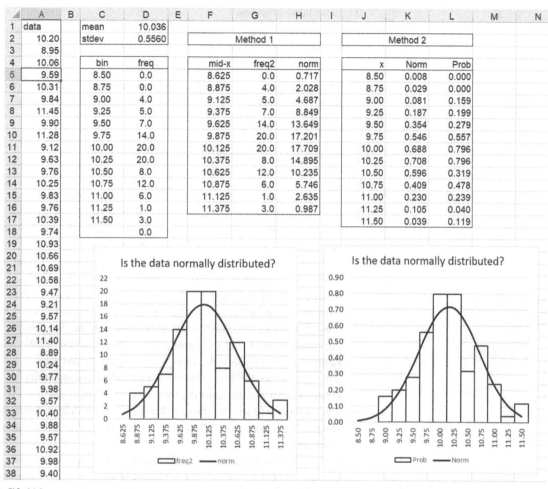

	A	B	C	D	E	F	G	H	I	J	K	L	M	N
1	data		mean	10.036										
2	10.20		stdev	0.5560			Method 1				Method 2			
3	8.95													
4	10.06		bin	freq		mid-x	freq2	norm		x	Norm	Prob		
5	9.59		8.50	0.0		8.625	0.0	0.717		8.50	0.008	0.000		
6	10.31		8.75	0.0		8.875	4.0	2.028		8.75	0.029	0.000		
7	9.84		9.00	4.0		9.125	5.0	4.687		9.00	0.081	0.159		
8	11.45		9.25	5.0		9.375	7.0	8.849		9.25	0.187	0.199		
9	9.90		9.50	7.0		9.625	14.0	13.649		9.50	0.354	0.279		
10	11.28		9.75	14.0		9.875	20.0	17.201		9.75	0.546	0.557		
11	9.12		10.00	20.0		10.125	20.0	17.709		10.00	0.688	0.796		
12	9.63		10.25	20.0		10.375	8.0	14.895		10.25	0.708	0.796		
13	9.76		10.50	8.0		10.625	12.0	10.235		10.50	0.596	0.319		
14	10.25		10.75	12.0		10.875	6.0	5.746		10.75	0.409	0.478		
15	9.83		11.00	6.0		11.125	1.0	2.635		11.00	0.230	0.239		
16	9.76		11.25	1.0		11.375	3.0	0.987		11.25	0.105	0.040		
17	10.39		11.50	3.0						11.50	0.039	0.119		
18	9.74			0.0										
19	10.93													
20	10.66					Is the data normally distributed?				Is the data normally distributed?				
21	10.69													
22	10.58													
23	9.47													
24	9.21													
25	9.57													
26	10.14													
27	11.40													
28	8.89													
29	10.24													
30	9.77													
31	9.98													
32	9.57													
33	10.40													
34	9.88													
35	9.57													
36	10.92													
37	9.98													
38	9.40													

■ FIG. 16.4

(e) Enter these formulas:

F5:   =(C5+C6)/2
G5   =D6
H5   =NORM.DIST(F5,mean,stdev,FALSE)*COUNT(data)*(C6-C5)

Select F5:H5 and drag the fill handle down to row 16.

In method 2 it is the frequency values that are scaled; they are converted from absolute values to probabilities. The normal distribution is shifted on the *x*-axis by a simple trick.

(f) Enter these formulas:

J5: `=C5`

K5 `=NORM.DIST(J5-0.125,mean,` (the 0.125 that accomplishes
    `stdev,FALSE)`   the *x*-axis shift.)

Select J5:K5 and drag the fill handle down to row 17. Name the range K5:K17 as *Norm*.

L5 `=D5*SUM(Norm)/100`

Select L5 and drag the fill handle down to row 17.

(g) Save the workbook.

The *x*-axis shift is fairly self-evident. Let's see if we can confirm that the scaling is done correctly. Anticipating the next part of the exercise, look at each chart. We can safely say that the area under the two data series should be more or less the same for a given chart. Here is how to check that. Select the data under *freq2* and observe the Sum in the status bar (exactly 100, as expected); repeat with the data under *norm* (approx. 100). Do the same test with the other method. Now we can make the charts. Excel has a slightly odd behavior when making anything but an XY chart. If the values in the first column are numbers then the chart engine assumes they represent a data series rather than the *x*-values. We can circumvent this behavior by making sure that the first column has no header, so temporarily delete cells F4 and J4, restoring them after the chart is made.

(h) Click somewhere within either the Method 1 or the Method 2 data and insert a clustered column chart—this is the first chart in the Column chart gallery.

On the chart, right-click the normal distribution data series, select *Change Series Chart Type*, and in the lower section of the resulting dialog change the Norm series to a Line type. Click OK.

Format the Norm data series specifying that the Line is to be smooth. (*Series Options / Markers / Border / Smoothed Line checkbox*) (On a Mac, there is a *Smoothed line* checkbox under *Format Data Series*.)

The reader will observe that, if the axis values are ignored, the two charts are more or less identical; this should give us confidence in the methods. But what would we do if we did not have the original data but only the grouped (frequency) data? We would need some way to compute both the mean and the standard deviation in order to use the NORM.DIST function. The mean is computed using the weighted average formula $\bar{x} = \dfrac{\sum x_i f_i}{\sum f_i}$ and the standard

deviation with $\sigma = \sqrt{\dfrac{\sum f x^2}{n} - \bar{x}^2}$.

These can be executed in Excel[4] with the formulas

Mean_g: `=SUMPRODUCT(mid_x,freq2)/SUM(freq2)`
Stdev_g: `=SQRT(SUMPRODUCT(freq2,mid_x,mid_x)/SUM`
`(freq2)-Mean_G^2)`

EXERCISE 3: CONFIDENCE LIMITS

It is often necessary to indicate the spread of the measurements. The most commonly used measure of spread in a dataset is the *standard deviation*. Statisticians speak of *population* and *sample* standard deviations, represented by σ and s, respectively. This quote[5] might help the reader: *Researchers and statisticians use the population and sample standard deviations in different situations. For example, if a teacher gives his students an exam and he wants to summarize their results, he uses the population standard deviation. He only wants his pupils' scores and not the scores of another class. If, for instance, a researcher investigates the relationship between middle-aged men, exercise, and cholesterol, he will use a sample standard deviation because he wants to apply his results to the entire population and not just the men who participated in his study*. The two Excel functions are STDEV.P and STDEV.S.

The data in column A of Fig. 16.4 might be reported as *the value of x was found to be 10.036±0.556 (n=100)*. In many cases, it would be appropriate to use only two decimal places since that was the precision of the raw data.

For various reasons we repeat experiments from which we obtain a sample mean ($\bar{x}$) while our goal is to determine the population (or the *true*) mean (μ). We would like some way of expressing how close we think our result is to the true value. If I wish to say *I have reason to believe with 90% confidence that $\mu=2.45\pm0.08$ (n=5)*, then the value 90% is referred to as the *confidence level* and 2.45 ± 0.08 is referred to as the *width of the confidence interval*. When computing confidence limits for the mean, we use the Student's t-statistic: $\mu=\bar{x}\pm\frac{ts}{\sqrt{n}}$ where t is the Student's t-value, s the standard deviation, and n the number of measurements. The value of s is found using the STDEV.S function. We determine t with the TINV function which has the syntax TINV(*probability, degrees of freedom*). The probability (α) equals

[4]By thinking about the mean, it should be clear that a better approximation is obtained with the mid-*x* values rather than the bin values. Also note that if we use the text *mid-x* to name a cell the result is the name *mid_x* with an underscore rather than a hyphen.

[5]http://smallbusiness.chron.com/difference-between-sample-population-standard-deviation-22639.html.

Throughout this chapter, we concentrate on a two-tailed test. This is appropriate when we compare two means to see whether or not they are unequal. A one-tailed test is appropriate when we need more than this and wish to know, for example, if one mean is really greater than another. A statistics text should be consulted for more information.

(1 − the confidence level). For repeated measurements of the same object, the degrees of freedom (f) is given by $n − 1$.

We begin by finding the mean and confidence limits of a set of seven measurements. At the end of the exercise, we will make the worksheet more flexible.

(a) On Sheet3 of Chap16.xlsx enter the text shown in columns A to D of Fig. 16.5. Ignore columns F to I temporarily. Enter the values in column A.

(b) The formulas in column D are:

D2:	=AVERAGE(A3:A9)	Mean
D3:	=STDEV.S(A3:A9)	Standard deviation
D4:	=COUNT(A3:A9)	Number of measurements
D5:	=D3/SQRT(D4)	Standard error of the mean
D6:	=95%	Required level (type 95% or 0.95 followed by % formatting)
D7:	=T.INV.2T (1-D6, D4-1)	Student's t-statistic (for $\alpha = 0.05$, $f = 6$)
D8:	=D7*D5	Confidence width
D11:	=D2	Mean, formatted to two places
D12:	=D8	Confidence limits[a]

[a]D12 is given a custom format of ±0.00. The ± symbol is produced with either $\boxed{\text{Alt}}$ + 0177 or $\boxed{\text{Alt}}$ + 241; the digits must be entered on the numeric keypad. Alternatively, you could insert the ± symbol in an empty cell and then copy and paste it into the custom format.

Some journals would require the data to be reported as having a mean of 1.95 and a standard error of 0.015 with $n = 7$. From this information, the

▲	A	B	C	D	E	F	G	H	I
1	Mean, standard error and confidence limits								
2	data		mean	1.9543		data		mean	1.9760
3	1.93		stdev	0.0408		1.94		stdev	0.0369
4	1.92		n	7		1.99		n	10
5	2.02		stderror	0.0154		1.98		stderror	0.0117
6	1.97		conf level	95.0%		2.03		conf level	95.0%
7	1.98		t	2.4469		2.03		t	2.2622
8	1.96		conf width	0.0377		1.96		conf width	0.0264
9	1.90					1.95			
10			rounded values			1.96		rounded values	
11			mean	1.95		1.92		mean	1.98
12			conf width	± 0.04		2.00		conf limits	± 0.03

■ FIG. 16.5

reader can compute the confidence limits for any required confidence level using the formula *confidence width* $= \pm t \times$ *standard error*. Our worksheet allows the same. You may change the confidence level value in D6 to say 95.5, to find new confidence limits.

If we wish, we can simply compute the confidence limit with one formula:

```
=T.INV.2T(5%,COUNT(A3:A9)-1)*STDEV.S(A3:A9)/SQRT
(COUNT(A3:A9).)
```

Complex formulas such as this, however, are error-prone.

Our worksheet would be useful for any experiment in which a measurement is repeated seven times or less. We can test this and at the same time double-check our worksheet.

(c) Use the *Descriptive Statistics* tool from the *Data Analysis* toolbox with the data in A3:A9. Do the values it reports for the mean, standard error, standard deviation, and confidence limits agree with your worksheet? Erase the values in A3:A9 and enter three new values. Use the *Descriptive Statistics* tool again (you will recall that its values are static and you must rerun the tool after the data changes) and check for agreement.

Our worksheet will not give correct results with more than seven data items unless we make appropriate changes to all the formulas in column D that reference A3:A9. We can, however, make the worksheet flexible.

(d) Copy A2:D12 to F2. Modify the formulas in column I to read:

```
I2:  =AVERAGE(F:F)
I3:  =STDEV.S(F:F)
I4:  =COUNT(F:F)
```

Recall that the range references to F:F may be interpreted as F1:F1048576. This means the worksheet will give the correct result no matter how many values are entered. The empty cell in F1 and the text in F2 have no effect. Enter the values shown in F1:F12 and use *Descriptive Statistics* to validate your worksheet results.

(e) Save the workbook.

In Exercise 1 we saw that the CONFIDENCE function result does not agree with the results reported by the *Descriptive Statistics* tool. This tool always uses a t-value for an infinite value of f, the degrees of freedom, that is, it uses z-values. Its results may be acceptable when n is very large, or when it is known that the sample standard deviation (s) for the n measurements is always close to the population standard deviation (σ).

EXERCISE 4: THE EXPERIMENTAL AND EXPECTED MEAN

A series of measurements may be made on a specimen where there is an expected result. A chemist may analyze a chemical sample thought to be compound X and compare the results with the known composition of X to determine if the chemical sample is pure X. An engineer may measure the thickness of a metal plate and compare the results with the known thickness to test a new measuring device. Statisticians speak of hypothesis testing in these cases. For the chemist, we have the null hypothesis H_0: *The sample is compound X.* For the engineer, we have H_0: *This new instrument is suitable for the task.* Both may be stated as H_0: *This measured average ($\bar{x}$) is the same as the expected average (μ).* There is the alternate hypothesis H_1: *This measured average ($\bar{x}$) differs from the expected average (μ).* Test the hypothesis on the mean by computing an experimental t-statistic and comparing it with the critical[6] value for a given confidence level. If the experimental t does not exceed the critical t, we dismiss the alternative hypothesis.

When testing hypotheses involving whether or not the experimental mean value is in agreement with a particular expected mean μ, the experimental t is computed using

$$t_{experimental} = \frac{|\bar{x} - \mu|}{s/\sqrt{n}}$$

where $\bar{x}$ is the mean and s is the standard deviation of the n measurements.

Scenario: To calibrate a packing machine, an engineer has made a series of measurements of the raisin content in boxes of breakfast cereal. The required value is 33%. To test the results, we will construct a worksheet similar to that in Fig. 16.6.

(a) On Sheet4, enter the text shown in A1:C11 of Fig. 16.6. For the time being, ignore the entries in columns F and G. Enter the experimental values in column A. Select A3:A13 and name A4:A13 as data.
    This will allow the worksheet to be used with up to 10 measurements, although we have only seven.

---

[6]Textbooks on statistics often use the terms t(calculated) and t(table). We use the terms t (experimental) and t(critical) to avoid confusion since we "calculate" both values. The Excel *Data Analysis Tool* uses the terms t(stat) and t(critical). Explanation: In the days before computers (BC), t-values had to be found from printed tables.

◢	A	B	C	D	E	F	G
1	Experimental and Expected Mean						
2							
3	data		Method 1			Method 2	
4	30.3		expected mean	33		expected mean	33
5	34.7		experiment mean	35.34		experiment mean	35.34
6	40.0		stdev	4.238		stdev	4.238
7	36.1		n	7		n	7
8	41.3		t expt	1.46		t expt	1.46
9	34.5		prob (required)	95%		alpha (required)	0.05
10	30.5		t(a, df)	2.45		prob (t, df)	0.19
11			Fail to reject Null hypothesis			Fail to reject Null hypothesis	

■ FIG. 16.6

(b) The entries in column D are as follows:

D4:	33	Required mean
D5:	=AVERAGE(data)	Calculated mean
D6:	=STDEV(data)	Standard deviation
D7:	=COUNT(data)	Number of measurements
D8:	=ABS(D5-D4)/(D6/SQRT(D7))	Experimental t-value
D9:	95%	Required level of confidence
D10:	=T.INV.2T(1 - D9, D7 - 1)	Critical *t*-value
C11:	=IF(D8>D10,"Reject Null hypothesis","Fail to reject Null hypothesis")	

The entry in C11 is centered across C11 and D11 using the *Merge and Center* tool.

In this case, the experimental *t*-value (1.46) does not exceed the critical value (2.45), so the null hypothesis is not rejected. With 95% certainty, we may say there is insufficient statistical evidence to believe the two values (the measured and the expected means) differ.

The alternative approach to this problem is to compute the probability that the mean value is statistically different from the expected value, that is, the *p*-value. We use T.DIST.2T to compute the *p*-value from the data and compare this to our required significance level (α), which is generally 0.05 or 5%.

(c) Enter the text in F3:F10 (you could copy from column C and edit). Copy the formulas D4:D10 to G4 and make these changes:

G9:	`=1 - D9`	The required α-value
G10:	`=T.DIST.2T(G8,G7 - 1)`	The computed P-value
F11:	`=IF(G9<G10,"Fail to reject Null hypothesis","Reject Null hypothesis")`	

We again fail to reject the null hypothesis that the two means are shown to be statistically the same since the calculated p-value (0.19) is greater than the stipulated α-value (0.05). We may interpret these results as saying that if the null hypothesis is true, there is a 19% probability that seven boxes taken at random will show a difference of 2.34 (=35.34–33.00) from the expected mean of 33. Note that we are saying that the difference between the found and expected means could occur by random errors. We are not necessarily saying this is an acceptable situation. The engineer may accept the accuracy of the machine but may decide to improve its precision in order to decrease the spread of the values.

In this problem, we used a two-tailed t-value since we were concerned with both positive and negative differences from the expected mean. Consider another scenario: The packing machine fills, on average, 50 boxes a minute. After modification, 10 trials were made, and the average filling rate was found to be 54.5 boxes/min with a standard deviation of 4.3. Has there been a statistically significant improvement in the machine? From these values, t $expt$ computes to 3.31 and a one-tailed p-value with formula `T.DIST.RT (3.31, 9)` is 0.0045 or 0.45%. So at the 5% level, there has been a significant improvement since the p-value is less than the α-value. In other words, we are not willing to accept that the difference in the means resulted from random measurement errors.

EXERCISE 5: POOLED STANDARD DEVIATION

This Exercise is primarily a prelude to the next one. The topic is repeated measurements on different specimens. We introduce the concept of the *pooled standard deviation* and the function DEVSQ. The pooled standard deviation is computed using the formula

$$s_{pooled} = \sqrt{\frac{\sum SSD_i}{\sum (r_i - 1)}} \text{ or } s_{pooled} = \sqrt{\frac{\sum SSD_i}{\sum r_i - n}}$$

	A	B	C	D	E	F	G	H	I	J
1	Pooled Standard Deviation									
2										
3	Sample	Replicates			Results				Mean	SSD
4	1	3	1.80	1.58	1.64				1.6733	0.02587
5	2	4	0.96	0.98	1.02	1.10			1.0150	0.01150
6	3	2	3.13	3.35					3.2400	0.02420
7	4	6	2.06	1.93	2.12	2.16	1.89	1.95	2.0183	0.06108
8	5	4	0.57	0.58	0.54	0.59			0.5700	0.00140
9	6	5	2.35	2.44	2.70	2.48	2.44		2.4820	0.06848
10	7	4	1.11	1.15	1.22	1.04			1.1300	0.01700
11	7	28							ss ->	0.20953
12										
13				mean of all measurements			1.67			
14				pooled standard deviation			0.10			

■ FIG. 16.7

where SSD_i is the sum of the squares of the deviations from the mean for the ith sample and r_i is the number of repeated measurements on the ith sample. The degrees of freedom are given in the divisors in these two equivalent formulas.

Scenario: A biologist has measured the mercury content of seven fish[7] taken from Lake Erie and obtained the results shown in C4:H10 of Fig. 16.7.

(a) On Sheet5 of Chap16.xlsx enter the text shown in the figure. Enter the values shown in A4:A10 and C4:H10.

(b) The number of samples n is found in A11 with the formula =COUNT(A4: A10). The number of repeated measurements for the first sample (r_1) is found in B4 with =COUNT(C4:H4) and this is copied down to B10. The total number of measurements ($\sum r_i$) is found in B11 with =SUM(B4:B10).

(c) In I4 enter =AVERAGE(C4:H4) to find the mean of the first sample. The sum of the squares of the deviations from this mean (SSD_1) is found in J4 with =DEVSQ(C4:H4). These formulas are copied down to row 10. The sum of the SSD values is computed in J11 with =SUM(J4:J10).

(d) The mean for all measurements is given in G13 by =AVERAGE(C4: H10). The formula in G14 for pooled standard deviation is =SQRT (J11/(B11 - A11)).

(e) Save the workbook.

---

[7]This data was taken from D. A. Skoog and M. W. West, *Analytical Chemistry*, 2nd ed., p40, New York: Holt, Reinhart and Winston, 1974.

EXERCISE 6: COMPARING PAIRED ARRAYS

In this exercise, we compare the mean between two sets of measurements made on a set of samples. Perhaps set A is the measurements using one technique, while set B was obtained from another. As in Exercise 4, we compute a standard deviation and use it to find a t-value. We compare the found t-value with the critical value computed for a specified α-value and the appropriate degrees of freedom. As before, we reject the null hypothesis for a two-tailed test if the experimental t-value is greater than the critical t-value.

For these circumstances (two sets of measurements on several different samples) the t-value is computed using $t_{\text{expt}} = \frac{\bar{d} - \mu_d}{s_d/\sqrt{n}}$ where $s_d = \sqrt{\frac{\sum(d_i - \bar{d})^2}{n-1}}$ (the standard deviation in the paired differences), n is the number of paired measurements, $\bar{d}$ is the average of the differences between the pairs, and μ_d is the expected average difference (usually 0).

In Exercise 4 we saw two methods to compare a measured mean with an expected value. We could call these the *t method* and the *p method*. We can also use a probability method for paired arrays of data. Without delving into the statistical theory, we will use the T.TEST function which has the syntax: TTEST(*array1*, *array2*, *tails*, *type*) where *tails* has the same meaning as before and *type* is given a value of 1 for paired arrays.

Scenario: There are two methods of making essentially the same measurement; we call them methods A and B. The experimenter has made six measurements using each method—see A3:C10 of Fig. 16.8. We wish to know if the results are statistically different.

(a) On Sheet6 of Chap16.xlsx, enter the text values shown in columns A: G of Fig. 16.8. For now, ignore columns I:K. Enter the values shown in A5:C10.
(b) In D5 enter =B5-C5 and copy it down to row 10.
(c) The formulas in column G are as follows:

G4:	=AVERAGE(D5:D10)	Computes $\bar{d}$
G5	0	The expected mean difference
G6:	=COUNT(A5:A10)	Computes n
G7:	=STDEV(D5:D10)	Computes s_d
G8:	=G4*SQRT(G6)/G7	Compute t(experimental)
G9:	0.05	The required α-value
G10:	=TINV(G9, G6-1)	Computes t(*critical*)
G15:	=TTEST(B5:B10,C5:C10,2,1)	Computes the p-value

◢	A	B	C	D	E	F	G	H	I	J	K
1	Paired sets										
2											
3		Data				Calculations			t-Test: Paired Two Sample for Means		
4	Specimen	A	B	diff		mean diff	0.16				
5	1	1.11	0.97	0.14		expected diff	0			A	B
6	2	3.77	4.33	-0.56		n	6		Mean	3.165	3.003
7	3	5.94	5.35	0.59		st dev	0.454		Variance	3.602	3.191
8	4	2.90	2.30	0.60		t expt	0.873		Observations	6	6
9	5	1.04	1.19	-0.15		alpha	0.05		Pearson Correlation	0.971	
10	6	4.23	3.88	0.35		t critical	2.571		Hypoth Mean Diff	0	
11									df	5	
12		Conclusion:				Fail to reject Null hypothesis			t Statistic	0.873	
13									P(T<=t) one-tail	0.211	
14						p approach			t Critical one-tail	2.015	
15						p from TTEST	0.423		P(T<=t) two-tail	0.423	
16		Conclusion:				Fail to reject Null hypothesis			t Critical two-tail	2.571	

■ FIG. 16.8

The formulas for the conclusions are as follows:

D12:   `=IF(G8<G10,"Fail to reject Null hypothesis",`
       `"Reject Null hypothesis")`

D16:   `=IF(G15>G9,"Fail to reject Null hypothesis",`
       `"Reject Null hypothesis")`

These are each centered over four cells.

(d) Save the workbook.

We are led to the conclusion that the two methods give the same mean (with an α-value of 0.05) since (i) $t(experimental)$ is less than $t(critical)$ and (ii) the p-value computed by TTEST is greater than the α-value of 0.05.

To round off this exercise, we use the *t-test: Paired Two Sample for Means* tool from the Data Analysis tool. The reader may wish to experiment or wait until the next exercise to see how to use this. Note that we should set the *Hypothesized mean difference* to 0 and the *alpha* value to 0.05 when completing the tool's dialog box. The results are shown in the figure. As expected, the results agree with our own calculations. The $t(experimental)$ values in G8 and J12 are the same, as are the p-values in G15 and J15. These serve as useful checks but recall that the results from the tool are static whereas our calculations will be updated if new experimental array values are entered.

EXERCISE 7: COMPARING REPEATED MEASUREMENTS

In the previous exercise, each specimen was measured once by each of two techniques. In this exercise, the same specimen is measured repeatedly by two techniques. Our task is the same: To determine if the means of the two sets of measurements are the same, assuming equal variances. Once again, we have two statistical methods we could use: the t and the p methods. For the former we compute a pooled standard deviation using the formula[8]:

$$s_p = \sqrt{\frac{\sum_{setA}(x_i - \bar{x}_A)^2 + \sum_{setB}(x_j - \bar{x}_B)^2}{n_1 + n_2 - 2}} = \sqrt{\frac{s_1^2(n_1 - 1) + s_2^2(n_2 - 1)}{n_1 + n_2 - 2}}$$

Using the result of this we compute $t(experimental)$ and compare it with t-$(critical)$. The experimental t-value is found using:

$$t_{experimental} = \frac{\bar{x}_1 - \bar{x}_2}{s_p}\sqrt{\frac{n_1 n_2}{n_1 + n_2}} = \frac{\bar{x}_1 - \bar{x}_2}{s_p\sqrt{(1/n_1 + 1/n_2)}}$$

For the p method, we will again use the Microsoft Excel functions T.DIST.2T or T.TEST to find a probability value, which we will compare to the required α-value. We will also use the Data Analysis tool *t-Test: Two-Sample Assuming Equal Variance* to check our results.

(a) On Sheet7 of Chap16.xlsx enter the text shown in A1:D19 of Fig. 16.9. Enter the experimental values in columns A and B. Name A5:A19 as A and B5:B19 as B. This will allow the worksheet to be used with up to 15 data points when it is used with other data.

(b) The formulas in columns E and F are as follows:

E5:	=AVERAGE(A)
E6:	=STDEV(A)
E7:	=DEVSQ(A)
E8:	=COUNT(A)
F5:	=AVERAGE(B)
F6:	=STDEV(B)
F7:	=DEVSQ(B)
F8:	=COUNT(B)
E9:	=SQRT((E7+F7)/(E8+F8-2))

---

[8]When the two datasets are of equal size, this reduces to $s_p = \sqrt{(s_1^2 + s_2^2)/2}$.

◢	A	B	C	D	E	F	G	H	I	J
1	Comparing repeated measurements									
2										
3	Data			Calculations				t-Test: Two-Sample Assuming Equal Variances		
4	A	B			A	B				
5	179.738	179.864		mean	179.7285	179.6645			A	B
6	179.707	179.611		st dev	0.016	0.144		Mean	179.7285	179.6645
7	179.731	179.537		ssd	1.852E-03	1.452E-01		Variance	0.000265	0.020747
8	179.722	179.903		n	8	8		Observations	8	8
9	179.745	179.543		s pooled	0.1025			Pooled Variance	0.010506	
10	179.731	179.661		t expt	1.249			Hypothesized Mean Diff	0	
11	179.749	179.544		prob	95%			df	14	
12	179.705	179.653		df	14			t Stat	1.248812	
13				t theory	2.145			P(T<=t) one-tail	0.116106	
14				outcome	Fail to reject Null Hypothesis			t Critical one-tail	1.76131	
15								P(T<=t) two-tail	0.232213	
16				p method				t Critical two-tail	2.144787	
17				p expt	0.232					
18				t-test	0.232					
19				alpha	0.05					
20				outcome	Fail to reject Null Hypothesis					

■ FIG. 16.9

E10: `=(ABS(E5 - F5)/E9) * SQRT ((E8*F8)/(E8+F8))`

E11: `95%`   The required confidence level

E12: `=E8+F8 - 2`   The degrees of freedom

E13: `=T.INV.2T(1-E11,E12)`

E14: `=IF(E10<E13,"Fail to reject Null Hypothesis", "Reject Null Hypothesis")`

Comparing the t(experimental) value of 1.249 in E10 with the t(critical) value of 2.145 in E13, we fail to reject the null hypothesis that the two means are statistically the same.

(c) For the p method, the formulas `=T.DIST.2T(E10,E12)` in E17 and is `=T.TEST(A, B, 2, 2)` in E18 are for a two-tailed test with sets having equal population variances. In E19 we use `=1-E11` to compute the required alpha. It is left to the reader to compose the formula in E20. The results here lead to the same conclusion: that the null hypothesis cannot be dismissed.

You may wonder why we used two formulas for the p method. The simple answer is that T.TEST is only of use when the two arrays are of equal size. The longer method, which involves computing a t-value from which to compute the p-value, is applicable when the sets are of unequal size.

(d) Use *Data / Analyze / Data* Analysis (*Data / Data* Analysis on a Mac) and select the tool *t-Test: Two-Sample Assuming Equal Variance.* Complete the dialog as shown in Fig. 16.10. The two t-statistics from the tool agree with our calculations, and so do the p-values. Unlike t-test, this tool may be used with arrays of unequal size. We can also test if the means differ by a specified nonzero amount by entering a value in the hypothetical mean difference box. If we wish to do a similar test with formulas, the t(experimental) value must be computed using:

$$t_{experimental} = \frac{(\bar{x}_1 - \bar{x}_2) - (\mu_1 - \mu_2)}{s_p} \sqrt{\frac{n_1 n_2}{n_1 + n_2}}$$

where $(\mu_1 - \mu_2)$ represents the hypothesized difference in the population means.

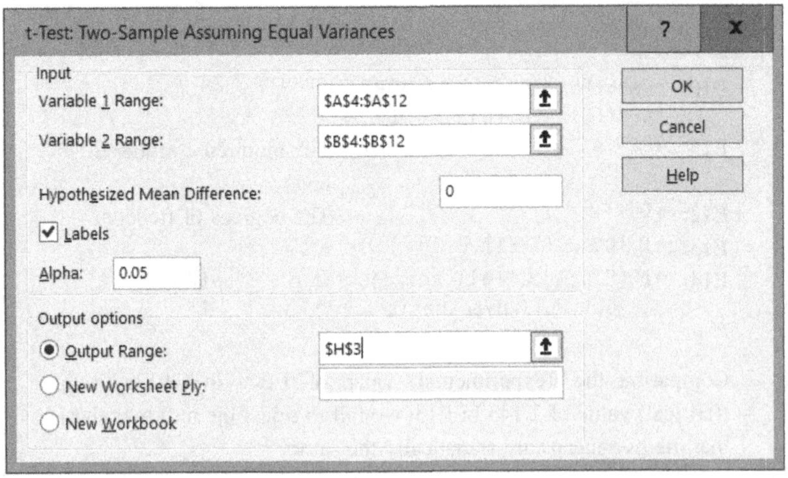

■ FIG. 16.10

EXERCISE 8: THE CALIBRATION CURVE REVISITED

In Chapter 8 we saw how to chart a calibration curve and add a trendline. We also used the functions SLOPE, INTERCEPT, and LINEST to find the slope and intercept of the line of best fit. This line, of course, has uncertainties associated with it. The LINEST function not only gives us the values for the slope and intercept, but it also gives the errors associated with them. Let s_b be the standard error (uncertainty) for the intercept b, s_m the standard error for the slope m, and s_y the standard error for the estimate of y. If y^* is the measured signal for an unknown, then the value of the unknown is computed using

$$x^* = \frac{y^*(\pm s_y) - b(\pm s_b)}{m(\pm s_m)}$$

In this exercise, we make a calibration curve and determine x^* for a measured y^* using the previous equation. The function LINEST is used to find the required parameters. We will see how a combination of INDEX and LINEST allows us to generate only those parameters that are necessary for the task. We shall need to recall that errors are combined using $e_3 = \sqrt{e_1^2 + e_2^2}$ and that for multiplication and division, we must work with percentage errors.

Scenario: A chemist has performed a calibration. Starting with five samples with known concentrations (x-values) she measured the corresponding responses from the instrument (the y-values); her data is in A3:B18 of Fig. 16.11. She then took a sample with an unknown concentration (x^*)

⬚	A	B	C	D	E	F
1	Uncertainty in a Calibration Curve					
2						
3	x	y			Linest	
4	1	2.86	slope	2.2880	0.5680	intercept
5	2	5.20	error in slope	0.0300	0.0996	error in intercept
6	3	7.40	R²	0.9995	0.0949	error in estimate of y
7	4	9.60	F statistic	5808	3	degrees of freedom
8	5	12.10	regression ss	52.3494	0.0270	residual sum of squares
9						
10						
11	Index with Linest			value	err	%err
12	m	2.288	y*	6.55		
13	b	0.568	numerator	5.98200	0.1376	2.30%
14	sm	0.0300	denominator	2.28800	0.0300	1.31%
15	sb	0.0996	x*	2.61451	0.0692	2.65%
16	sy	0.0949	reported x*	2.61 ± 0.07		

■ FIG. 16.11

and measured its response value (y^*). The task is to use the calibration data to find x^*.

(a) On Sheet8 of Chap16.xlsx enter the text shown in Fig. 16.11.
(b) Enter the calibration data in A4:B8. Name the columns as x and y, respectively.
(c) Select D4:E8, enter the formula =LINEST(y, x, TRUE, TRUE), and press Ctrl + ⇧ Shift + Enter↵ to complete the array formula. The entry will appear in the formula bar surrounded by braces { } because it is an array formula.
(d) To see how we may obtain individual parameters from the LINEST function, enter the formulas shown here. These are not array formulas, so complete them normally.

B12: =INDEX(LINEST(y, x, TRUE, TRUE), 1,1)
B13: =INDEX(LINEST(y, x, TRUE, TRUE), 1,2)
B14: =INDEX(LINEST(y, x, TRUE, TRUE), 2,1)
B15: =INDEX(LINEST(y, x, TRUE, TRUE), 2,2)
B16: =INDEX(LINEST(y, x, TRUE, TRUE), 3,2)

The first formula returns the LINEST value that would normally be in the first row and first column, that is, the slope of the line of best fit. Likewise, the second gives us the intercept, which is in row 1, column 2, of the LINEST array.

Name the cells in B12:B16 with the text to their left. This will make it easier to understand the formulas that follow.

We now compute the unknown concentration (x^*) using the formula given at the start of the exercise.

(e) For the purpose of the Exercise, assume our measured signal had a value of 6.55. Enter this value in D12. Enter the following formulas:

D13: =D12 - b	The numerator ($y^* - b$) in the equation
E13: =SQRT(sy`2+sb`2)	The error in the nominator
F13: =E13/D13	The percentage error in the nominator
D14: =m	The denominator m
E14: =sm	The error in the denominator
F14: =E14/D14	The percentage error in the denominator
D15: =D13/D14	The value $x^* = (y^* - b)/m$
E15: =D15*F15	The error in x^*. This will mean nothing until F15 is computed

F15: =SQRT(F13 2   The percentage error in x^*
    +F14 2)
D17: =ROUND(D15,2) &" ± "& ROUND(E15,2)

When using a spreadsheet (or a calculator) to do such computations, we let it use its full precision. We may wish to format the cells to show a limited number of digits. We must round off the values when reporting the results; our y^* value had three significant figures. We would report x^* as $2.61_5 \pm 0.07_0$ or $2.61_5 \pm 2._7\%$. The formula in D17 is cheating a bit; it uses rounding rather than true significant figures.

EXERCISE 9: MORE ON THE CALIBRATION CURVE

The statistical analysis in the previous exercise ignores the fact that the estimations of the slope and intercept are interdependent. A full treatment of the alternative approach is beyond the scope of this book. The interested reader may wish to consult an advanced statistics book.[9] We will take a more pragmatic approach and give the *how* without the *why*. The author is indebted to the Royal Society of Chemistry for permission to quote from one of its technical briefs.[10] Be aware that this brief follows the convention using $y = a + bx$ as the equation of a straight line; we shall stay with the North American convention of $y = mx + c$.

The regression line has an associated confidence interval. The center line in Fig. 16.12 is the linear line of best fit with the 95% confidence limits shown above and below. The data used had an R^2 value of only 0.58; a poor fit was used to enable us to clearly see the three lines. The expression for the confidence interval for the computed y- values is as follows:

$$CI(y_i) = \pm t(\alpha, df) \cdot S_{yx} \cdot \sqrt{\frac{1}{n} + \frac{(x_i - \bar{x})^2}{S_{xx}}}$$

This is for a set of n data pairs having an average x-value of $\bar{x}$. We will see later how S_{yx} and S_{xx} may be computed in Excel.

When we use calibration data we reverse the procedure: we measure y^* in order to estimate x^*. This type of analysis is sometimes referred to as an

---

[9]For example, P. C. Meier and R. E. Zünd, *Statistical Methods in Analytical Chemistry*, Wiley, New York, 1993.

[10]AMC Technical Brief No. 22, ed. M. Thompson, March 2006, Royal Society of Chemistry, London (http://www.rsc.org/images/Brief22_tcm18-51117.pdf). Reproduced by permission of the Royal Society of Chemistry.

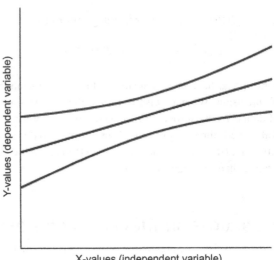

■ FIG. 16.12

inverse regression. We may use either of the following equation to compute the confidence limits for inverse regression:

$$CI(x^*) = \pm t(\alpha, df) \cdot \frac{S_{yx}}{|m|} \cdot \sqrt{\frac{1}{n} + \frac{1}{k} + \frac{(y^* - \bar{y})^2}{m^2 \cdot S_{xx}}}$$

$$CI(x^*) = \pm t(\alpha, df) \cdot \frac{S_{yx}}{|m|} \cdot \sqrt{\frac{1}{n} + \frac{1}{k} + \frac{(x^* - \bar{x})^2}{S_{xx}}}$$

In these equations, m is the slope of the fit and n the number of points in the regression data, while k is the number of duplicates in the y^* measurements. The quantity S_{yx} (the standard error of the estimate) may be found with the Excel STEYX function while S_{xx} (sum of the squares of x deviations) is computed with the DEVSQ function. The approximations inherent in these equations are valid only when $\frac{t(\alpha, df) \cdot S_{yx}^2}{m^2 \cdot S_{xx}} \leq 0.05$

or our example, we have calibrations consisting of five data pairs (A5:B9 in Fig. 16.13). In F4:F13 we find the slope and intercept of the line of best fit together with various quantities needed to compute the confidence levels. In B11:C11 we test to see if our data meets the criterion (previous equation) to allow us to use the CL formulas. Then in F15:F20 we do the actual CL calculation; note how the two formulas give the same results. The reader may decide not to enter the documentation in column G and rows 22:23. Because we have somewhat involved formulas, this is a most appropriate time for

	A	B	C	D	E	F	G
1	Calibration Curve Uncertainty (Confidence Limits)						
2							
3	Calibration data				Parameters		Formulas
4	x	y			m	2.353	=SLOPE(y,x)
5	1.12	2.86			c	0.289	=INTERCEPT(y,x)
6	2.05	5.20			n	5.000	=COUNT(x)
7	2.99	7.40			df	3	=n-2
8	4.03	9.60			Syx	0.131	=STEYX(y,x)
9	4.99	12.10			SSx	9.452	=DEVSQ(x)
10					avgx	3.036	=AVERAGE(x)
11	Test	0.0033	TRUE		avgy	7.432	=AVERAGE(y)
12					p	95%	number
13					t	3.182	=T.INV.2T(1-p,df)
14							
15							
16	Measured y values for unknown				k	5	=COUNT(YY)
17	y*				avgY*	6.550	=AVERAGE(YY)
18	6.55				compX*	2.661	=(avgYY-c_)/m
19	6.47				CL X*	0.114	=t*(Syx/ABS(m))*SQRT(1/n+1/k+(avgYY-avgy)^2/(m^2*SSx))
20	6.56				CL X*	0.114	=t*(Syx/ABS(m))*SQRT(1/n+1/k+(compXX-avgx)^2/SSx)
21	6.57				X*	2.66 ± 0.11	=ROUND(compXX,2)&" ± "&ROUND(F19,2)
22	6.60						

■ FIG. 16.13

using named cells. As we have planned the work in advance we will actually name the cells before they contain any data or formulas.

(a) Open Chap16.xlsx and on Sheet 9 enter the text shown for A1:F20 in Fig. 16.13.

(b) Name A5:A9 as X and B5:B9 as Y.

(c) Select E4:F17 and use the naming tool to name the cells F4:F17 with the text to the left.

(d) Using the information shown in column G in the figure, enter formulas in F4:F13. If you use the pointing method, Excel will use the cell names in the formulas. This will make it much easier to check that your formulas are correct.

(e) Now we will see if we satisfy the criterion that allows the CL values to be computed with the approximate equations. In B11 enter =t^2*Syx^2/(m^2*SSx) and in C11 enter =B11<0.05. The result of TRUE tells us we may proceed.

We now have a calibration and are ready to enter the data for the "unknown." For this demonstration, we assume the experimenter has repeated his measurements five times on the same sample.

(f) Enter the five readings in A16:A20 and give the range the name YY.

The formula in F15 gives us the x^*-value. In F16 and F17 we compute CL using each of the equations shown at the start of the exercise. Finally, in F20 we use a formula to neatly display the x^*-value with its confidence limits.

The companion website contains an Excel file called CalibrationCurve.xlsx, which has worksheets showing the use of Excel with the data from the Royal Society of Chemistry technical brief and a demonstration of the use of a spinner coupled to a calibration curve.

PROBLEMS

1. The thickness of two paper samples[11] was measured four times for each sample. The results for Sample A were 772, 759, 795, and 790 (the units are inches $\times 10^{-4}$). For Sample B they were 765, 750, 724, and 753. Does this data suggest the samples have the same or different thickness?

2. *Is there a statistical difference in datasets A and B?

A	B
2.31017	2.30143
2.30986	2.29890
2.31010	2.29816
2.31001	2.30182
2.31024	2.29869
2.31010	2.29940
2.31028	2.29849
	2.29889

3. *To test for any difference in wear, 10 tires (5 of type A and 5 of type B) were randomly placed on the front rims of 5 cars and wear measurements taken after the cars had been driven a set number of miles; see the following table.[12] Use a *Data Analysis* tool to statistically decide if the two means of the two tire wear values differ.

---

[11] W. J. Youden, *Experimentation and Measurement*, U.S. Department of Commerce, 1984.

[12] W. Mendelhall et al., *Statistics for Management and Economics*, Duxbury, Belmont, CA, 1993.

Automobile	Tire Type	
	A	B
1	10.6	10.2
2	9.8	9.4
3	12.3	11.8
4	9.7	9.1
5	8.8	8.3

4. Grissom and Sara have analyzed samples from the same crime scene. Each has reported the mean values of their tests. Also, in accord with CIS policy, they have reported how many tests were made and the SSD (sum of squares of deviations from the mean) values.

	Grissom	Sara
Mean	59.15	59.62
Number of analyses	6	4
SSD	0.0214	0.0295

From the data in the following table, can you state there is a statistical difference in their results?

5. *A widget manufacturer measured the lifetime of 5000 of his product. The data is approximately normally distributed with an average of 585 h with a standard deviation of 89 h. If he plans to sell two million and promises to give buyers a free widget if theirs lasts less than 750 h, how many should be made? How many are expected to last between 600 and 700 h?

6. The lifetimes[13] for Acme Premium tires are approximately normally distributed with a mean of 45,000 miles and a standard deviation of 2500 miles. The tires have a 40,000-mile warranty. (i) What percentage of the tires will fail before the warranty expires? (ii) What percentage of the tires will fail within 1000 miles of the warranty expiration?

7. Apply the method of Exercise 9 to the data in Exercise 8 to see the difference in the computed confidence intervals.

8. The accompanying table shows the calibration data for the chemical analysis of silica; X is the known amount of silica and Y is the absorbance of the solution. An unknown sample had an absorbance value of 0.242.

[13]G. Keller et al. *Statistics for Management and Economics*, Duxbury, Belmont, CA, 1984.

Find the amount of silica in the unknown and give the 95% confidence limits.

X	0.000	0.020	0.040	0.060	0.080	0.100	0.120
Y	0.032	0.135	0.185	0.268	0.359	0.435	0.511

9. We have not had time to consider the F-distribution. Using information from a textbook and/or the Internet, look again at the data in Problem 4 and determine if there is any statistical reason to think there is a difference in the precision of the two agents.

10. Write a UDF to compare the mean of two ranges at the 0.05 significance level and return either Fail to reject Null hypothesis or Reject Null hypothesis, depending on the relative values of t-expt and t-critical. The UDF should be called with something like =TwoMeans (A1:A10, B1:B15).

11. Write a UDF to compare the mean of a range with an expected value and return either Fail to reject Null hypothesis or Reject Null hypothesis depending on how the computed p-value compares to 0.05. The UDF should be called with something like =ExpectedMean(A2: A11, B2, 0.05). Test your function by comparing this data with an expected value of 59.3 at a significance of 0.05.

59.09	59.17	59.27	59.13	59.10	59.14

APPENDIX: SUPPLEMENTARY MATERIAL

Supplementary material related to this chapter can be found on the accompanying CD or online at https://doi.org/10.1016/B978-0-12-818249-9.00016-9.

Index

Note: Page numbers followed by *f* indicate figures, *b* indicate boxes, and *np* indicate footnotes.

Iteration, temperature profile using, 361–363, 361–362f

K

Kirchhoff's voltage law, 275–276, 285, 285f
Kutta-Simpson formula, 340np

L

Ladder down the mine, 276–278, 277f
LCM function, 70
Least-squares analysis, 188
Legend box, 136f, 137
Limits, 11
Linearization methods, 305–307
  Eadie-Hofstee equation, 305–307
  Lineweaver-Burk equation, 305–307
  Michaelis-Menten equation, 305–307
Linearization vs. solver, 305–307, 306f
Line charts, 134, 134–135f
LINEST function, 193–194, 193f, 302, 389–390
Lineweaver-Burk equation, 305–307
Logarithmic fit (LOGEST), 199–200, 199f
Logical comparison operators, 86
LOOKUP function, 97–99, 98f
Looping structure, 226, 230

M

Macro, 245
  adding control, 260–261, 260f
  computation, 249–252, 250f
  private, 251–252
  public, 251–252
  recording, 246–248, 246–247f
Macro Security tool, 218, 219f
Map chart, 169–170, 170f
Margin settings, 49, 53
MATCH function, 97
Mathematical limitations
  copy command, 17
  displayed values, 21–22
  integer values, 37
  real numbers, 37–38
Mathematical operators, 18–21
Matrix algebra, 358–361, 358f, 360f
Matrix functions, 71–72
Mean, 376
Merge and center, 60
Michaelis-Menten equation, 305–307
Minimization problem, 307, 307–308f
Modeling
  ammonia VP, 271–273

circuit analysis, 275–276, 276f
four-bar crank, 355–358, 356–357f
population model, 269–271, 270–272f
structure member force analysis, 273–274, 274–275f
tank empting model, 364f, 365–368, 366f
temperature profile using iteration, 361–363, 361–362f
temperature profile using solver, 363–365, 363–364f
Modular, 324–326, 325f, 327f
Module, 245, 248, 281
Monte Carlo technique, 331–332, 333f
MROUND function, 69
Multilinear regression, 195–197, 197f
MUNIT function, 360np

N

#N/A, 21
#NAME?, 21
Name Box, 5
Naming a cell or range, 27–29, 28f
Nested IF formulas, 89–90
Nesting, 58
Newton–Raphson methods, 283
Newton's method, 267, 268f
NIST, 301
Normal distribution, 372–374, 376
NPER function, 77
#NULL!, 59
#NUM!, 21
Numerical integration
  accuracy, 319
  adding flexibility, 323–324
  Gaussian integration, 328–331, 330f, 332f
  modular, 324–326, 325f, 327f
  Monte Carlo technique, 331–332, 333f
  RAND function, 331
  Romberg integration, 326–327
  Simpson's 1/3 rule, 318, 321–323, 322f, 326
  Simpson's 3/8 rule, 318, 326
  tabular data, 326–328, 327f
  trapezoid rule, 317–321, 320f
  UDF, 323–326

O

ODD function, 69
OED, 352
Operations
  arithmetic operators, 16–21

copy command, 17
displayed values, 21–22
Evaluate Formula tool, 34–35
formatting numbers, 21–22
fractions, 22–23, 23f
mathematical limitations, 37–38
mathematical operators, 18–21
naming method, 27–29, 28f
range finders, 15, 18, 34
round-off errors, 37–38
stored values, 21–22
subscripts and superscripts, 35–36
symbols, 35–36
& operator, 106, 113
Optimization problem, 307–310, 309f
  constants, 308
  dependent variables, 308
  independent variables, 308
  sand suppliers, 307–308
  worksheet, 309f
Orientation, 51–54, 53f

P

Page break, 54–56
Page Layout tab, 51–54, 53f
Page View, 6
Parametric equations, 155, 156f
Pareto chart, 166, 166f
Paste command, 17
Pie charts, 135
PI() function, 70
PivotTable feature, 102, 121–123, 122f
Plotting functions, 141–144, 142–143f
Pointing method, 16, 17f
Polar (radar) chart, 156–158, 157f
Polygon, centroid of, 280–283, 282f
Polynomial expression, 273
Polynomial fit, 198–199, 198f
Pooled standard deviation, 382–383, 383f
Population model, 269–271, 270–272f
POWER function, 70
PPMT, 77
Predictor/explanatory variable, 187–188
Primary axis, 136
Print area, 52
Print dialog, 48
Printing process
  documentation and cell formulas, 56, 56f
  Header/Footer, 54–56, 55f
  margins, 49, 53
  page break, 54–56
  Page Layout tab, 51–54, 53f

Printed in the United States
By Bookmasters